Prime Mover

ALSO BY STEVEN VOGEL

Cats' Paws and Catapults: Mechanical Worlds of Nature and People

Vital Circuits: On Pumps, Pipes, and the Workings of Circulatory Systems

Life's Devices: The Physical World of Animals and Plants

Life in Moving Fluids: The Physical Biology of Flow

Prime Mover

A Natural History of Muscle

STEVEN VOGEL

Illustrated by Annette deFerrari and the author

W. W. NORTON & COMPANY

New York • *London*

For information about permission to reproduce selections from this book, write
to Permissions, W. W. Norton & Company, Inc., 500 Fifth Avenue,
New York, NY 10110

Manufactured by Maple-Vail Book Manufacturing Group
Book design by Brooke Koven
Production manager: Julia Druskin

LIBRARY OF CONGRESS CATALOGING-IN-PUBLICATION DATA
Vogel, Steven, 1940–
Prime mover : a natural history of muscle / Steven Vogel ; illustrated by
Annette deFerrari and the author.
p. cm.
Includes bibliographical references and index.
ISBN 0-393-02126-2
1. Muscles. I. Title.

QP321 . V64 2002
612.7'4—dc21   2001044842

ISBN 0-393-32463-X pbk.

W. W. Norton & Company, Inc., 500 Fifth Avenue, New York, N.Y. 10110
www.wwnorton.com

W. W. Norton & Company Ltd., Castle House, 75/76 Wells Street, London
W1T 3QT

1 2 3 4 5 6 7 8 9 0

to Roger, Frances, Booth, and Avery, who, wearing my genes, must now take charge of my fitness

Contents

Preface

We humans remain animals. Biology talks about us all, something that can create a lot of uneasiness—or worse. It implies constraints on human aspirations, constraints that may loom large or small, depending on one's political and social outlook. On the one hand, we worry about how much our genes might set our individual and collective courses; on the other, we use nonhuman animals as surrogates for investigating our illnesses. For the biologist, our animal nature is second nature, even an opportunity. It provides a larger arena in which to explore how we do what we do, with the "we" then transcending the temporal and structural boundaries of our species.

That's the intent here. But you'll encounter nothing so controversial as arguments about inherited behavioral predilections. No, I just want to take a piece of our biological nature, the "flesh" of "flesh and blood," and explore how it works and how we work with it. I mean to argue the general message that because we are animals, a biologically based view can shed light on our human world. My last book, *Cats' Paws and Catapults*, argued the same point from a different vantage point. It used biomechanics as a base line from which to explore why we make the kinds of things we do. Here I ask what muscle physiology may reveal about human history, prehistory, and culture.

Muscle has been our sole engine for most of our time on earth. Our muscle differs only a little from that found elsewhere in the animal kingdom. The same stuff propels water flea and whale. With it we walk and run and climb and swim; we lift heavy weights and manipulate tiny screws; we bend things, twist things, kick and bite things. But we don't fly—nor does any other animal as heavy as a human—at least without an extraordinary prosthesis. Nor can we burrow like an earthworm or jet like a squid or jellyfish, even if we're near the top of the heap for the overall diversity of our muscle-powered activities.

Our muscles aren't the most forceful or the hardest working, but on both accounts they come close to nature's best, and we're well endowed, devoting almost half our body mass to muscle. We can paddle boats, pull carts, push wheelbarrows, pedal bicycles, carry each other on litters and sedan chairs. In return for food and breeding opportunities, our bovids, equids, and camelids contribute their muscular efforts to tasks we choose. Both the pyramids of Egypt and the Great Wall of China represent muscular accomplishments as much as they do social and technological ones. In both cases human muscle did most of the job.

Thus muscle provides a rich and multifaceted tale. I begin by explaining what we know about how it works and how we've come to know what we know. The remainder of the book ties the operation of this engine to the tasks it does, looking first at how the basic engine can be hooked onto skeletons to perform diverse activities. It then moves farther afield to consider such things as the design of tools, the harnessing of beasts of burden, the limits to our performance, even our predilection for eating muscle. I hope to make the case that biology, physiology, biomechanics—my interests—matter in contexts far beyond their immediate scientific domains. Writing the predecessor of this book hooked me on the seductive pleasure of moving out of my scientific box. I nourish the hope that the results of the activity may prove enjoyable and enlightening for others.

Muscle has not yielded its secrets easily. Over the years I've met quite a few people actively investigating how muscle works at every level from molecules to animals. I can report that without exception they uphold the best of the scientific tradition—from pure intellectual quality to their attitudes toward their subject and one another. They're made in different molds—flamboyant (Szent-Györgyi), noisy (Dick Taylor), quiet (Tom McMahon), and so forth—to pick people who can no longer take issue with my characterizations. Lesson? Between our interest in people and the tough job of explaining science, perhaps we pay too much attention to disputes and rivalries and areas of scientific work in which they're prevalent. My forty-year experience as a scientist suggests that the muscle people are closer to the norm.

A large number of people have generously advised me at one point or another in my getting together the present material, explaining things, suggesting sources of information and useful examples, extracting my foot from my mouth. Several afternoon tea drinkers helped throughout the whole enterprise: John Gregg, Daniel Livingstone, and Fred Nijhout, and Knut Schmidt-Nielsen, in particular. Mike Reedy, in whose company I regularly work my muscles and cardiovascular equipment, survived my repeated questions about muscle microstructure and the people working on muscle. More specific items came from Al Buehler, Peter Burian, Marjorie Dosik, Bob Full, Jeff Goldman, Lew Greenwald, Barbara Grubb, Bill Holley, Abdul Lateef, George Newton, Rad Radhakrishnan, Kalman Schulgasser, Anne Scott, Lloyd Trefethen, Steve Wainwright, and Martha Weiss, although not all these folks realized just what they were aiding.

In particular, I owe much to my anthropological colleague, Steve Churchill, and my ever-tolerant spouse, Jane, each of whom annotated drafts of the entire manuscript. And to my editor at W. W. Norton, Edwin Barber—instigator, adviser, and one of my three best writing teachers. If, as he says, this manuscript came in cleaner and clearer than my last, he should take much—maybe most—of the credit.

Figure 3.2, from Borelli, appears by courtesy of the Trent Collection of the Duke Medical Center Library. William Kier kindly permitted me to trace illustrations from his Ph.D. thesis (Duke Zoology) for Figure 6.12. Permission to use material for Figure 4.2 was granted by the *Journal of Experimental Biology* (copyright 1954) and for Figure 9.1 by *Nature* (copyright 1986, Macmillan Magazines Limited). I am especially indebted to Mary Reedy, who contributed all the electron micrographs.

Steven Vogel
Durham, North Carolina

Prime Mover

CHAPTER 1

Body Work

O F THE WEIGHT OF A HUMAN IN DECENT SHAPE—ALL TOO few of us—muscle makes up fully 40 percent. Not blood, bone, brain, or liver contributes as much; only together do they add up to our weight of muscle. Moreover, all that muscle does just one task: It makes the chemical fuel that originally came from our food produce force and motion. It does neither more nor less than what we ask of the combustion engines of our cars and airplanes. As in the engines of our technology, its imperfect efficiency makes it get warm, a nuisance when we work hard in warm weather but nice enough as we jump around or shiver to offset the cold.

In power efficiency—how much work it can do for a given amount of fuel—muscle differs little from those combustion engines. In weight efficiency—how much work a given weight of muscle can do in a given time—it compares well with automobile engines but suffers badly when put up against a good jet turbine. Still, this is one remarkable device. Nature perfected it around a billion years ago (give or take a few hundred million), launching multicellular animals on our glorious trajectory. It powers ant and elephant alike, so alike that only a trained eye can see the subtle differences between their muscles when bits are viewed under a microscope. Flies fly with it; clams clam up with it.

For over a million years, we humans (or our putative ancestors) have taken bits and pieces of nonliving material—sticks and stones, old bones and pieces of metal—and fashioned devices that apply our old engine to new tasks. Stone axes, spears, bows and arrows, fire drills, knives, saws, rasps, atlatls, and boomerangs; rowboats and bicycles; hand-cranked centrifuges and meat grinders and can openers: None does more than harness our various muscles. For ten thousand years or so we've pressed nonhuman muscles into our service as sources of power. We discovered that bovids, equids, camelids, and others could be persuaded to trade mechanical work for vegetation, mostly vegetation we found indigestible.

When the day's work is done, what then? We can eat the engine, either casually or ceremonially. Livers, kidneys, and eggs yield only a few calories next to that muscle. Most cultures account animal muscle as the best of foods, and rich ones rear animals solely to put muscle on the table. We even eat our own muscle—beyond episodes of cannibalism. In extreme starvation, when our bodies run out of fat, muscle gets mobilized by our metabolic machinery.

The urbane urbanite uses verbal tricks (making pig into pork, calf into veal, cow into beef) to disguise that link between barnyard and kitchen, at least for English speakers and their mammals. In odd contrast, we take our fowl at face value even when we're affluent enough to put two chickens in the fleshpot. Farm families, though, know quite a lot about that which they eat. My son, at age eight, remarked on the bad taste shown by the North Carolina State Fair, where, in full view of the pigs, hung a large poster illustrating the various cuts of pork. Some farmers at least distinguish between animals personal and animals victual: "Don't name it if you mean to eat it."

A decade or so ago I taught a course for adult nonscientists that looked at various mechanical aspects of existence. In one session I took apart the hind leg of a lamb, purchased at the nearest supermarket, to show how its diverse bones, muscles, tendons, and joints enabled the lamb to gambol about. Midway through the demonstration, one student, a person of some substance in our local computer industry, blossomed out with that look of epiphany that brings joy to any teacher. I paused for some comment on the subtle biomechanical role of the kneecap. Instead what I heard was "Muscle—you mean that that's what meat is?"

And that's what this book is about. Muscle, the prime mover of human history, aka meat: how it works, how we work it, and how the nature of this engine has shaped our history, our cultures, and our technologies. We're animals. Whether we're more than animals turns on one's theology, but we're at least animals. It would be strange, indeed, if our animal nature made no difference to history, culture, technology.

Nothing is more animal than movement—animation, in a word—and underlying our every movement are our muscles.

LET'S TALK NOT about some substance called muscle but about a muscle in particular, even if we pick a generic and paradigmatic example. Removing the rest of the animal, then, what have we left? A flabby, spindle-shaped bit of flesh, thick in the middle and tapering off into the shiny tendons at either end, as in Figure 1.1. If this happens in an undergraduate physiology lab, the muscle is most likely the great calf muscle of a frog, its gastrocnemius (yes, the *c* is pronounced, as a *k*). Extending the foot rapidly and forcefully matters for both jumping and froggy swimming, so frogs have fine calves. Since some of us occasionally eat the leg muscles of frogs, the truly compulsive reader should be able to get a look at this particular bit of anatomy with no more than a trip to the gourmet meat market.

Muscle acts simply. Appropriately stimulated by nerves, it draws its ends closer together, or at least it pulls on its ends, trying to bring them

FIGURE 1.1. The gastrocnemius muscle of a frog, as it jumps, and of a human. Notice that the fibers in the human muscle run obliquely; this is important to our story.

close. In the case of both the frog's calf and our own, the lower tendon runs over the heel, so that action makes the foot protrude. A frog jumps; a person's heel (and body) get pulled up, leaving only the toes for support. Still, no muscle can make ends meet. It ordinarily shortens by about 20 percent of its length or, under unusual conditions and at lower power and efficiency, by 30 or 40 percent. Some muscles of notably high effectiveness barely shorten at all; the flight muscles of many insects contract less than 5 percent.[1] So it's a short-stroke engine that usually needs help—levers and such—for most tasks. Nor can muscle lengthen without assistance. So it needs yet more help in the form of springs, other muscles, or hydraulic devices if it's to make more than one contraction per lifetime. Not that such behavior characterizes muscle alone. Burning fuel drives a piston of an automobile only a relatively short distance in one direction, and the piston needs either a flywheel or the action of other pistons to complete its cycle, not to mention a transmission to convert its motion into something appropriate for driving a set of wheels.

Not even its associated levers and such suit muscle for all tasks. How hard can you hit? You can hit harder if you first grab a weighted stick, a hammer. How fast can you send off a projectile? You can propel it faster if you use a golf club, baseball bat, or spear-thrower. Nor did muscle-amplifying machinery originate with humans. Galileo first showed that, neglecting air resistance, all muscle-powered animals, whether fleas, grasshoppers, galagos, or gazelles, should be able to jump to the same height, roughly three feet. But the flea, being so small, suffers mightily from air resistance, which ought to reduce its height over 80 percent. Undeterred, it remains fully competitive in the high jump. It offsets air resistance by using a cunning bit of elastic and a trigger to store up and then suddenly to release a burst of muscular energy, doing just what the ancient mariner did with his crossbow and the vaulter does with the pole.[2]

Even levers, elastic elements, and triggers can't circumvent every limitation of the basic machine. Consider what happens when you lift something: You do some work, an amount equal to the weight of what you lift times the height to which you lift it. If that's what the physical scientist means by work, then when you simply hold something aloft, you do no actual work. So that task should take no energy, if we define energy (as we do) as the capacity for doing work. Is something wrong here? Since you get tired doing it, you know that holding something up takes energy. Your muscles clearly work even though you do no work in a strictly physical sense. The problem, not merely a definitional one, comes down to a distinct drawback of a muscular engine. A rope can hold something aloft indefinitely without expending any energy. For all its billion years of experience and perfection, muscle hasn't figured out how to lock up at some shortened length so as to hold that length without further work. Indeed, the counterintuitive feel of the physicist's definition of work comes from just that need to use energy when your muscles do no more than hold a weight aloft.

Oh, yes, while we're on the queerness of this engine, here's another, one, though, of language and habit. We speak of muscle "contraction," but does muscle actually "contract"? A simple test shows that it doesn't. Immerse yourself in a pool or lake, retaining just enough air in your respiratory system so you barely float. Now squeeze as many of your muscles as you can. If they really contracted, you'd get a little smaller. Your mass being unchanged, you'd be denser. If you were denser, you'd sink. But you don't sink when you "contract" your muscles because they don't really contract. A muscle may pull its ends together, but in doing so, it gets thicker, enough so that its volume remains the same. For that matter, a "contracting" muscle may not even get shorter, as when you hold up a weight or push on a wall. It can even get longer, as when you try to stop something, such as a baseball, that's approaching you at high speed.

A few more items at the start to give the flavor of what will follow:

- Recently I spent a few days brainstorming with kindred eclectic souls for a company intent on producing some appliance that would transform humans (ideally a large and profitable number of humans) into efficient swimmers. We do swim, but by the standards of any salmon or seal even the most highly trained and talented among us is laughably bad. In cardiovascular equipment we rank among the better-endowed mammals, and we're not seriously short of muscle. The main troubles trace to our nonstreamlined shape and the location of all that muscle; fundamentally we're terrestrial walkers and runners, not aquatic swimmers. So what would be required to make proper swimmers out of us? What kind of snap-on or strap-on gear might harness our muscles to the task?

    The specific questions may be of immediate and commercial interest, but they exemplify a problem far older than recorded history. We want to do more than directly grab, pull, bite, run, and lift. So we make tools, from levers and cranks to vaulting poles and micromanipulators, designing each to couple effectively to our muscular engines. We want to harness the power of domestic animals, so we make plows and windlasses, even treadmills and carriages. Again coupling proves critical, and according to well-regarded historians, improvements in coupling mark turning points in human history.

- Usually biology draws on physics, but on rare occasions biology gives something in return. Muscle has made at least two contributions to the physical sciences. Among the key people in the development of the principle of energy conservation in the mid-nineteenth century a pair came from biology. Julius Robert von Mayer was a physician, and Hermann von Helmholtz began as a physiologist. Muscle played at least an indirect role in getting Mayer on the right track, while for Helmholtz the relationship between muscle's heat production and metabolism was central.[3] We can't any longer imagine doing physics—indeed science of almost any kind—without that energy conservation principle, what we call the first law of thermodynamics. The other occasion came earlier, in the late eighteenth century. Luigi Galvani found that he could use the twitch of an exposed frog muscle to detect electricity, as what we now term a galvanometer. Fortunately for the frogs, this shocking bioassay for electricity soon passed from the scene; voltage and other electrical phenomena, as emphasized by his contemporary Alessandro Volta, transcended the world of organisms. Incidentally, Galvani used just that calf muscle mentioned a few pages back. Some experimental systems long retain their utility!

  For that matter, the muscle needn't be exposed. I used to ask students in a biology lab to move small pairs of electrodes around on their forearms, giving themselves a mild electric shock at each location and noting the particular finger movements that resulted. With patience (and a little fortitude) each could thus map the locations and determine the actions of the main muscles of the forearm. A look at the anatomical drawings of the great Vesalius (1514–1564)[4] then confirmed the result. But that stimulating exercise required a less sensitive (sorry, again) era.

- How hard can a muscle pull? Or, to put it another way, with how much force would we have to stretch a muscle just to counteract its contractile force? That depends a little bit on which muscle we look at. A force of 30 is typical, although the range extends from about 15 to 140 pounds per square inch (100 to 1,000 kilopascals, or kPa, in standard scientific notation).[5] As rope muscle is wimpy. A steel cable with a cross section of a square inch can withstand a pull of over 50,000 pounds (350 megapascals, or MPa), around a thousand times more. But the cable of course can't contract and act as an engine. Of more relevance here, we can compare the tensile strength of muscles with that of tendons, the things to which muscles ordinarily attach. Tendons will withstand pulls of 15,000 pounds per square inch (100 MPa) before they break, over a hundred times what a muscle can manage. Figure 1.2 puts these comparisons in graphic form. Appropriately enough, that calf muscle we considered earlier has a belly that's a few hundred times thicker (in cross section) than the tendons on its ends. So those spindle-shaped ends reflect underlying mechanical behavior, as does so much of our anatomy. As Hamlet put it, "There's a divinity that shapes our ends. . . ."[6]

- How fast can a muscle shorten? That depends on the particular muscle, on both its behavior and its size. Size comes into the picture in a way that's perhaps clearest if one views a muscle as a long series of people linked hand to hand (perhaps on frictionless ice

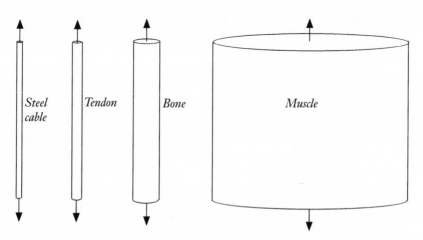

FIGURE 1.2. Cross sections of steel cable, tendon, bone, and muscle that can withstand the same weight or, more generally, the same tensile force.

skates). If all of them simultaneously flex their arms, the line will get shorter. If another series of people, twice as long as the first, flex their arms in just the same way, that longer line will shorten twice as fast. So we need to consider an intrinsic speed that adjusts for length. A muscle, then, shortens by about 10 percent of its length in something between about a hundredth and a tenth of a second. To put it another way, a muscle that's a foot long shortens at between ten and a hundred feet per second, or between seven and seventy miles per hour. As a general rule, the faster it shortens, the less forcefully it pulls.[7]

- Work gets done when a muscle shortens against some force that would rather it didn't do so. How much work can a muscle do? That depends on how long we give it to work. So the question is better posed in terms of the power a muscle can deliver, with power defined as the rate at which work gets done. The best muscular systems can sustain power outputs of around 90 watts per pound (200 W/kg). That suggests that a 150-pound mammal (one of us, perhaps) with 40 percent muscle ought to put out over 5,000 watts of power or about 7.5 horsepower! Well, we can't and don't. We just can't supply fuel and oxygen or remove wastes and heat at anything close to the requisite rate. How much power you can put out turns out to depend strongly on how long you're asked to sustain it, even more than on your age and physical condition, neither of these being negligible, by the way. The longer the run, the slower the average speeds. The issue becomes critical when you design human-powered aircraft, which have high minimum power requirements.

  As terrestrial mammals go, we humans go decently enough, up there with dogs and horses among the sustained runners. What distances people, dogs, and horses from less competent cats and cows isn't so much our muscles, though, as our cardiovascular equipment. Oxygen supply defines what an animal can do, and we do pretty well on that score. Still, runners of any species finish behind fliers—birds, bats, and insects—and sustained swimmers: billfish, seals, whales. The difference may, once again, come back to muscle. The complexity of the motions involved in running with legs may preclude equally effective use of large masses of muscle.

- While muscle neither looks nor works at all like the engines of our technology, we've now set the stage for the comparison alluded to at the start. What factor should we consider? Fuel efficiency would be nice, but differences in kinds of fuels and in the cost of getting fuel to the engines complicate the issue. Probably power output relative to weight gives the easiest common reference, as in the table below.[8]

Engine	*Power Output, Watts per Pound*
Muscle	90
Early steam pump	50
Electric motor	100
Automobile engine	200
Motorcycle engine	500
Aircraft engine, piston	700
Aircraft engine, turbine	2,500

• Flex your arm at the elbow while tensing its muscles, especially those just below your shoulder; "make a muscle," as we say. The biceps on the inner surface of your upper arm forms a bump that moves up and down as you change the angle of your elbow. Some imaginative ancient saw that bump as a subcutaneous mouse; in the mysterious alchemy that underlies language, the word for mouse generated a word for muscle. In Greek, *mys* serves for both, and our "myosin," "myology," "myoglobin," and "myocardium" preserve that perception. Latin does likewise but adds a flourish of its own. "Mus" is a generic mouse, while "musculus" specifies a little mouse. "Musculus" means muscle as well, a second meaning distinct enough to be unambiguous in context. Linnaeus borrowed the pair for genus and species, *Mus musculus*, of our house mouse.

CHAPTER 2

How Muscle Works

Two bicycles were arranged in opposition; one subject pedaled forward, the other resisted by back-pedaling. The speed had to be the same for both, and (apart from some minor loss through friction) the forces exerted were the same. All the work done by one subject was absorbed by the other; there was no other significant resistance. The main result was evident at once, without analysis: the subject pedaling forward became fatigued, while the other remained fresh. . . . The experiment was shown in 1952 at a conversazione of the Royal Society in London and was enthusiastically received, particularly because a young lady doing the negative work was able quickly, without much effort, to reduce a young man doing the positive work to exhaustion.[1]

A. V. HILL, muscle physiologist

A LIMP BIT OF FLESH WAS LEFT HANGING MIDWAY THROUGH the last chapter. We now return to it, asking how it converts fuel into tension, shortening, work. As it happens, we have a good idea of how its bits and pieces interact to enable a muscle to work (in both vernacular and physical senses). We'll defer to the next chapter the tale of how we came to know what we know.

First we need to dispose of the stuff that holds a muscle together and connects its ends to bones, stuff called unambiguously, if unimaginatively, connective tissue. Tendon makes up most of this component, but connective tissue most often gets thoroughly interspersed throughout the contractile material as well. This connective tissue presents a problem for the investigator, not just for the explainer, but the problem can be minimized by our picking a muscle with a minimum of the stuff. Nothing new about that; much of what we know of biology has come from shrewd choices of experimental material. Think, for instance, of the special contribution of fruit flies to genetics or (if less generally appreciated) of squid to neurobiology.

Here's how one choice came about, as told by the person who made it, the late Albert Szent-Györgyi. The light dawned in a fancy restaurant in Budapest in the late 1930s:

> What we need is a muscle with long fibers running parallel, possibly with little connective tissue. It is easy to find the muscle which contains the least connective tissue, for it is the quantity of connective tissue which decides the culinary value of a muscle, to which it is inversely proportional. The culinary value is expressed numerically on the right side of the menu. If this column of numbers is studied, a significant difference is always found at the level of filet mignon, scientifically, *Musculus psoas*. The reason is easy to understand. The contractile matter is a semi-solid gel which could easily be damaged mechanically. The protection it needs depends on its location. If much exposed, as is the case in muscles of the extremities, it is embedded into heavy sheets of connective tissue, the fasciae. The most protected major muscle of the body is the *Musculus psoas* lying in the depth of the abdominal cavity, protected from the back by the vertebral column and in front by the varying mass of the abdomen.[2]

(The history of research on muscle involves no more colorful figure than Szent-Györgyi [1893–1986]. He came to the subject after his classic work, which identified ascorbic acid with vitamin C and showed its metabolic function, work that brought him a Nobel Prize. He decided that movement defined animal life, that muscle enabled movement, and thus, by working on muscle, he could come closest to the essence of life. He defined a spectrum of scientists from Apollonians to Dionysians— rationalists to romantics—and unhesitatingly offered himself as paradigm for the latter. He disdained authority, initially in Hungary, where during and just after World War II he managed to become persona non grata to both Germans and Russians.[3] At Woods Hole, Massachusetts, after the war, he was within the same year first cited as the outstanding foreign-born citizen of the state and then charged with tax evasion for the way he ran his research institute. A lecture by Szent-Györgyi was always a show; at the end of one I recall his rolling up his sleeves to show that nothing remained hidden.)

So filet mignon—psoas muscle, with a silent p—played (and continues to play) a major role in muscle research. Of course one doesn't want to kill a cow to get an ounce of muscle for the lab. Besides, big mammals such as cows put more of their fat in among the muscle fibers (giving additional gustatory appeal) whereas small ones put more of their fat just beneath the skin and leave their muscles leaner. So what's needed is a small mammal with a relatively well-developed psoas, one, furthermore, that's easy to raise in captivity. On that rationale, rabbit psoas became a standard material. As a fringe benefit, the investigator could eat the rest of its muscles—at least before animal care regulations put an end to such sensible recycling.

The Basic Machine

A muscle is made of muscle fibers, which run parallel to each other. Our calf muscle (the gastrocnemius, mentioned earlier) contains about a million of them. Each of these in turn is made up of a thousand or so smaller parallel units, the so-called myofibrils of Figure 2.1. Thus the force generated by a myofibril times the number of myofibrils in the bundled bundles that make up a muscle gives the force produced by the muscle as a whole. Force is multiplied just as it would be if when a group of people pull something, each person is tugging on a separate rope.

Still smaller units line up, end to end, in each myofibril. These are the so-called sarcomeres. In our gastrocnemius muscle, about fifteen thousand make up each chain.[4] Since they're serially arranged, they pull on each other, and only the ones on the ends of a myofibril pull against the load itself. So the force produced by a myofibril differs in no way from the pull of any one of its sarcomeres. Two things do go up as the number of sarcomeres in the chain increases: the distance the free end of a muscle can move and the speed of the movement. Distance and speed are multiplied as they would be if each person in a group that was joined arm to arm (as mentioned above, perhaps on frictionless ice skates) simultaneously bent both elbows.

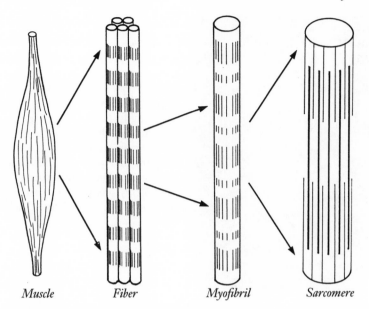

Muscle   *Fiber*   *Myofibril*   *Sarcomere*

FIGURE 2.1. A muscle is made up of fibers, which in turn are made up of myofibrils, which are formed of lengthwise arrays of sarcomeres and their associated machinery.

(A note on the jargon of the trade. "Myo-" and "sarco-" will tediously recur in this account. The former comes from the Greek word for muscle—and mouse—and the latter from the Greek word for flesh. No special significance attaches to the use of one as opposed to the other. While both reflect a certain pretentiousness, they do provide an unambiguous allusion to muscle. At least we're so far spared upscale meat markets that call themselves sarcoporia.)

With the sarcomeres, we've descended to the level at which the basic contractile mechanism can be described. That a muscle has such units can be dimly perceived with ordinary light microscopes. One can use techniques for preserving, slicing, and staining muscle that have been available for more than a century. Or one can quickly freeze muscle, slice it thinly, and look at the slices between a pair of polarizing filters. But tantalizingly close as that comes to the operating elements, it isn't close enough. Pushing an electron microscope to its practical limit just suffices to get one down to the level that matters—or up to the necessary resolution, to put it another way. Individual protein molecules now

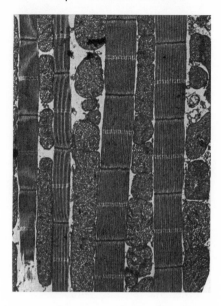

FIGURE 2.2. An electron micrograph of a lengthwise slice of a muscle, in particular of the main flight muscle of a fruit fly. Compare this with the diagram in Figure 2.3. The bodies between the fibrils are mitochondria, the energy converters needed for sustaining intense activities, such as flight.

become detectable, as in Figure 2.2, at least if the proteins are large and not dissolved in the liquid of the cell.

Of the various proteins in a sarcomere, two kinds make up most of it and matter most. They're arranged in a specific pattern that varies only minimally from muscle to muscle, as in Figure 2.3. Myosin makes up the

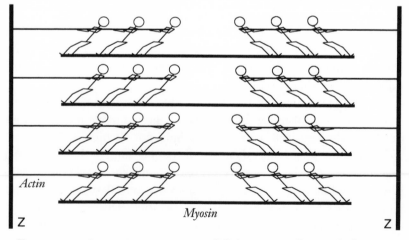

FIGURE 2.3. A sarcomere, represented diagrammatically, an idealization based on many pictures like Figure 2.2.

thick filaments in the middle of each sarcomere. Each myosin molecule weighs about half a million times more than a hydrogen atom, so these are large proteins. Actin molecules are much smaller, less than a tenth of that size, but in muscle they attach to one another in long, doubly helical chains. They form the largest part of each thin filament that runs inward from a sarcomere's ends.

Neither the myosin nor the actin filaments ordinarily contract. Instead contraction comes from interdigitation of myosin and actin, as you might do by sliding the fingers of one hand between the fingers of the other. "Sliding filament" has become the usual term for what's now the accepted model for muscle contraction. What happens is that the filaments ratchet (rapidly!) along each other. The molecular interactions between myosin and actin correspond most agreeably with the behavior of an intact muscle. Thus (Figure 2.4) the force a muscle can produce

State of the sarcomere　　　　　　　　*Maximum tension it can produce*

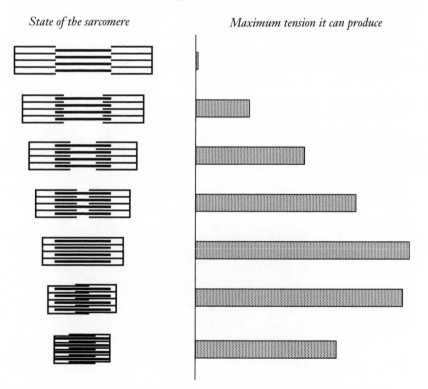

FIGURE 2.4. The force a muscle can exert relative to the degree of interdigitation of the filaments of its sarcomeres.

depends on the degree of overlap between thick and thin filaments.[5] If one stretches a muscle so overlap is reduced and then stimulates it to pull as hard as it can (against an immovable load), the force with which it pulls decreases. That maximum pulling force drops to zero at just the muscle length at which the fibers stop overlapping. Allow a muscle to shorten and then ask it to pull against a similarly immovable load; the pulling force begins to drop off at the point where the thin filaments bump into each other. But force continues to be generated, if at ever-decreasing levels, as the thin filaments slide along one another. Force reaches zero again when each thin filament hits the opposite end of the sarcomere.

While the exact mechanism involved in the ratcheting action remains elusive, the general scheme seems clear enough. Hauling something toward you by pulling, arm over arm, on a long rope provides an adequate analogy. The "arms" in muscle are protrusions from the thick myosin filaments. Each myosin molecule has one on its end, so with several hundred myosin molecules in a thick filament, that many arms reach across toward the thin actin filaments. As in Figure 2.5, these cross-

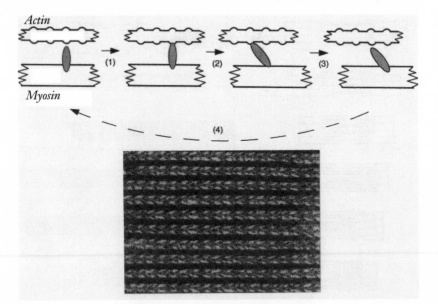

FIGURE 2.5. The current interpretation of the action of the cross-bridges and what filaments and bridges look like in a very high-magnification electron micrograph of the flight muscle of a giant water bug.

bridges between the filaments attach, swing through an arc, let loose, and then reattach farther along the thin filaments. The swinging movement while they are attached accomplishes contraction. Not that we truly understand what sounds so straightforward when it's summarized in a few sentences. At this writing, the mechanism by which these crossbridges do their job constitutes the frontier, with a host of bright people and active laboratories working on what has turned out to be a surprisingly tough nut to crack.[6]

One more piece of this initial, simple look at contraction: when and where does the energy get put in? Muscle, as an engine, moves things that would not otherwise move or at least would move along different paths. The energy coming in is not kinetic (as in a flywheel), electric (as with wires from a battery), or thermal (as in the expansion of hot gases). Chemical energy runs muscle directly, with no other form of energy intervening. Still, "directly" misleads at least a little; you can get energy from olive oil by burning it in a lamp, but to make it run muscle, a whole series of transformations are needed. At the end of the transformations (in their various versions), molecules of adenosine triphosphate (henceforth called ATP) are generated in the muscle. They're not synthesized from scratch but instead get made by attaching a third phosphate to adenosine diphosphate (ADP). In the process, energy must get fed in, or the last phosphate would not join the outfit. When thin and thick filaments ratchet along each other, ATP gets broken down into ADP and phosphate again. The scheme amounts to a kind of short-term rechargeable battery.

Oddly enough, at no time does a muscle contain much ATP. Running at top speed on ATP alone, a human would run out, having converted it all to ADP, in two to four seconds.[7] So we can't go far without reconstituting ATP from ADP. Another chemical, creatine phosphate, provides the immediate source of both energy and phosphate; splitting off a phosphate from creatine provides the wherewithal to make ATP from ADP. But we haven't much creatine phosphate either, not even enough for twenty seconds of hard running. For a longer run, we bring into use the fat and starchy stuff (glycogen, essentially a chain of sugars) in the muscle. That can keep a runner going for an hour or so. For still longer bouts of activity, fat and carbohydrate have to be moved to the muscles from the liver and the body's fatty tissue.

As an analogy to the hierarchical energy supply for muscular work, consider how you might eat and make bread. From time to time you eat a few slices of bread (ATP). When the bread runs low, you get out the flour (glycogen) and bake a few loaves, putting most (creatine phosphate) in the freezer in case houseguests suddenly descend. Still less often, you buy flour (from liver and fat depots). The depletion of bread stimulates baking, which in turn stimulates marketing, with the stuff in the freezer as a reserve for sudden demand.

The Nature of the Natural Engine

Breaking ATP into ADP and phosphate ordinarily requires an enzyme, one called a phosphatase in the jargon. When a muscle pulls, the protein myosin acts as the phosphatase. Without a supply of ATP, the filaments won't ratchet their way along each other. So muscle is an engine that takes energy in a nonmechanical form—in its case in the form of chemical bonds. From that input it produces mechanical energy, acting as both enzyme and engine. In some ways muscle is ordinary: in its efficiency, in the power it produces relative to its weight; in other ways it's quite a remarkable engine: soft, wet, and contractile.

Muscle, in dramatic contrast with electric motor or turbojet, depends on a molecular motor. For organisms, though, the scheme represents nothing radical but rather a modest modification of what happens in all cells all the time. A molecule attaches to a second and exerts a force on it by changing from one shape (configuration) to another. As a result, the second molecule changes its shape (or splits or joins another), whereupon the first detaches itself and goes back to its original shape, ready to repeat the process. The process often changes the form of still other molecules, which by their changes feed energy into the system. A person who has taken a biology course in the past few decades will recognize that we're talking here about the way an enzyme works. A big molecule changes back and forth from one form to another and, in the process, alters some other molecules, one at a time. Most enzymes do such chemistry as an end in itself, as we convert, for instance, what we eat into what we're built of. Muscle, by contrast, puts this change in shape to direct, large-scale, mechanical use. They may be small—even big molecules are still small—but muscle's molecules are lined up, as end-to-end sequences of sarcomeres, thus multiplying the distances and speeds they achieve. The sarcomeres lie in side-by-side parallel arrays, thereby multiplying the forces they generate. Operating in tight coordination, these reversible, enzymatic changes in configuration on a molecular level add up to movements useful on our personal scale of operation.

So right down to the molecular level the engine remains a mechanical device. By our technological yardstick, cross-bridges between protein molecules are fabulously tiny engines. By the yardstick of nature, what's fabulous is the way these engines, harnessed in long and broad arrays, produce large-scale, forceful motion. Otherwise they remain chemomechanical devices like all other enzymes. Humans happen to build few chemomechanical engines, and none yet graces our everyday lives. Chemomechanical motors may have no obvious advantage—or, for that matter, disadvantage—over our familiar electromechanical and thermomechanical devices. But familiar they are not. "Enzymatic" still rhymes with "enigmatic."

A Few Other Critical Pieces

A look at your car's engine reveals a lot more than pistons and cylinders. Similarly actin, myosin, and ATP don't make up the whole story of muscle; much more machinery goes into making it a functional motor. An electron micrograph of a slice of muscle (Figure 2.6) shows a variety

FIGURE 2.6. An electron micrograph, this one of a leg muscle (the quadriceps) of a mouse, at a lower magnification than that of Figure 2.2. Between the myofibrils, you can see mitochondria and the bits and pieces of other membranes. The edge of the cell is at the upper left. Some myofibrils appear discontinuous, but that's just an accident of the way they run in and out of this flat, thin section.

of other structures, and chemical analyses turn up many more things than make themselves visible in a picture. By and large, though, we now have a fair idea of what each of these items contributes to the enterprise.

One might wonder, for instance, what keeps the thick filaments in place rather than creeping or slipping toward one or the other end of the sarcomere. Another protein, called titin, does the job. Its very thin filaments extend from the thick myosin elements toward each end of the sarcomere. Titin seems to fold or contract when the sarcomere shortens, although it contributes little in the way of force to the process. In addition, at least half a dozen other proteins link up with the actin and myosin, each playing its separate role.

And what of those end plates that provide anchorage for both the thin filaments of actin and the thinner ones of titin, the plates that connect adjacent sarcomeres? These Z-discs are complicated structures made up of at least four other kinds of protein. They contribute strands that run crosswise and keep adjacent sarcomeres aligned and bound to the outer membrane of the muscle fiber.

More yet. Specific structures between the myofibrils generate most of the ATP that powers contraction. Since we feed carbohydrate, sugar, or fat, not ATP, into the contractile system, muscle needs something analogous to the transformers with which we domesticate the electricity that comes out of high-voltage transmission lines. So-called mitochondria play the principal role in that process. These appear in electron micrographs as round or oval bodies, and they're often described as just that in textbooks. But in preparation for electron microscopy, muscle must be sliced into extremely thin sheets, and the resulting two-dimensional views may mislead us. A thin slice of a mass of cooked spaghetti would show a bunch of round and oval bodies, not long and tortuous cylinders. Working with thicker slices of tissue, the old microscopists made no mistake when they picked a name, mitochondria, that meant filamentous bodies.

A muscle fiber must contract when its animal wants it to, and all the sarcomeres in a series must simultaneously feel the urge to pull. That demands a triggering system, the equivalent of coil, distributor, and spark plugs in the ignition systems of the simpler automobiles of yore. Cells ordinarily carry an electrical charge across their surface membranes, negative on the inside and positive on the outside, in no fundamental way different from the poles of a battery. In this respect muscle fibers are quite ordinary cells. Simplifying shamelessly, we might say that the arrival of a nerve impulse at the surface of a muscle fiber causes a short circuit across its membrane. That eliminates the difference in charge across the membrane where the nerve makes contact. That chargeless region then spreads rapidly across the surface of the fiber. An impulse travels down a nerve by just such propagation of a short circuit.

Muscle, though, takes the process a step fancier, as in Figure 2.7. The surface has deep indentations that run like pipes far into the interior of the fiber, so no part of the fiber is at all far from some bit of its external membrane. (Perhaps we shouldn't say external membrane since it protrudes well inside!) These pipes extend inward opposite the Z-discs in many muscles, elsewhere in others. Knowledge of the existence of the pipes (the T-system, for transverse tubules) preceded recognition that they represented extensions of the outer membrane. Proof of the connec-

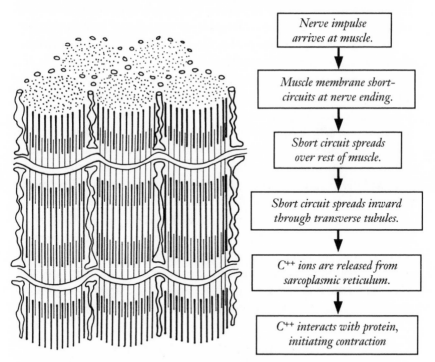

Nerve impulse
arrives at muscle.

Muscle membrane short-
circuits at nerve ending.

Short circuit spreads
over rest of muscle.

Short circuit spreads inward
through transverse tubules.

C^{++} ions are released from
sarcoplasmic reticulum.

C^{++} interacts with protein,
initiating contraction

FIGURE 2.7. Triggering muscle contraction. Notice, on the left, the way the T-system penetrates across the muscle between the myofibrils, and note also the sarcoplasmic reticulum that runs lengthwise between elements of the T-system. On the right is the sequence of events involved from arrival of nerve impulse to contraction.

tion rests on observations that large (comparatively!) particles that can't penetrate cell membranes nonetheless find their way into the system.

Dropping the electrical charge across the surface (wherever that might be) doesn't in itself stimulate contraction. Instead it affects another set of membranes entirely within the fiber, the so-called sarcoplasmic reticulum. When the charge drops, this system of irregularly shaped bodies releases calcium ions (atoms of calcium carrying positive charges), which diffuse into the sarcomeres (finally back to sarcomeres!). The arrival of calcium makes one of the auxiliary proteins change shape and initiate actual contraction. That the final, diffusive link involves an atom rather than even a molecule of modest size carries significance. Diffusion speed depends on size; smaller atoms and molecules diffuse faster. Distances around the fibrils may be small, but the contraction of all sarcomeres must approach perfect simultaneity. A larger molecule, to do the job, would have to be released in still more intimate contact with the fibrils.

(A parenthetical note on names. We name the parts of the machines we make according to their functions, combining logic and convenience. But what have we here? Sarcomeres, mitochondria, sarcoplasmic reticulum, T-system, even thick and thin filaments—no term carries any immediate implication of function. For that matter, the mitochondria of muscles were called sarcosomes, meaning "bodies within muscle," until their identification in the late 1940s with what were already known as mitochondria in other tissues. We glimpse a peculiarity of both living systems and the history of science. Daunting complexity at every level of size and organization faces the biologist, and description of structure almost always precedes analysis of function. So we name the parts on the bases of appearance rather than function, and the names stick. Sometimes we get into a strange situation in which we know how a system works and what parts it contains, yet we can't put the two together. How cilia beat, how cells divide, and how fish swim provide decent—or embarrassing—examples. The terminology for each memorializes our mystification.)

Putting the Pieces Together

So far everything has turned on small units—fibers, myofibrils, sarcomeres. How can we reassemble the components into a functional muscle? Or do we deceive ourselves when we identify a muscle, dissected out and named by an anatomist, as the functional unit? Yes, a muscle ordinarily works as a unit, but a smaller piece of the system might be more basic. Viewed functionally, a muscle represents a confederation of what have come to be called (perhaps because anatomists didn't recognize them) motor units.

A gastrocnemius muscle of one of us contains, as noted earlier, about a million muscle fibers. But we don't send a million nerve cells into each gastrocnemius, only a little over five hundred. A typical nerve cell, arriving at the muscle, thus connects up with almost two thousand fibers, which must therefore operate as a group, as a motor unit. To put it another way, the gastrocnemius contains over five hundred motor units. That's not atypical, with motor units containing from a few hundred to a few thousand fibers in vertebrates such as us. (Crustaceans and insects, among other nonvertebrates, use a different arrangement for triggering even if their muscles look similar to ours, so this discussion doesn't apply.) That the anatomists missed these functional elements should raise no eyebrows (an action for which we use muscles with especially large numbers of motor units). The muscle fibers innervated by branches of a given nerve cell form no anatomically recognizable bundle but instead are distributed through much (or all) of the muscle as a whole.

In short, the nerve noticed by the anatomist that runs to the gastrocnemius muscle comprises a bundle of five hundred nerve cells (neurons) running parallel to one another. An impulse arriving at one of its motor units from a nerve cell triggers a brief contraction of the fibers of that unit; the unit gives a twitch of a standard strength. We vary the strength of contraction of a muscle in two ways: by adjusting the rate at which impulses arrive and by changing the number of motor units active at any time. We reduce the overall twitchiness of the contraction by making the muscle's different motor units twitch at different times. We have other tricks. Even within a single muscle, motor units vary in size (mainly in number of fibers) ten- or a hundredfold. For minimally twitchy fine movements against low loads, we use the smallest motor units. For sustained and forceful output, we send impulses to our motor units frequently enough—forty or fifty times each second—so the muscle fibers don't have time to get stretched out again between triggerings. Thus their twitches merge, and each pulls steadily. For severe loads, motor units still do synchronize somewhat, producing the slight tremor you may notice when you're trying to lift or hold something that's about as heavy as you can manage.

Finally, what about cells? We're taught that cells are the basic biological building block, the basic unit of structure and function, the minimal "living" unit, and so forth. Where do they come into this account? The issue can't be dismissed as trivial. Even in tissues where cells don't divide from time to time, cells still do cellular things. In particular, they periodically renew all their internal proteins, and that demands the full machinery for protein synthesis, starting with the recipe in some DNA in some nucleus. In mammals, only the red blood cells lack that full machinery, and we discard them after they have given a few months of service.[8]

Cells, albeit pretty strange ones, do make up muscle. A muscle fiber equals a cell, with a proper cell membrane on the outside (sarcolemma) and with proper cytoplasm in the inside (sarcoplasm). But it differs dramatically in size from the cellular norm. Most of our cells are about a hundredth of a millimeter across in any direction. Nerve cells provide the textbook exceptions; one from your spinal cord to your big toe extends several feet. But nerve cells are skinny, so overall, they're not unusually massive. By contrast, muscle fibers measure between a hundredth and a tenth of a millimeter across, ten to a hundred times thicker than the long part of a nerve cell. At the same time, they extend almost as far as nerve cells, up to a few feet in a big animal. We have in fact no cells that are more massive than muscle cells. For so much cell, one nucleus apparently won't do, and muscle cells most often have many of them. In our muscles they're arranged around the periphery of the fiber, along with the other bits of cellular machinery needed to replace or enlarge things. That's not the only way to do the job; many insect muscles, for instance, have a core of nuclei running down their middles. Either way, muscle cells don't do much dividing when a muscle enlarges, perhaps at the instigation of some fitness program. Instead the cells enlarge by making more of all the proteins involved in contraction.

And Making It Perform

In the last chapter we put a few numbers on what muscles could do, in particular giving maxima for force and power. We add two provisos: first, that the muscles that produce the most force don't give the most power, and second, that for a given muscle, force and power hit their peaks under different conditions. You can't maximize both simultaneously. And we noted a few pages ago that a muscle produces its greatest force when operating near its normal, resting length. We need a few more items to link the mechanism of contraction with the way we use muscle as engine.

Early in the game (at least back as far as the early 1920s) muscle physiologists realized that a muscle produces its greatest force when it isn't changing length at all—that is, when the load is too great to be moved at all. Under the opposite condition, an insignificant load, a muscle hits its top speed of shortening. That's about how an old-fashioned steam engine operates, but it's unlike the performance of ordinary automobile engines or electric motors. An automobile engine produces no force when not turning, and it needs a second engine to start it. Electric motors manage to start themselves, but they often use some special parts to do so. Getting peak force at zero speed turns out to be a general feature of muscles, with only the specific numbers for maximum speed and maximum force differing from muscle to muscle. That kind of performance should surprise no one since that's how an intact animal works. You can pull or lift with the greatest force if you don't have to move or raise the load, and you can move (not just accelerate) a lighter load faster than a heavier one.

Figure 2.8 gives a typical curve for force (or load) against speed. This one comes not from an isolated muscle but from a person (the physiolo-

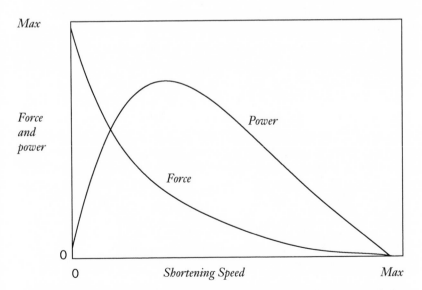

FIGURE 2.8. How the force and power a muscle can produce vary with the speed at which it shortens.

gist Douglas Wilkie, I think) bending an elbow and flexing a forearm against different loads, picked to emphasize the direct link between muscle physiology and animal performance.[9] Such curves have nice, tidy extreme values for force and speed; the mathematically savvy reader should note that they're only superficially hyperbolic.

Now force times speed gives power. So the same data can tell us how much power a particular muscle (or this particular forearm) might put out. Of equal interest, it also tells us the combination of load and speed that gives that maximum power. The power curve, as in the figure, has a peak, since neither zero speed nor zero load yields any power output at all. That peak at intermediate force and speed turns out to be general as well, if once again the particular numbers differ muscle to muscle and animal to animal.

A Final Physiological Foible and Fable

Let's return to that meeting of the Royal Society in London early in the 1950s described by A. V. Hill, the one in which two bicycles were hitched together head to head with a chain connecting their cranks and pedals. When the rider of the first pedaled normally, the feet of the rider of the second were driven backward, as in Figure 2.9. The second rider then attempted to prevent that backward motion. The two riders thus

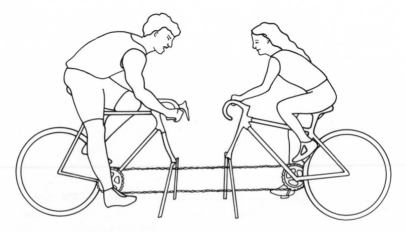

FIGURE 2.9. The arrangement of bicycles and people for the comparison of negative and positive work.

used the same muscles, the first trying to make them shorten more rapidly, the second trying to keep them from lengthening as fast. So the muscles of the second rider developed their tension while extending rather than while shortening. In other words, they worked as brakes, trying to stop some motion rather than instigate it.[10]

What happened? The second rider had no trouble driving the first rider to exhaustion, or, perhaps more accurately, making the first rider drive himself to exhaustion. Working in the laboratory rather than at a soiree, the first rider used 3.7 times as much oxygen as the second. Something distinctly odd was going on, something that demands explanation and that says a lot about what muscle can do.

First, the physical picture. Work is defined as the force applied times the distance something moves *in the direction of the force*. The first rider clearly did work, pushing forward on the pedals, which resisted but ultimately moved in the direction they were pushed. But the second rider pushed against the pedals one way while they persisted in moving the other way. Taking the definition of work at face value implies that the second rider must have done negative work! While that sounds contradictory, it merely states that the muscles of the second rider absorbed rather than produced work. Negative work, then, does no violence to our basic physics; work simply goes into the machine rather than comes out.

We can do the same thing with a pair of electric motors. If we drive the second motor with the first, the second becomes a generator. I once put the phenomenon to use to monitor the speed of a turning shaft; I just connected a tiny motor to the shaft and ran the wires from the motor to a voltmeter. Still, the second rider's muscles didn't quite match the generator, since the work done on them appeared as heat, the least useful form of energy, rather than as electricity or new chemical bonds. Muscle action doesn't reverse in such an accommodating manner. But then neither does an automobile engine. We can use the motor as a brake when going downhill or slowing up, at least in cars with manual transmissions, but no gas flows back into the tank.

The real peculiarity of muscle turns on the relative strength and efficiency of negative work. Consider three ways of using a muscle. It might develop tension as it shortens. Or it might develop tension against an immovable load, one against which it isn't strong enough to shorten. The third and oddest possibility consists of its developing tension while it gets stretched by the load.[11] That's what happened to the second rider in the demonstration. We noted earlier that muscle can develop greater tension when not actually shortening than when it does. What we note now is that muscle develops even greater tension when it tries to shorten during a stretch. Moreover, while the energy absorbed (the negative work) does little that's of much use, at least it gets absorbed with little metabolic cost. In that sense, negative work is more negatively efficient than positive work is positively efficient. The rider who was braking not only exerted more force but consumed less oxygen in the process; the first rider was left gasping.

All this sounds as if we ought to rearrange our muscle-powered machines to use lengthening rather than shortening "contractions" of muscle. Unfortunately that won't do. The efficiencies may be higher in absolute terms, but negative remains negative, and work goes in rather than comes out. You can walk uphill with an efficiency of 25 percent; you gain a quarter as much gravitational potential energy as the metabolic energy you put in. You walk downhill (in one experiment) with an efficiency of -118 percent. The number may impress, but it just says that the amount of energy you spend to absorb energy is a little less (by 18 percent) than the energy you absorb. Absorbing energy still extracts a cost rather than pays a dividend.[12] Brakes on bicycles or automobiles do far better; before power brakes became standard on cars, even the least of us could abruptly stop a car weighing between one and two tons without raising a sweat. Systems such as electrical generators or regenerative brakes that can recycle the energy do better still.

One might guess that the whole business of negative work remains an academic exercise—down to the slight pun of that last word. No, it matters with every step you take. When either walking or running, you limit the forward swing of your legs by contracting your gluteal (buttock) muscles while those muscles are lengthening. Going uphill, you do almost exclusively positive work; going downhill, you do almost entirely negative work. On the level, you do both. You use your biceps not only to lift weights but also to lower them. We're so accustomed to it that we rarely think about the fact that lowering is easier than raising. But consider: if both raising and lowering happen at steady rates, exactly the same force must be exerted. Yet lowering, with negative work, feels easier and, as we've seen, costs less in an energetic or physiological sense. These are just two examples of something ubiquitous. Almost every time one of your muscles shortens, some antagonistic muscle lengthens while actively exerting tension. The long and short of it is that you're forever doing negative work with your muscles.

Nor does concern end here. A special hazard attends making a muscle contract while it lengthens and thereby having it develop greater force with less expenditure of energy. Such contractions, especially if repeated, can cause a lot of damage to muscle fibers. Worse, the damage—inflammation and degeneration—doesn't appear (or make you feel bad) for about a day. "Delayed onset muscle soreness" (DOMS) labels it but says even less than do the preceding sentences.[13] It intrudes most noticeably when a sufferer does anything that again asks the same muscles to develop tension while lengthening. I once descended from a small mountain on an especially steep trail with a somewhat overweight companion; both of us carried packs of perhaps thirty pounds. The descent seemed easy, but it was evident the next day (and for some days thereafter) that my companion had badly injured her muscles; while most activities caused no trouble, descending a flight of stairs (now without a pack) involved a remarkable level of pain. Fortunately, training can minimize the problem, and the afflicted muscles do heal completely.

CHAPTER 3

And How We Found Out

Theories of muscle contraction are available on the market by the dozen. The author himself is responsible for a few of them. Most of these theories were fatally hurt in the impact between physics, chemistry, physiology, and electron microscopy, fitting only the requirements of the science of their author but being incompatible with that of others. The situation with muscle is at present similar to that of the holy elephant, which had ninety-nine names, the real one being the hundredth, known only to the elephant himself.[1]

ALBERT SZENT-GYÖRGYI, muscle biochemist

THAT'S HOW SZENT-GYÖRGYI, THE GOURMAND WHO PUT THE rabbit's psoas muscle on the map, described the state of the art in a 1957 book. Fortunately, things have improved since then. A pair of short papers published in 1954 completely changed the landscape, as was evident almost immediately after their appearance.[2] So when written, the statement must have been on the very cusp of anachronism. But we're getting ahead of ourselves.

Before Microscopes Helped Much

Often one hears that Western science began with Aristotle (384–322 B.C.E.). My own view, perhaps reflecting the bias of a biologist, is that Western science instead represents a reaction to Aristotle. When considering organisms, he got so little right that one suspects mere accident when he was on target. Unfortunately, Aristotle's writings were read with a dogmatic rigidity far beyond what he could reasonably have intended and then remained canonical for nearly two millennia. His *De motu animalium*—why a Greek work now takes a Latin title I do not know—touches on muscular action. To quote:

> And it [muscle] is obviously well disposed by nature to impart movement and supply strength. Now the functions of movement are pushing and pulling, so the tool of movement has to be capable of expanding and contracting. And this is just the nature of the *pneuma*. For it contracts and expands without constraint, and is able to pull and push for the same reasons; and it has weight by comparison with the fiery and lightness by comparison with its opposite. Whatever is going to impart motion without undergoing alteration must be of this kind. For the natural bodies overcome one another according to their predominance: the light is overcome and kept down by the heavier, and the heavy kept up by the lighter.[3]

Aristotle appears trapped by his conceptual framework, a paradigm into which he has to force everything. All is made of earth, fire, air, and water—hard, hot, windy, and wet. Air is the pneuma, "the breath of life."[4] Not that he's the only scientist or philosopher ever stuck with a poor paradigm, but the lack of contemporary competition left no one to rescue him from his cul-de-sac. For that matter, we're undoubtedly still mired in many such misguided paradigms, but one knows for sure only post hoc—when a better picture emerges. A look at our recent past should dispel any smugness. In the 1940s suggestions of how proteins could encode inheritance were shown wide of the mark by the demonstration that nucleic acid did the job instead. In the 1960s geology and paleontology had to shed some peculiar notions when continental drift became accepted. In the 1970s our earlier explanations of how flagella propelled bacteria took a fatal hit from the discovery that they involved a true rotary engine.

Beyond Aristotle's peculiar paradigm lie two other oddities: the pneumatic model and the notion that muscle must both expand and contract actively. The first flies in the face of muscle's obvious gaslessness. The second ignores the way muscles so often oppose the action of other muscles. It also squares poorly with the obvious lack of stiffness of non-pulling muscle; one can't get far by pushing on a rope. These disparities must reflect the absence of an experimental or even observational tradition. Szent-Györgyi referred to Aristotle as among the greatest prescientific thinkers, noting: "Aristotle said that a big stone falls faster than a small one. The interesting point about this statement is not that it was wrong, but that it never occurred to Aristotle that he could try an experiment to test his ideas."[5]

I find equally odd, and of greater general significance, the disparity between the steady technological progress of the Middle Ages—devices for farming, mining, building, ship construction, time measurement, warfare, and other endeavors—and the retarded development of science. The birth of science, even of sciences relevant to practical matters of mechanics and human health, awaited the Renaissance. For some reason, the experimental, trial and error tradition of technology provided no procedural model. Perhaps science was a retarded child because its parent was philosophy rather than engineering, because, we might say, it put Aristotle above Archimedes.

Two people who lived a century apart figured largest in breaking through the old traditions of Aristotle and his intellectual successor the Roman-era anatomist Galen (129–199 C.E.). The first, Andreas Vesalius (1514–1564), ranks first among anatomists, ever. Flemish by birth (even now a Belgian university bears his name), he had a peripatetic career, studying or working at Louvain, Paris, Padua, Bologna, Venice, Mainz, Brussels, and Madrid and dying of illness while returning from a pilgrimage to Jerusalem. In 1543, at the age of twenty-eight, he published (in Basel, Switzerland) his main work (I am tempted to say "corpus"), a seven-volume compendium on human anatomy, from which Figure 3.1 has been taken. This enormous body of text and figures (the latter from the studio of the great Renaissance painter Titian) broke both with the anatomy of Galen, which had been derived from nonhuman mammals, and with notional stuff like pneuma. Not only did Vesalius get the pieces in their proper places (finally!), but his view of what muscle does needs no revision after nearly half a millennium.

Nor do I with Plato and Aristotle (who did not at all understand the nature of muscle) attribute to the flesh so slight a duty as to serve, after the fashion of fat or grease or some sort of clothing, the purpose of lessening the effects of heat in summer and of cold in winter. On the contrary, I am persuaded that the flesh of muscles,

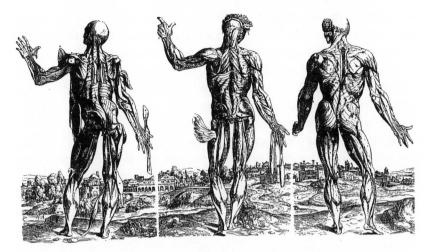

FIGURE 3.1. Several of the illustrations from Vesalius's great human anatomy. Viewed together, the background of the drawings shows an area a short distance southwest of Padua.

which is different from everything else in the whole body, is the chief agent, by aid of which (the nerves, the messengers of the animal spirits not being wanting) the muscle becomes thicker, shortens and gathers itself together, and so draws to itself and moves the part to which it is attached, and by the help of which it again relaxes and extends, and so lets go again the part which it had so drawn.[6]

He had the right idea. Muscle is the active engine, shortening is active while reextension is passive, and shortening happens when nerves so direct. But for a little backsliding when Fabricius (1537–1619, William Harvey's teacher) attributed active contraction to tendon, not muscle, Vesalius's view still stands.

The second great figure, although mentioned in all the standard sources, remains far less famous. The great book of Vesalius has been translated, reprinted, and continuously valued. Most of the great book of Giovanni Alfonso Borelli (1608–1679), published posthumously in 1680 and 1681, didn't get translated from Latin into English until 1989.[7] The part that did, the section on bird flight, may be his weakest; one couldn't possibly get off the ground decades before Jacob Bernoulli established the relationship between flow speed and pressure. Not that the book, titled, like Aristotle's, *De motu animalium*, disappeared or remains all that rare and valuable—my university, for instance, owns two different editions—but Borelli hasn't become a household name except among those of us who call ourselves biomechanics.[8] Perhaps physiology dates more quickly than gross anatomy. One can still use Vesalius as a dissection guide; using Borelli as a textbook would be like using Caesar's *Gallic Wars* as a travel guide for modern France.

Since the great *Principia* didn't appear until 1687, Borelli didn't have Newton at his side to straighten out the interrelationships of force, speed, acceleration, and work. But he got a lot right. He understood statics and applied the principles effectively. He was sound on pulls and pushes, both individually and in combinations, and on the action of levers. Consequently he worked out the integrated actions of the muscles and bones that Vesalius and others had by that point described, and in the process he straightened out all kinds of misconceptions about how both humans and other animals comport themselves. He explained how muscles could produce much greater forces than we exert with our intact limbs, how the ways they were hooked to bones traded their strong, short movements for less strong, longer movements. Most of his contemporaries viewed that trade-off of force for distance as something paradoxical since it does the opposite of what most hand tools accomplish; he showed it to be just ordinary leverage in reverse. He showed how the force of a muscle depended on its thickness, its cross-sectional area. He also explained the way muscles with oblique fibers (pennate muscles) generated even greater force, if at the cost of giving still shorter movements. He drew on nonmammalian animals, from birds and fish to insects and earthworms, to make general arguments about muscle, support, and locomotion. Figure 3.2, a composite of figures from *De motu*, gives some sense of how he made his points.

But if he understood what muscle did, he was far off base on how it did it. Contraction, he correctly asserted, didn't depend on forcible injection of blood, although he had nearly the right idea on how a penis used blood pressure to become erect.[9] He correctly associated hardening and contraction of a muscle. But he incorrectly guessed that contraction happened when blood brought in one substance, nerves supplied another, and the mixing of the two, acid and alkali, produced an effervescence (sometimes "ebullition" to the translator) analogous to that of fer-

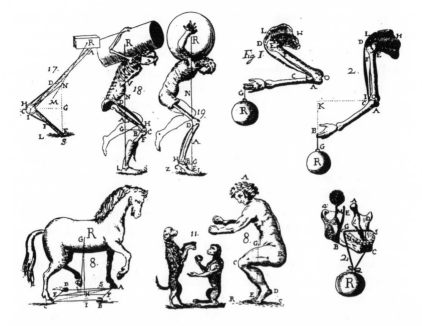

FIGURE 3.2. A composite of drawings from the jumbled-looking plates of Borelli's *De motu animalium*.

mentation. The resulting swelling in girth, in Borelli's view, caused the decrease in length.

Worse than this forgivable error, considering the science of the time, was his assertion that an increase in total volume accompanied muscle shortening, admittedly a logical consequence of his notion of effervescence. Not that such a system is impossible. If, for instance, we attach a set of nonstretchy cords lengthwise along an expandable balloon, further inflation of the balloon will shorten it, essentially what Borelli argued. One wrong guess, effervescence, instigates a sequence of tortured rationalizations. That's something all too common in the history of science. We've had our porous ventricular septum of the heart, our phlogiston coming off in combustion, and our luminiferous ether pervading the cosmos, to spread blame among biology, chemistry, and physics. We're all too prone to shoehorn ostensibly contradictory evidence into an established model.

Even given what was known, his claim that muscle expands in volume when it contracts in length comes as a surprise. Muscle contains no obvious gas after contracting, and a person doesn't float higher in the water when tensing muscles. Still, maybe water moves into muscle from blood or lymph during contraction—not exactly Borelli's idea, but something ruled out by neither of these arguments. Also, a mechanical model made of diagonally crossed elements of constant length, as those Borelli used in his diagrams (what we'd call a pantograph), will change volume as its length and width interchange.

Reflecting the interest of the issue, a lovely demonstration of muscle's constant volume was done in about 1663—antedating the publication (but probably not the writing) of Borelli's book by seventeen years—by a Dutch investigator, Jan Swammerdam (1637–1680). Swammerdam, better known for extensive work on insect anatomy, managed to stimulate a frog gastrocnemius while it was completely contained in a vessel filled with water. From the top of the vessel a capillary tube protruded, so even a tiny increase in volume would have moved the water level upward in the capillary. But the level didn't rise when the muscle contracted, and it may even have dropped ever so slightly. While the experiment wasn't properly published for about sixty years, the result seems to have been known to some of Swammerdam's contemporaries.[10]

Borelli's century had seen William Harvey's work on circulation (1628), Marcello Malpighi's discovery of capillaries (1661), the introduction of microscopy by Robert Hooke in his *Micrographia* (1665), and the use of microscopes to look at living organisms by Antonie van Leeuwenhoek (from about 1675). Seventeenth-century life scientists performed experiments and reasoned quantitatively, following in Harvey's footsteps.[11] Borelli was just such a quantitative experimentalist and a lot more besides. He was a late-Renaissance Renaissance man, a mathematician and astronomer who, among other things, first suggested that comets travel along parabolic paths, no small accomplishment without the explicit aid of Newton's laws. Harvey and Malpighi retained the concerns of physicians; Borelli treated physiology and animal mechanics (iatromechanics the area was sometimes called) as an integral part of science rather than as a branch of medicine.

After the Microscope Came into the Picture

Malpighi, Hooke, Leeuwenhoek, and Swammerdam made good use of microscopes, and Borelli himself had a look at muscle through one.

Still, seventeenth-century microscopy contributed next to nothing to the study of muscle. While the instruments were crude and clumsy, the real trouble lay elsewhere. A light microscope can be a frustrating device—as I found out when using an old but competent one that belonged to my father—for looking at anything not intrinsically tiny or thin. Light must pass up through the object on its way to the eye of the viewer. If the object isn't extremely thin, one sees only a silhouette. A brighter light reveals merely the blurs of its scattering rays. Fortunately, many tiny organisms are sufficiently thin, as are blood cells, the membranes of lungs, and the webbing between a frog's toes. But a muscle thin enough to transmit light isn't a muscle that's capable of doing much. One might look at muscles from very tiny animals or tease apart fibers from loosely bound muscles (like that psoas or the flight muscles of insects), but the exposed fibers look like little more than long, thin fibers. Maybe you'd learn that the fibers run lengthwise, but every butcher and cook could tell you as much.

Biologically informative microscopy on a large scale (if I can so term it) begins in the nineteenth century, not the seventeenth—with no offense to the pioneers just mentioned. Yes, the instruments became better; indeed by the end of the nineteenth century they had approached the theoretical resolution limit set by the wavelength of visible light. But even more important was the evolution of sophisticated preparative techniques. Microscopists learned to preserve material so microscopic structure, not just gross form, didn't deteriorate. They learned to slice material into sheets sufficiently thin that enough light would go through without too much scatter. This required them to learn how to invest preserved tissue with materials, such as wax, that wouldn't just collapse under the knife. Slices about a hundredth of a millimeter thick could be glued to glass and the wax removed; with paper that thin, an inch-thick book would have had about five thousand pages. Such thin slices, though, turned out to be nearly invisible under the microscope. So a large armamentarium of intense stains came into use, some from natural sources and some from the rapidly developing technology of artificial dyes for fabrics and coatings. Random splotches of bright colors came to mark the lab coats of histologists, as the fix-embed-slice-stain people came to be known.

A complex and sophisticated analytical technology thus joined microscopy's underlying optical technology. Stains not only differed in color but attached themselves to different structures within the slices and thereby provided information about the chemistry of those structures. Solvents could be applied to the glued slices, the subsequent examination of which would tell what components had disappeared. The overall effects of such changes could be evaluated if one used the same solvents to extract chemicals from larger pieces of the same material. By such indirect approaches, information could be gleaned about structural elements at and even below the limit of resolution of the microscope.

One technique, used by mid-century, was polarization microscopy. If light passes through a polarizing filter (made of a so-called Nicol prism at that time) oriented one way and then through another filter oriented at right angles to the first, the light is blocked, and the field appears dark. If a bit of material between the two filters changes the polarization of the light that passed through the first filter, some of that light can pass through the second, and the observer will see a bright image of the material against a dark background. In a polarizing microscope the first filter lies below, and the second above, the specimen on its stage. For biological material, a change of polarization usually indicates parallel sheets or bundles of fibers, all running in the same direction and individually thinner than the microscope can resolve. Bundles of parallel fibers of course bring us back to muscle.

Leeuwenhoek had noticed that muscles had crosswise bands, what we now recognize as the Z lines or Z-discs (Z for *zwischen*, meaning "between") that separate the sequential sarcomeres. The polarization microscope revealed much more structure. Several people, most notably Ernst Wilhelm von Brücke, in 1857, saw, midway between Z lines, a broad band that shone brightly, which they correctly interpreted as a bundle of longitudinally aligned rodlets (Figure 3.3). This region, now called the A band (for "anisotropic"), behaved—or failed to behave—in an interesting manner when muscle was stretched or stimulated to shorten. It didn't get longer, nor did it get brighter, meaning that these suspected rodlets weren't themselves stretching or contracting. It also didn't get dimmer, implying that the rods weren't folding. What changed were the relatively dark areas between A bands and Z lines, now called the I bands (for "isotropic"), which broadened when the muscle was stretched and narrowed almost to disappearance when it shortened.[12] In hindsight (also recall the last chapter), these investigators were observing the interdigitation of myosin and actin filaments, with

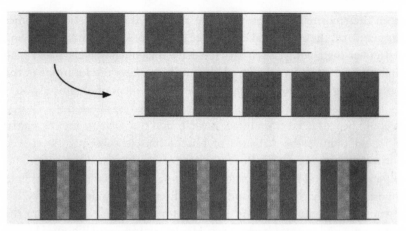

FIGURE 3.3. Top: How muscle, relaxed and contracted, looks when viewed between crossed polarizing filters, with the bright A bands shaded here. Bottom: How muscle looked in histology books early in the twentieth century.

myosin (making up the A band) affecting the passing light more than actin did.

Other pieces of the picture were already in place. Back in 1835, Theodor Schwann (1810–1882) had shown that a muscle exerted its greatest force when contracting at or near its resting length. That ruled out all manner of theories based on magnetlike attraction, which would have given maximum force at minimum length. Curiously, Schwann, best known for microscopical work, made his contribution to the mechanical rather than to the microscopic aspect of the muscle matter. Schwann, as you may dimly recall from a biology course, was, with Matthias Schleiden, the first to recognize the fundamental importance of cells. They argued not just that organisms were made of cells but that a cell formed a basic functional unit, a far more radical notion. Schwann had to make the greater leap. Schleiden's plant cells were much larger—typically ten times longer or a thousand times greater in volume—and were surrounded by easily visible wooden (cellulose) walls. Schwann's animal cells lack such walls; indeed the cell membrane separating cells is too thin to be resolved by a light microscope. What Schwann (and the rest of us who've peered down light microscopes) call the membrane may be just an artifact of preparation. Imperfect fixation pulls cells away from their normal contacts with one another. If such fixation becomes the standard, the artifact takes its place in textbook and course, achieving a peculiar metareality. Such artifacts have unavoidably bedeviled a field that looks at deposits of colored chemicals marking the locations of the parts of long-dead and elaborately treated thin slices of organisms.

Back to muscle. Myosin itself was recognized by about mid-century. It could be extracted from intact muscle, and the solvents that extracted it would remove the A bands of bits of muscle subsequently viewed between polarizing filters. But for some reason the discovery was gradually forgotten and had to be refound in 1953. Nor was the localization of myosin the only thing lost and then rediscovered. Andrew Huxley has written a fascinating account of the history of muscle research in the nineteenth and twentieth centuries. Huxley, grandson of Thomas Henry Huxley, Darwin's great defender, was one of the two people who in 1953 proposed the sliding filament model. He wonders (at some length) why still other critical parts of the muscle story either failed to be appreciated or subsequently passed from active consideration. Part of the problem must have been the difficulty, just mentioned, of telling what was real from what was just an accident of specimen preparation, pieces of microscopy's haunting legacy of artifacts. Part must have come from the sheer number of little facts that must fit into a single, coherent picture. How can one know which should carry weight and which matter little? Part might just be some accident of history, the humors of her muse.

Along with Andrew Huxley, though, we're still surprised. With so many theories proposed, someone should have guessed correctly! Perhaps we (and Huxley especially) ought to be glad that no one did make the right guess. What could be more frustrating than to make a monumental discovery and then have to yield pride of primacy to some predecessor who happened to have made a lucky guess? On more than one occasion we've chosen a correct explanation for what in retrospect was an indefensible reason or on inadequate evidence.

More People and Pieces

While the first part of the twentieth century saw lots of progress on a variety of fronts, a coherent overall picture still resisted focus. One can read into this history any of a variety of mutually incompatible lessons. Steady onward and upward progress can be argued if one picks, with the wisdom of hindsight, the right experiments. Steady, if not consistently upward, progress can be illustrated by picking others. Still other scenarios are defensible; the historian of science can grind almost any ax. Erroneous theoretical notions may have distorted interpretations of experiments.[13] Critical microanatomical information may have been increasingly disregarded as the field passed into the domains of physiologists and biochemists.[14] Synthesis may have become harder as the volume and diversity of information led to intellectual fragmentation. I wonder about this last, at least prospectively: whether a field can reach a point where too much investment actually impedes conceptual progress, whatever its yield of data. A friend who works on the relationship between the microstructure and the operation of muscle tells me that he can keep up with only a small subset of new papers on the subject, so active is the area at present.

Early in the twentieth century we became aware of an important way in which muscles resembled nerves. Individual fibers could either contract or not, just the way a nerve either conducted or did not conduct an impulse. Frequency of contraction, like impulse frequency, could be controlled, but the strength of a contraction depended solely on the load faced by a muscle fiber or motor unit. An animal had to increase the output of its muscle in other ways. It could bring more motor units into play, or it could make the fibers twitch more frequently, or it could make them pull a little harder by sending nerve impulses at rates high enough so fibers couldn't reextend between twitches.

The power source for muscle contraction gradually revealed itself, although not without lots of wrong guesses and other difficulties. Using a muscle led to the production of lactic acid, something that one feels after a brief burst of activity but that disappears (both the lactic acid and the crampy feeling) after a bit of hard breathing. Was lactic acid critical to contraction? By 1930 we knew it wasn't, since poisoning the lactic acid producer didn't stop contraction, at least immediately. About the same time the liberation of phosphate during contraction became established, along with the discovery of ATP and creatine phosphate in muscle. Muscle wasn't the only target of this work on metabolic pathways. The study of intermediary metabolism intended to reveal, and succeeded in showing, the general sequences of chemical reactions by which cells break down big molecules of starch, sugar, and fat to produce carbon dioxide, ATP, and such other things as alcohol and lactic acid. Muscle gave as much as it got, providing some of the best experimental material for these biochemical investigations. Muscles capable of especially high power output drew especial attention; the biochemists preferred insect flight muscle and pigeon breast muscle to the muscle physiologists' frog gastrocnemius or (because fibers could be teased apart and sarcomeres were long) insect leg muscle.

The principle of energy conservation dates from the mid-nineteenth century, and efforts to use it to reveal something about muscle contraction started almost as early. According to the principle, the energy liberated when a muscle contracts ought to appear as the sum of the work it does and the heat it produces. The twitch of a muscle against an immovable load liberates heat. No great surprise. Paradoxically, though, more heat comes out when muscle is allowed to shorten against a load and do mechanical work—recall the distinction made back in the first chapter between force and work—but (as Andrew Huxley notes) the phenomenon was forgotten and had to be rediscovered in the 1920s.

One person in particular focused on the way muscle liberated heat and on the relationship among heat production, work done, and contraction speed. That was a Briton, Archibald Vivian Hill (1886–1977), better known as A. V. Hill. He faced formidable technological problems. First, one doesn't ordinarily measure heat production itself but rather the resulting increase in temperature; conversion from temperature back to heat poses a few practical problems. Furthermore, the twitch of a muscle liberates very little heat, so the muscle warms only a tiny fraction of a degree. Finally, things happen fast, and Hill wanted to find out not only how much heat comes out but when during contraction the heat appears.

While ordinary thermometers are neither sensitive nor fast enough, an old device can be persuaded, with much coaxing and ingenuity, to give useful results. Many combinations of liquid or solid conductors can yield measurable voltages, as attested to by all of our batteries (good) and a lot of awkward corrosion of metals in seawater (bad). Some combinations turn out to be usefully temperature-sensitive. The usual way to capitalize on that behavior involves a pair of junctions between dissimilar metal wires, as in Figure 3.4. The back to back pairing normally cancels out their junction voltage—unless the two junctions are exposed to different temperatures. So the voltage detected indicates the difference in temperature between the junctions, one of which is held at some

THERMOCOUPLE THERMOPILE

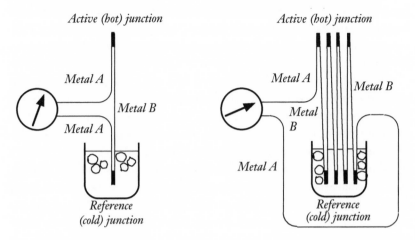

FIGURE 3.4. The paired junctions of a thermocouple and the more sensitive multiple junctions of a thermopile.

known reference temperature. The scheme is simple, and junctions can be made so tiny they heat in a fraction of a second. More sensitivity can be had by using a series of such junctions, exposing alternate ones (those with the same orientation of the two metals) to the heat source. The single device is called a thermocouple; the serial junction version is a thermopile.[15]

(Scatological digression. An ice-water bath provides a common temperature reference for thermocouples, although electronic substitutes for these baths can be purchased. According to legend, a graduate student once approached his sponsor for such an electronic fix so he could measure body temperatures of animals in the field, as was the custom in the old "noose-'em-and-goose-'em" school of physiology. The sponsor told him where he could shove his reference junction, which he did with, in the end, excellent results.)

Hill and his collaborators found that the heat produced during contraction didn't at all depend on whether oxygen was available; oxygen must serve an indirect and long-term role rather than form an immediate part of the contractile process. In nice congruence, they found that heat continues to emerge for minutes after contraction, marking the chemical transformations that recharge the batteries. That rationalizes our sense that breathing matters only after the first few seconds of a run, that it's irrelevant to the performance of a good runner in a hundred-yard dash. It also squares with our sense of getting warm after vigorous exercise—although other mechanisms contribute as much to the latter.

Hill's laboratory also rediscovered and put into good quantitative terms the earlier observation that a muscle that is allowed to shorten liberates more heat than one trying to move an immovable load. The amount of extra heat depends on the amount of shortening and on the load against which the muscle shortens, but it doesn't depend on the speed of shortening. So doing work costs more than just pulling, and viscous resistance to internal motion, long presumed important, matters little. Any acceptable theory about how muscle works had to be consistent with these and their other results, done with great care over many years.

Albert Szent-Györgyi and his psoas muscle enter at this point in the story. During the late 1930s and the 1940s he (and his coworkers) focused their attention on the proteins in muscle rather than on the energy-supplying metabolism, the microscopic structure, or the mechanics and thermal aspects of contraction. A good test for a chemist like Szent-Györgyi is whether he can take something apart, identify the parts, and then put it back together and have it function. Chemists thus speak of the definitive synthesis of a complex molecule, meaning a completely unambiguous synthesis that gives a substance with the same properties as the unknown original. Szent-Györgyi came remarkably close to accomplishing just that with the contractile machine. He first showed that a muscle that had been soaked in glycerin at very low temperature lost (by dissolution) almost everything except its actin and myosin. Yet what was left of the muscle would contract when subsequently bathed in a solution containing ATP. That squeezed out a whole bunch of contraction theories.

ATP had other interesting effects. When it was added to a solution of myosin, the viscosity dropped; the solution flowed more readily. When it was added to threads of extracted myosin, they softened and stretched more easily. Most exciting of all, under the right conditions actin and myosin could be combined and extruded as a filament; if ATP was added, the filament then shortened. (When I was a teaching assistant in undergraduate laboratories, in the early 1960s, we used to do this as an exercise.) ATP was specifically required to do all this magic, and ADP was released in most of the manipulations. No doubt at all, this test tube and beaker work sounded prescient.

Breakthrough

By 1950 muscle contraction was receiving the attention of some people of truly stellar scientific credentials, including two past and one future Nobel laureates. Hill had received (with Otto Meyerhof, of Germany) the Nobel Prize back in 1922 for his work on muscle, and time had not dimmed his skill and perspicacity. Szent-Györgyi had gotten the prize in 1937 for discovering vitamin C, after which he switched to working exclusively on muscle. And Andrew Huxley entered the field. Huxley and Alan Hodgkin had figured out how nerves conduct impulses, certainly the most important discovery in neurobiology of the century, for which they (with John Eccles) were to receive the 1963 Nobel Prize.

A remarkable amount of information had been accumulated, even if as yet it had failed to fit any single model. Some powerful new techniques—in particular, X-ray diffraction and electron microscopy—were being brought into action. Retrospectively, one might claim that the time was ripe for a breakthrough, but one should be deeply suspicious of such whiggish rearward, or forward, glances. Since my graduate student days I've been hearing, for instance, that developmental biology stands poised at the very threshold of revolutionary synthesis. It may well be true, of course.

(For that matter, one ought to be suspicious of any so-called breakthrough, especially of claims contemporary with the event. While scientists rarely use the word, science journalists suffer from breakthrough addiction. Bear in mind both their mission—the selling of excitement—and the implied lack of temporal perspective of the "jour" in "journalism," its allusion to events of the day.)

Skepticism aside, our understanding of muscle contraction was transformed in the early 1950s; it took its greatest upward leap ever. I recall a splendid lecture in my undergraduate physiology course that began with the statement that both the textbook and last year's story now seemed confused and misguided, that this year we could at last enjoy a coherent and defensible account. Two people played the principal roles in the transformation. Both arrived at the same basic model (if using slightly different arguments), both were British but worked independently, and by wild coincidence, these two unrelated people, Hugh E. and Andrew F., shared the same surname, Huxley. History must indeed have a perverse sense of humor.

Simultaneous discovery happens from time to time in science. We take this as a sign of an open community in which data and ideas circulate freely, one in which from time to time the time gets ripe for some major advance or synthesis. Leibniz, in Germany, and Newton, in England, simultaneously invented the calculus, empowering physics but generating a bad interpersonal relationship. Alfred Russel Wallace and Charles Darwin came up with the basic mechanism of evolution, natural selection, at about the same time; by contrast, they interacted with grace and civility.

But simultaneous discovery by people with the same surname? Awkward, at the least, and in this instance several things compounded the awkwardness. As with Wallace and Darwin, the two Huxleys were at different stages of their careers and enjoyed different levels of social and scientific recognition: Hugh had just finished his doctoral work in 1952, by which time Andrew's name formed half of the great Hodgkin-Huxley mechanism for nerve conduction. Andrew was one of the great Huxleys of science; Hugh was the scion of a working-class family from the provinces. I've heard it suggested that their model would have earned a joint Nobel Prize had Andrew not shared one for his earlier work on nerves and had the people who decided on the awards been less reluctant to recognize a person twice. The final trick of fate: Andrew (A. F.) precedes Hugh (H. E.) in the alphabetized bibliography of every account. I do not believe, though, that either Hugh or his work has lacked appreciation by the scientific community. Still, we talk of the Huxley rather than the Huxley-Huxley model.

What they proposed of course was the sliding filament model described in the last chapter. In adjacent papers in a single issue of the journal *Nature*, in 1954, they presented arguments and evidence that muscle consisted of overlapping, interdigitating filaments; that contraction involved the sliding of one set of filaments relative to the other; and that no change in length of either kind of filament accompanied the process.[16] Three different techniques provided a satisfyingly consistent picture.

Hugh Huxley had applied a new microfocusing X-ray generator to a living muscle from the inner thigh of a frog. The X-ray diffraction images showed smudges indicating a hexagonal latticework of filaments. The same latticework was inferred from diffraction images from a contracted muscle, using a Szent-Györgyi type of glycerinated preparation to get rid of every-thing but the actin and myosin. But the intensities of the smudges differed, suggesting that while material was redistributed, the form was the same.[17] How X-ray diffraction yields such information resists easy explanation, but the underlying phenomenon of diffraction can be observed by shining a point source light (as from a slide projector or unfrosted light bulb) through round holes of decreasing diameters. Moderate-size holes (as done with a paper punch) produce sharp shadows, but light spreads out—dif-fracts—as it passes through tiny pinholes. While the technique wasn't new, applying it to material other than dry crystals involved novel problems. Incidentally, X-ray diffraction data obtained at almost the same place and time, by Rosalind Franklin and Maurice Wilkins, provided the basic evi-dence for the Watson-Crick model of the DNA double helix.

Andrew Huxley (with Rolf Niedergerke) used a technique called interference microscopy to show that when living muscle shortened, the width of the A band didn't change, something known a century earlier but forgotten. Interference microscopy is one of several techniques that permit examination of transparent material without the necessity of staining and thus usually killing it. Variation in light-transmitting char-acteristics other than color or optical density gets converted to color or light intensity differences that are visible to the microscopist. Hugh Huxley confirmed their finding with a slightly different technique.

The pictures from electron microscopy proved most persuasive—how we like direct images!—but involved the greatest risk of deceptive artifacts. An ordinary light microscope faces a resolution limit set by the wavelength of light, even if the presence of structures smaller than that limit (as with crossed polarizers) can sometimes be inferred. For visible light, the practical limit is about 0.1 micrometers, or a ten-thousandth of a millimeter. That limit can be reduced by using radiation of shorter wavelengths. For instance, electrons accelerated by a potential of sixty thousand volts have a wavelength a hundred thousand times shorter than that of visible light. So, beginning in the 1930s, microscopes using such electrons were built: no mean trick. Magnetic lenses have to replace optical ones, the beam of electrons must travel to and through the spec-imen in a vacuum, the specimen must be sliced at least ten times thinner than for a light microscope, and contrast must be gained by "staining" with such heavy metals as lead and osmium. Wax embedding no longer suffices; plastic of the right hardness must be infiltrated and polymerized through the specimen. Metal knives no longer work adequately, so sections have to be cut on cunningly designed slicing machines with freshly broken glass edges or diamond knives. Sections can no longer be viewed through glass, so not only must they endure high vacuum, but these ultrathin sheets must stretch between the filaments of a wire gridwork. One doesn't just pop a piece of muscle into the EM!

None of this was at all routine in 1953, but Hugh Huxley (with Jean Hanson and now at MIT) managed to make remarkable electron micrographs, with usable magnifications not of the order of a thousand (as in light microscopes) or ten thousand (as done by the few other people turning the electron microscope to biology) but of a hundred thousand. Slicing lengthwise, they saw the thick filaments known to be myosin in the middle of each sarcomere. They also saw the thin filaments known to be actin, projecting outward from the Z lines, running between the thick filaments, and nearly but not quite reaching the centers of the sarcomeres. Slicing crosswise, they saw the hexagonal latticework of filaments indicated by X-ray diffraction. Where the crosswise slice passed through the center of the sarcomere, they saw only thick filaments; where it passed through the rest of the A band, they saw both thick and thin filaments (as in Figure 3.5, newer and clearer than what they had to work

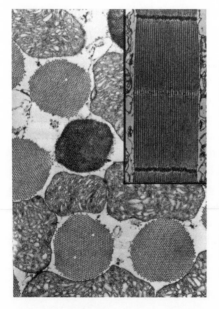

FIGURE 3.5. Electron micrographs of the flight muscle of a fruit fly—lengthwise (inset) and crosswise. In the crosswise section the myosin and actin filaments appear as a hexagonal array of thick and thin dots respectively. The section crosses a Z line in the middle, which appears darker than the filament cross sections. Between the myofibrils are of course mitochondria. In the flight muscles of insects, which contract by only a few percent, the area of overlap between thick and thin filaments is especially great.

with); where it was adjacent to the Z line, they saw only thin ones. An unchanging distance between Z line and H zone made it clear that the thin filaments didn't themselves shorten.

Another piece fell into place when Andrew Huxley, with several collaborators, showed how the force that a muscle could generate at different lengths—stretched, normal, and contracted—correlated nicely with the degree of overlap between thick myosin and thin actin filaments. Conversely, the speed of shortening for a minimally loaded muscle didn't depend at all on the degree of overlap. These and other observations, some old and some new, pointed to some interaction between thick and thin filaments as the place where energy got put in, where ATP was broken down into ADP.

That, together with even better images showing cross-bridges between the filaments and lots more analyses of the molecules involved, has led to the present picture of swinging heads protruding from myosin molecules. The heads extend forward, attach to the actin, swing backward, giving the crucial push, and then detach. Holding position (pulling against an immovable load) takes energy, since producing force requires that bridges still be made and unmade. But making filaments move across one another takes more bridge action and thus more energy; Hill's measurements fit nicely. The independent action of the heads explains the smooth action of even the smallest bit of muscle, in much the same way that multiple cylinders operating out of synchrony smooth the rotation of an automobile engine. Cross-bridges may operate in oarlike fashion, but the oars don't pull together like those of a racing crew.

Does everything now fit a seamless, satisfying model? Does it ever? The sliding filament model with force generators that ratchet along independently of one another has stood remarkably well, the first part for nearly half a century. In a sense, though, it just shifts the focus of our ignorance to another level. How cross-bridges work has commanded great attention in recent years. The nature of the myosin heads, how they interact with actin, how the ATP fits into the picture, how calcium starts the process—all remain cloudy. As uncertain is whether some specific breakthrough will be needed or these issues will be settled by incremental progress.[18] History affords examples of both.

The history of muscle work parallels that of other areas of biology (and perhaps science as a whole) in a curious way that's perhaps not politically correct to notice. In the sixteenth and seventeenth centuries

Italy, particularly northern Italy, provided the center of activity. Not that only natives participated; Vesalius, after all, came from Belgium, and people such as William Harvey were educated in Italy. The eighteenth century showed no such centralization, with most of Western Europe making contributions. German scientists dominated the nineteenth century but played a smaller role in the twentieth, even before the Second World War. In the twentieth century the British and Scandinavians contributed disproportionately. North America may now dominate many areas, but perhaps this simply reflects the uniquely vast size of our scientific establishment. I may have been born and trained in North America, but two of the four people who have most influenced my scientific outlook were of European birth and education.

One other historical note. Why was the mechanism of muscle contraction so difficult to unravel? I touched on one possibility earlier; it's worth a few words of reiteration now. Elucidating the mechanism has required a convergence of work on structure, on force and motion, on thermal behavior and energetics, and on biochemistry. Each of these areas had (and largely still has) separate techniques and even terminology. Worse, people working in these various areas differ in training, experience, habit, and maybe even how they use their brains. Some biologists—morphologists—care mainly about structure with little interest in exactly what the structure may accomplish for a creature. Structure may imply important things about ancestry, or it may aid in identifying and naming the creature, or its geometry or complexity may fascinate the morphologists. Other biologists—physiologists—care more about how the parts of an organism work and less about their appearance. Biochemists have still other concerns.

Tying genetics to the structure of DNA took a similar convergence, but the structural complexity of muscle dwarfs that of DNA. For DNA, the central but by no means sole empowerment was X-ray diffraction; for muscle, electron microscopy played a similar role but did so less completely. In each case the key elements were revealed almost as soon—but not at all easily—as the tools were put to use. Moreover, for both DNA and muscle, revelation of the structure provided the key to understanding how it worked. Both A. V. Hill and Albert Szent-Györgyi, great physiologist and biochemist respectively, were active at the time of the breakthrough and had been working on muscle for many years. But neither had in him much of the morphologist, and that may have proved the decisive limitation. I don't deny the importance of either of their contributions by pointing out that neither ended up playing much role in the final showdown. Indeed the most persuasive thing about the Huxley model may have been the way it squared with what both had found earlier.

Making Molecules Move Macroscopically

At this point we might put aside history and take a general look at this peculiar engine, a look from a slightly different perspective from that taken in the last chapter. Muscle is an enzymatic engine. The forces with which it can pull ought to be of the order of the forces that hold solids together, large forces, relative to the dimensions involved. But, as Andrew Huxley has reminded us,[19] those forces can't move things fast or far. So how does muscle manage them? Several tricks hold particular interest for the present story.

Cross-bridges, again, seem to work like oars. Consider a long oar. A smaller, slower, but forceful movement of the arms of the rower gets transformed into a larger, faster, but less forceful movement of the end of the oar. So by the way it protrudes, a cross-bridge can amplify speed and distance over those of ordinary enzymatic reactions.

More speed comes from the way sarcomeres line up end to end. Take a 10-inch-long muscle with an intrinsic speed of ten lengths per second—one that shortens by 10 percent of its length in a hundredth of a second or 100 inches per second. Both the whole muscle and each of its sarcomeres contract with that speed relative to length. The real speeds, though, measured as distance per time, differ dramatically. A human sarcomere is about 2½-thousandths of a millimeter long, the whole muscle 250 millimeters long. To put it another way, the muscle consists of a hundred thousand sarcomeres, end to end. (As well as a lot of sarcomeres across, but that's not important here.) So the muscle, shortening at 100 inches per second, contracts a hundred thousand times as fast as each of its sarcomeres; the latter move a thousandth of an inch per second. If each sarcomere of your arm muscles ran all the way from one tendon to the other, it would take you thirsty hours to raise mug from table to lips.

Still more speed and distance come from the way we hook muscles to bones, the subject of another chapter. Here let's just note that muscles attach so as to do the opposite of what we ordinarily refer to as leverage, as Borelli recognized. With a pair of pliers or a can opener you move your fingers a long way but with a small force, while the business end of the device moves less far and fast but more forcibly. By contrast, the biceps muscle of your upper arm contracts by only an inch or so, but it makes your hand swing through an arc of several feet. Stage by stage, a slow and short-range but powerful molecular force generator becomes transformed into a fast, large-scale mover.

CHAPTER 4

Flying High, Making Noise, and Clamming Up

Today I had about the duddest experiment imaginable. Some time ago William told me he had some "African frogs" and I could have them if I liked, so I thought I would try one at 0°C. Do you know the beasts? First you have to catch them—they are very smooth, slimy animals and dart about like fishes in the water. As most of them were rather large I wanted to catch about the smallest—that took a good quarter of an hour—and they are by no means easy to hold when you have caught them. . . . I write this to give you fair warning against these horrible beasts—they look it too: they have wicked little eyes on the top of a curiously thin flat head, and no doubt they spend most of their life at the bottom of the water. . . . I hope I shall never see another of them![1]

WILLIAM HARTREE, engineer, to A. V. Hill

THE CROSS-STRIATED MUSCLE USED FOR JUMPING, SWIMMING, walking, and so forth appears almost everywhere in the animal kingdom. Only a few kinds of creatures, such as sponges and single-celled animals, lack the engine. So nature probably invented it only once. Not that this basic engine hasn't been retuned for its different tasks; in this respect, its natural history parallels the technological history of the various engines of human ingenuity. Both lawn mower and automobile use four-cycle, gasoline-powered, spark-ignited, internal-combustion engines, not two cycles, or diesels, or steam engines, or turbines. On the one hand, the engines of lawn mower and automobile differ in details extending well beyond size and number of cylinders; on the other, they're still the same kind of motor. That generic similarity without specific identity links all the muscles we know.

We vertebrates, though, belong to a conservative lineage. If we have appendages, they're four in number. To make a wing or a hand, a leg inevitably is sacrificed. We use the same protein to make scales, horns, and hair. All of us have recognizably similar livers, kidneys, hearts, eyes, brains, and so on, not just functionally equivalent organs. Skeletal muscle is skeletal muscle. Our sarcomeres, for instance, come in a basic pattern and a basic size: about 2.5 micrometers long, four hundred to the millimeter, ten thousand to the inch. Nonvertebrates—insects, mollusks, worms—vary a lot more. A single creature, for example, may use sarcomeres of different lengths in different muscles. But even among vertebrates, indeed even in a single blue jay or bluefish, muscles get tuned to their tasks.

Why Color Matters

We tinker and tamper with just about every aspect of our muscles other than the length and arrangement of the sarcomeres. You know that vertebrate muscle is diverse stuff; it looks far from uniform as it sits behind glass or beneath plastic wrap in the food store. Cow muscle comes in red, veal and pork are pink, while skinned chickens are still paler pink. (The variation of skin color in chickens comes from the amount of stuff like marigolds that are fed to the birds.) Few stores sell it, but horsemeat looks even redder than beef, and whale steaks (occasionally available in the United States when I was young) are darker yet.

With the range of color runs a spectrum of functional properties. Two extremes define the range. Twitch fibers, the lightest in color, receive their stimulation from large nerve cells that conduct impulses rapidly. Stimulated, they give quick, sharp, maximally strong contractions—twitches—almost immediately after they've been triggered. While they can twitch many times each second, most of them can't maintain the performance for very many seconds; they fatigue. By contrast, thinner, slower nerve cells supply tonic fibers, darker-colored. These contract more slowly, and in response to a series of stimuli they contract farther and don't fully relax in between each stimulus. They resist fatigue much better. For frogs, in which the distinction was first recognized, twitch fibers do the jumping and tonic fibers handle postural matters. As a rule, twitch fibers can exert a lot more force—up to ten times as much—than can tonic fibers. But, again, most fatigue quickly.

Sometimes different muscle types occur in different parts of a body. That's the case in many fish. They cruise with tonic muscle, the redder meat located just beneath the skin or in against the backbone, halfway from the top (back) to the bottom (belly). The light muscle, which makes up most of the body of a fish, comes into play during quick starts and bursts of speed, when a fish tries to catch something or to avoid being caught. Not only do the two look different, but their textures are different, and they taste different. A fish that does little cruising, such as a flounder, has little other than light muscle. A real cruiser like a mackerel has a lot more dark meat, and even the tuna's light meat is relatively dark.

Birds make an analogous distinction. Nothing taxes a bird's muscles as much as flying, and the biggest are the breast muscles that move its wings downward to generate most of its lift. Sustained fliers, such as ducks and geese, have dark breast muscles with which they fly over mountains without fatigue. I was once involved in an experiment in which a gull flew for more than twelve hours in a wind tunnel with only a brief pause for a drink of water each hour. The end came when the gull simply refused to go on. Examination showed that it had used up its fuel supply completely; it no longer had any fat that we could find. Nothing about it suggested that it suffered our kind of short-term muscle fatigue. By contrast, domesticated chickens and turkeys, which are incapable of sustained flight, have breasts of light meat. For some reason duck and goose command higher prices than chicken and turkey even though more people seem to prefer light meat to dark. A glance at the color of the breast muscle will tell you whether you're being served squab (pigeon, a good flier) or "rock Cornish game hen" (some kind of bland midget chicken), irrespective of the menu's promise.

Sometimes individual muscles contain a mix of fiber or motor unit types, and the particular mix may vary from muscle to muscle. That's how we're arranged. Nor can all fibers be divided into two tidy categories. For instance, some twitch fibers turn out to resist fatigue quite well. Even within a single muscle, fibers vary greatly in the time elapsing from the beginning of tension to its peak, the contraction time. In a muscle just in front of your ankle, one with which you raise your big toe,[2] some fibers have contraction times of as little as a fiftieth of a second while others take over a tenth of a second. We use our soleus muscle, located in the calf between the gastrocnemius and the bone, when we maintain a standing posture, and we shift to the gastrocnemius when we walk or run. The soleus has an overall contraction time of an eighth of a second, and 80 percent of its fibers are dark, fatigue-resistant tonic ones. The muscles with which we move the eyeballs in their sockets as we shift our gaze have contraction times around three times as fast, and only 15 percent of their fibers are the tonic type.[3] Clearly nature values versatility and flexibility over ease of summarization!

Oh, yes, the color. Why is beefsteak red? If you think that the color comes from blood, you're just slightly wrong. The red stuff in blood isn't proper hemoglobin but myoglobin, sometimes called muscle hemoglobin. A myoglobin molecule amounts to a quarter of a hemoglobin molecule, or to put it the other way around and more formally, hemoglobin is a tetramer (quadruplet) of myoglobin. Hemoglobin, as you probably know, takes on oxygen in the lungs and carries the oxygen to the rest of the body. Oxygen diffuses out across the capillary walls and is used to break down carbohydrates and fats; the energy released in the process eventually appears in the form of ATP. Myoglobin's role lies in that "eventually." Myoglobin attracts oxygen more strongly than does hemoglobin, so given the opportunity, oxygen moves from hemoglobin in the blood to myoglobin in the muscles.

According to the classical picture, myoglobin maintains an oxygen reserve in muscle, in much the same way that creatine phosphate maintains a reserve energy store of bound phosphate to regenerate ATP. The reserve should help the system make a smooth transition from a high-power anaerobic dash to a sustainable level of effort. Muscles with lots of myoglobin are those with lots of tonic, fatigue-resistant motor units, so the picture makes good sense. It tells us why muscles in diving birds and mammals have more myoglobin and a darker color than the muscles of other vertebrates. Diving works an animal hard while depriving it of atmospheric oxygen. In air-breathing divers, such as whales and penguins, myoglobin concentrations can reach thirty times the levels of their nondiving relatives. As a result, they may store more oxygen in the myoglobin of their muscles than in all the hemoglobin of their blood.[4] That's despite their having blood unusually rich in hemoglobin.

Nonetheless, various forms of hemoglobin (including myoglobin) occur in places where neither circulatory transport nor storage looks important. Sometimes they turn up in organisms without circulatory systems. Sometimes they appear in situations in which oxygen, even if scarce, is used so steadily that a reserve shouldn't matter. Some parasitic roundworms, root nodules of some plants, and some worms that live on the bottoms of lakes have forms of hemoglobin, odd occurrences not easily explained as simple accidents of ancestry. A particularly creative physiologist, Per Scholander, proposed an ingenious explanation back in 1960, but his idea remains at most a footnote in textbooks.

To be used, oxygen has to diffuse from red blood cells in capillaries to the mitochondria of muscle. Distance is daunting to diffusion. Double the distance from one point to another, and you'll need twice as long to walk, run, or drive it. Double the distance for diffusive transport, and not twice but four times as long will be needed.[5] A tenfold increase in distance means a hundredfold slowdown in delivery. Thus we noted that the final diffusive link in triggering a contraction was the release of calcium right next to the myofibrils—small (therefore relatively speedy) molecules and very short distances. Scholander, calling the process facilitated diffusion, showed that myoglobin speeds up the diffusive transport of oxygen. Exactly how this happens has yet to be fully explained, but facilitated diffusion seems to work as a kind of bucket brigade, handing off oxygen molecules from one myoglobin molecule to another. The lower the overall oxygen concentration, the more effective the process becomes. Nothing trivial, it can increase the oxygen transport rate as much as eightfold.

This phenomenon of facilitated diffusion may provide the key to the peculiar distribution of hemoglobin and myoglobin among organisms. Augmentation of diffusive oxygen movement may have been the original use of these substances, which is what it does for root nodules and roundworms. Oxygen transport in a moving bloodstream would then have put a bit of preexisting machinery to a new use, as evolution often does, the way we use our forelegs for manipulation instead of locomotion.[6]

A few words about Per F. ("Pete") Scholander, who gave us both the idea and the initial demonstration of facilitated diffusion. Scholander (1905–1980) ranks with the most versatile physiologists ever. Originally trained in Norway in medicine and botany, he did most of his work in the United States, first at Woods Hole and then at the Scripps Institute, two of the best places in the country for marine biology. However, marine work made up only a small part of his marvelously diverse accomplishments. He devised the instrument, still used, to measure the extreme tensions in the sap of trees. He identified the basic heat exchange mechanism used by many mammals and birds to run cold-blooded appendages on warm-blooded circulatory systems. He showed how an analogous exchanger helps deep-sea fish drive scarce oxygen into their swim bladders and keep it from diffusing back out into the ocean. He figured out all sorts of important aspects of thermal physiology, from how fur functions to how Australian aborigines can sleep in the cold to how Korean pearl-diving women keep from getting fatally chilled. He showed how fish could live at the bottoms of fjords where the seawater was below the freezing point of their blood. He identified the basic respiratory and circulatory adjustments in diving mammals—just to touch on his endeavors.

Scholander deserves better recognition than we've given him. Perhaps he spread himself too thin. People who know of his work in one field rarely know of his work in any other! But that argument has merit only if recognition rather than accomplishment is the objective. His papers stand as classics in their separate domains; he was no dilettante. While he may not have had the last word on anything, he certainly instigated a host of productive lines of work and stimulated all who came in contact with him. Besides, he did what he wanted to do and where he wanted to do it, testing his ideas from Australia and New Guinea to the high Arctic and, by all accounts, including his own, having a great deal of fun. I recall with pleasure an informal dinner at which he entertained the diners by recounting his greatest scientific blunders, including sins of both commission and omission.[7]

The fact remains, though, that we remember the person, John Kendrew, who figured out the structure of myoglobin, for which he received a well-deserved Nobel Prize in 1962, but we forget the person, Scholander, who figured out what myoglobin most likely does for us. My own graduate adviser, the late Carroll M. Williams, used to complain about the practice. As he put it, identify and show how to isolate a hormone, and you're a footnote; get its chemical formula, and you're a hero. I suspect a connection with commercial value. A formula excites the people interested in chemotherapy, if the hormone is one of ours, or extermination, if the hormone belongs to an insect. Chemistry pays better than physiology.

Back to red and white muscles. In noting earlier how much power muscle could put out and how much force it could exert, I commented that the same muscle could not be at the top of the heap in both. Red muscle puts out power; white muscle puts out force. Power takes more energy, so red muscle has a better oxygen supply, with more capillaries, far more myoglobin, and more mitochondria. Force depends on the aggregate cross-sectional area of contractile machinery, so white muscle does better by dispensing with some capillaries and mitochondria. For some tasks, one matters; for other tasks, the other. An archer needs great force to draw a heavy bow; a rower needs power even if at some expense in maximal force. A cat needs force to accelerate as it lunges at its startled prey; a dog needs power to chase down a meal that's running ahead. The cat has small lungs and exhausts quickly; the dog has large lungs and can keep the pace. Horsemeat is darker than beef. The distinction (or spectrum) runs from an animal's activities down to the finest structural details of its muscles. Ecology orders; physiology delivers.

Powering Flying Insects

Insects began to fly back in the Carboniferous, the first animals to leave the ground under their own power. After that auspicious launch they've never looked back, most of them at least. A few groups of insects, such as the silverfish that eat books and wallpaper, have found perpetually flightless prosperity. Some inhabitants of oceanic islands have discarded flight; blame reduced predation and the danger of being blown out to sea. Some, such as many cockroaches, make out quite well without bothering to fly. The rest represent life's greatest diversity of muscle-powered flying machines. Being smaller than bats and birds helps in one way. No insect can fall fast enough to injure itself when dropping from any height. Never mind Galileo; in practice bigger things do fall faster since in practice air resistance matters. But that same heightened air resistance makes it harder to get anywhere with any speed, and low speed puts a flier at the mercy of even modest breezes. So whether an animal is big or small, active flying from place to place (not just gliding) takes a powerful engine.

In most lineages of animals, larger forms resemble the ancestral versions less than smaller ones—that is to say, ancestors tend to be relatively small. Early dinosaurs, early mammals, early primates weren't notably large. Of course we may be taking a biased look at the record, as my colleague Dan McShea reminds me; if an early form is of average size, descendants will be both larger and smaller. If the larger ones fossilize better or stick out in the digs or just get better press, then we'll *think* that size has increased with time.[8] But even if (to take Dan's point at full value) average size changes little in most lineages, the major groups of fliers have clearly minified. In birds, bats, and flying insects, early forms were large. Some present forms may still be large, but they've been joined by smaller fliers. Among birds, the passerines (perchers) and hummingbirds are relative newcomers. Among bats, the recent arrivals are the tiny insectivorous ones most familiar to people outside the tropics, not the large fruit-eating flying foxes. Among flying insects, the flies, wasps, and beetles are the least ancient groups and include many of the smallest species.

How should a small flier differ from a large one? Small size presents a curious problem. A machine that flies forward by blowing air backward can achieve no forward speed greater than the rearward speed of the air it blows back.[9] Other things being equal, the smaller flier will have shorter wings, and the tip speeds of these wings will be lower. So the smaller flier will blow air backward more slowly and will therefore fly forward more slowly. That means it won't get as far in a given time, it will be more vulnerable to predators, and—perhaps worst—even slight breezes will completely determine where it ends up. The obvious solution is for it to beat the wings more frequently, often enough so their tips will move about as fast as the wing tips of larger fliers. Because smaller wings will be lighter, making them reverse direction more often should present no special trouble.

Muscle, though, runs into trouble when this miniaturization gets pushed down to the size of small flying insects. Each upstroke and each downstroke of each wing require that a muscle contract. Earlier a contraction time of a fiftieth of a second was declared fast, and that was just the time from the start of contraction to maximum force or minimum length. So with any allowance for reextension and recovery, wingbeat frequencies above about twenty-five per second look problematic. But tiny insects, ignorant of muscle physiology, persist in beating their wings up to as many as a thousand times a second. Don't malign the mosquito; as it makes its soprano buzz, it beats its wings a remarkable three to four hundred times per second. Even the basso bumblebee can exceed a hundred. Insect wings reach rates far exceeding those of any other muscle-driven appendages with which animals move around.

How do insects manage such rates and at the same time work at power levels as high as anything anywhere? As a first trick, they don't shorten their flight muscles much—by about 2 to 5 percent rather than the more typical 10 or 20 percent.[10] That drops contraction time. It also allows the use of muscles of slightly different structures, ones with greater overlap between thick and thin filaments relative to the length of their sarcomeres. More overlap means more cross-bridges, which mean an increase in the force a contraction can generate. The shorter contraction distance does have a downside, however. The work done equals the pulling force times the distance that the pulling force moves a load. Less distance means less work for a given force. So reducing contraction distance may push the frequency limit higher and increase force, but at the same time it decreases the work per stroke. What comes to the rescue turns out to be frequency itself: The higher the frequency, the more often each muscle gets used, and the higher its power output. Power, once again, is the rate at which work is done, work per unit time. When all the variables combine, power output goes up as frequency increases.[11]

Most flying insects that beat their wings at rates of over a hundred per second—as well as a lot that beat at lower rates—use another, yet more radical trick. Their scheme involves the most drastic modification of muscle operation we've yet encountered. Normally a nerve impulse arrives at a muscle and triggers a contraction. More impulses either give more contractions or prolong contraction. When a locust flies, one or two impulses stimulate the muscles that force its wings upward, and then one or two impulses stimulate the antagonistic muscles that force its wings downward. When a wasp flies, something different happens.

With no nerve impulses, the wasp's flight motor, as expected, does nothing. But impulses don't determine how often the wings will beat; they only keep the motor turned on. The mass and elasticity of the system set wingbeat frequency, with nothing as fancy as reflex activity. Shorten a pendulum, and it swings more often; tighten a guitar string, and it vibrates more frequently. If mounted so their feet dangle, wasps (or many other insects) immediately start beating their wings. Measuring frequency is easy; one need just adjust a strobe light to match its flashing rate (shown on its dial) with that of the wings. That's the point at which the wings don't appear to beat at all. If one cuts a bit from the end of each wing (taking care not to be stung by the abdomen's tip!) and again allows the feet to dangle, the wingbeat frequency rises. Another bout of snipping, and the frequency goes up still farther. The rise closely follows what one calculates for a purely physical system based on the masses of the cut pieces and their distances from the base of the wings. I did this back in my final undergraduate year, when I was first trying my hand at delicate experimentation.

Monitoring the nerve impulses presents no great difficulty either; I've done it as a classroom demonstration. You just mount the insect on the end of a stiff wire (warmed wax makes a quick glue) so its feet dangle. You then stick a pair of insulated wires with bared tips in through little holes in the insect's thorax. The other ends of the wires go to a suitably sensitive amplifier and from there to a loudspeaker (to hear impulses as pops) or an oscilloscope (to see impulses as vertical lines on a screen). With its feet off the ground (and perhaps with a little wind on the head) the insect will, in favorable cases, continue to beat its wings for hours. Audibly or visually, locust and wasp differ dramatically. A locust (or a dragonfly) gives a steady deep buzz of impulses or a regular train of vertical lines; a wasp (or a fly) just produces unattractive noise or blips on the screen with little evident pattern.

That takes care of the electrical activity of nerves and muscles. To hear or display wingbeat frequency takes only the addition of a mechanical transducer. Wonderfully sensitive ones come in the guise of phonograph cartridges; microgroove records don't make a needle wiggle very much. It doesn't much matter where the cartridge contacts the insect, since everything shakes in tune with the beating wings. Both signals display nicely on an oscilloscope, as shown in Figure 4.1.

Making a muscle operate in this autonomous (asynchronous) mode takes little modification of its basic mechanism. Calcium, instead of being periodically released and then sequestered again, remains free to stimulate the system. Nothing fancier than a mechanical stretch instigates contraction, so contraction of the muscles that pull the wings down stretches and thus stimulates the muscles that pull the wings back up. In turn, contraction of these up-pullers stimulates the down-pullers, and so on. The muscles must be fairly stiff, but the stiffness of insect muscle normally exceeds that of vertebrate muscle. Furthermore (to be technical about it), the load on the muscle must have a large inertial character with low damping—like a swing that keeps moving after you push rather than like an oar that goes nowhere unless you keep pulling.

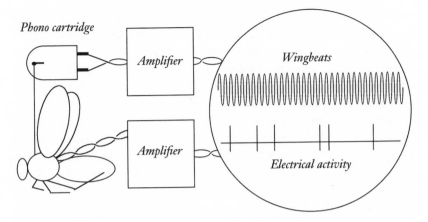

FIGURE 4.1. An insect mounted for monitoring its wingbeats with a phonograph cartridge and its neuromuscular electrical activity with electrodes.

Flies and mosquitoes, beetles, wasps and bees, thrips, book lice, and true bugs plus a scattering of other insects use such asynchronous muscle. While they share the same muscle structure, we're sure that they don't share an ancestor that did the trick. In other words, to judge from what we know of the phylogenetic tree of insect evolution, this way to use muscle appeared independently in a number of lineages.[12] Getting smaller puts a premium on having high wingbeat frequencies, and that premium must have been worth the loss of direct nervous control of frequency. It's a persuasive story of convergent evolution, with different lineages facing similar selective forces and each making similar adjustments to its existing machinery.

Still, it may not be the whole story. Robert Josephson, who works on invertebrate muscle at the University of California, Irvine, suggests another advantage of asynchronous operation. Josephson is one of those people who, when you mention something you're up to, inevitably respond with some insightful point that you should have thought of yourself. Ever iconoclastic, he points out that we now know of muscles with normal neural triggering that contract several hundred times each second; whiteflies (which are not true flies), for instance, beat their wings with normal muscles about 180 times per second.[13] High frequencies can thus be achieved without becoming asynchronous. Josephson reminds us, though, that the thorax of an insect is none too roomy, and good flying machines always avoid excess weight. Insect flight muscle can (and indeed must) use oxygen to generate ATP at a rate unsurpassed elsewhere in the animal kingdom. That takes machinery, which takes space and weight. Mitochondria unavoidably occupy as much as 30 or 40 percent of the space in a flight muscle fiber. (It was work on insect flight muscle, in the late 1940s, that showed that the sarcosomes of muscle, bodies then of unknown function, were in fact mitochondria.) Rapid triggering also takes machinery, and that membrane system (T-system plus sarcoplasmic reticulum) takes up as much as 20 percent of the flight muscle of insects that use nerves normally. The insects that have asynchronous muscle dispense almost completely with such membranes, freeing space in their muscles for more myofibrils. That translates into more power relative to volume or weight.

In any case, asynchronous insects may still be able to adjust the frequencies at which they beat their wings. When I was a graduate student working on fly flight, something caught my eye in several papers on the muscles of their thoraces. As well as bearing one or two pairs of wings, an insect's thorax accommodates all six of its legs. In flies, one of the muscles that goes from each middle leg up into the thorax doesn't simply attach near the leg itself on the bottom of the thorax. Instead the muscle broadens out and runs clear up to the top. As a leg muscle it helps a fly jump, to judge from its position and attachments. I found that for a fraction of a second at the very start of flight the wingbeat frequency was abnormally high, either when a nonattached fly took off or when I pulled the support out from under a mounted fly. If I unhooked that odd muscle by cutting off the middle legs, that initial extra–high-frequency buzz disappeared. Cutting front and hind legs, which lack the muscle, had no effect. No great surprise: Squeezing the thorax should, like tightening a guitar string, increase frequency. That ought to speed takeoff, and rapid takeoff must be a good thing in a world that wants to swat you or eat you. How clever of the fly to use the same muscle for jumping up and for raising wingbeat frequency! The general point: even without the direct nervous triggering of each beat, frequency can be controlled. If mechanical resonance sets it, then just alter the resonant frequency by adjusting the stiffness of the thoracic box.[14]

You might expect insect flight muscle to be the darkest of dark muscle. In reality, it's an undistinguished tan and has no myoglobin at all. For that matter, most insects have no hemoglobin, no red blood cells, puny hearts, no lungs, and relatively simple circulatory systems. Not that diffusion makes up the difference; for that, insects would have to be far smaller. Instead of using hemoglobin to transport oxygen in blood, insects have completely separate oxygen delivery systems. These pipes (tracheae) get air from ports along their sides and then branch and rebranch down to subcellular dimensions as they ramify through the muscle fibers. The pipes bring air to within about one two-hundredth of a millimeter (five micrometers) of the mitochondria. In the finer branches, diffusion (much more rapid in a gas than in a liquid) moves the oxygen. In the larger branches, insects can do a lot of pumping, moving abdominal segments in and out like accordions and taking advantage of the pipe-squeezing effect of flight muscle contraction itself.

Making a Lot of Noise

Many animals produce sound, and they do so in more ways than the instruments of an orchestra. Many expend remarkable amounts of energy to get lots of sound out of the tiniest of instruments. Just think of the noise made by crickets or cicadas or frogs. Anyone who refers to the still of night cannot be living in a warm, wet place! Relatively speaking, we higher vertebrates—birds and mammals—have it easy. We pump large volumes of air in and out of our lungs for other reasons, and converting our air pipes to biological analogues of organ pipes, flutes, or clarinets takes little additional machinery. In turn the machinery takes only a little energy to run—howler monkeys, kookaburras, and Wagnerian sopranos notwithstanding—next to such activities as keeping warm. Moreover, in sound intensity relative to body size, only small birds are big league anyway. We may make more complex sounds that carry more information, but for sheer volume insects put us to shame. You don't need eyes to detect a mass emergence of periodical cicadas.

Producing sound poses a basic problem much like that of beating tiny wings. The higher the frequency, the less energy it takes to make sound of a given perceptual intensity. Treble comes cheap while bass costs, as those of us who've built hi-fi systems learned early in the game. Tweeters are naturally tiny; the art lies in building a nonmonstrous woofer. Even now, fancy stereo systems use extra amplifiers for their subwoofers. If you want to attract females or to announce your presence and territory, you do best with frequencies above at least 100 hertz (vibrations per second; middle C on a piano has a fundamental frequency of 263 hertz). But that pushes the envelope for rates at which muscles contract, implying that producing sound (birds and mammals aside), like beating tiny wings, must be a tough job, a collection of extreme cases.

We have two reasons to treasure nature's extreme cases. The first is that they tell us how far natural design can be pushed, showing us what nature has to modify to make an ordinary design deal with extraordinary demands. That in turn helps us understand how organisms in general—not just the extreme cases—work. How can a warm-blooded duck afford to walk around in ice-cold water while pumping blood in and out of its feet? The answer helps us understand when and why our own appendages are colder than our heads and torsos. Scholander of course was the grand master of that approach. He didn't invent it, though. Borelli certainly used it, and William Harvey, in the book that established the circulation of the blood in 1628, makes the case for the comparative method in no uncertain terms, asserting: "If only anatomists were as familiar with the dissection of lower animals as with that of the human body, all these perplexing difficulties would, in my opinion, be cleared up."[15]

In the twentieth century the approach was used with especial effectiveness by a lineage of physiologists that began with Christian Bohr (father of the famous physicist Niels Bohr) and was continued by Bohr's student, the Nobel laureate August Krogh and by Krogh's student Knut Schmidt-Nielsen, and, of course, the students of all three.[16]

The second reason turns on the way similar extremes often occur in separate lineages; we call the phenomenon convergence. Tall, woody, treelike plants have evolved on a number of occasions from short, herbaceous plants. The features that these tall plants share tell us a lot about the underlying logic of making a tree from the basic materials available to higher plants. They do so the more clearly because those common features cannot represent mere accidents of common ancestry.[17] To state it another way, if natural selection drives evolution, then features that converge must be those that matter to the success of an organism. Therefore, convergence points to functional significance; it's nature's way of telling us what aspects of her designs she has had to worry about.

Back to sound. How can a little animal be a loud presence? As you might guess, one approach consists of making a big enough muscle contract at a high enough frequency. We're talking about contraction rates as much as ten times higher than anything a human muscle ever manages. That requires rapid triggering and rapid cycling of calcium; calcium must be released and reabsorbed every few milliseconds. Not surprisingly, these rapid muscles have in common very elaborate and voluminous internal membrane systems—T-systems and sarcoplasmic reticula. Indeed in the early 1960s the fact that these membranes were most elaborate in superfast muscles provided an important clue to what they normally accomplished.[18]

Oh, yes, whom are we talking about? One such noisy creature is the katydid, or the long-horned grasshopper. Katydids, like crickets, sing by rubbing their forewings together, and they do it by using the same muscles that flap their wings in flight. Curiously, to fly, they beat their wings and contract these muscles about twenty times each second, whereas to sing, they contract the muscles at up to two hundred times per second. But they trigger their two hundred separate contractions per second with nerve impulses arriving at the same impressive frequency.[19]

Lobsters too make noise, although few of us have ever heard one. When disturbed, a Maine lobster voices its displeasure by vibrating its antennae and producing a loud, buzzing underwater sound. The muscles that move the antennae may have the most extensive of all sarcoplasmic reticula. Myofibrils occupy only a quarter of their volume, mitochondria are scarce, and sarcoplasmic reticulum takes up over half the muscle's volume.[20]

Echolocating bats make ultrasonic pulses loud enough to hear and find the echo returning from an individual insect. Muscles don't hit ultrasonic frequencies (that's a purely acoustic business), but they do determine the form of the individual pulses of ultrasound, which can be as short as a millisecond each and as frequent as two hundred per second. Again the crucial muscle is special, with very slender myofibrils wrapped and even penetrated by a particularly extensive sarcoplasmic reticulum.[21]

What else? Toadfish (Figure 4.2) wrap their air-filled swim bladders with another of these superfast muscles with lots of sarcoplasmic reticulum. They emit their call, often described as a boat whistle of an intensity comparable to a loud automobile horn, by making about two hundred muscular pulses per second. Rattlesnakes rattle rapidly, with the tail shaker shaken up to ninety times each second, using superfast mus-

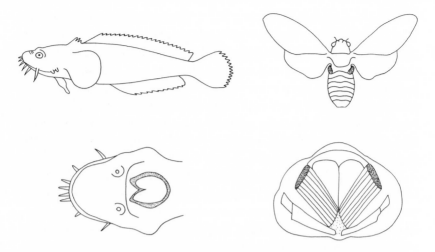

FIGURE 4.2. On the left, a toadfish and its sound-producing air bladder; the bladder with its muscular wall actually lies within the fish. On the right, a cicada with its timbals and an internal view across its abdomen to show the diagonal muscles beneath the timbals.

cles of similar construction. Rattlesnakes, at least western diamondbacks, do something even more impressive: They can rattle away for hours without interruption (no need to take a breath!). So, besides all that sarcoplasmic reticulum, the rattling muscle has to have a lot of mitochondria to supply energy. But profligate they are not; the tail shaker muscle has the lowest cost per twitch of any known muscle.[22]

A final sound story puts the pieces together nicely. A cicada (Figure 4.2) doesn't so much sing as scream. It does that with a pair of muscles attached to a pair of timbals (or tymbals), one on each side of the abdomen. Contraction of the muscle pops the stiff, domed timbal inward; when the muscle relaxes, the timbal pops out again. John Pringle, then at Cambridge University, who had discovered asynchronous flight muscle in the late 1940s, found that cicadas operate their timbal muscles the same way.[23] The timbal's outward pop stimulates the muscle to pull, and the resulting inward pop causes the muscle to relax again, allowing the timbal to pop out once more. So the cicada screams.

But not all cicadas use such asynchronous muscle, and Bob Josephson, whom we've already met, used a pair of species to compare the conventional synchronous triggering system with the radical asynchronous triggering system. In one respect, in twitch duration when isolated from the animal, the asynchronous one was the more ordinary. Paradoxically perhaps, it wasn't a superfast muscle in the normal sense, but otherwise it was satisfyingly special. Screaming costs, and both devoted about the same large volume, about 40 percent, to mitochondria. But the super-fast synchronously triggered muscle devoted about 12 percent more volume to sarcoplasmic reticulum and gave a corresponding 12 percent lower volume to myofibrils—shades of bats, katydids, toadfish, and rattlesnakes. So the asynchronous muscle was stronger, in fact by a greater than expected twofold. That brings us back to Josephson's original point about flight. Achieving high frequency may not be the primary or original benefit of the asynchronous arrangement; the payoff that matters (or mattered, as these things evolved) more likely turns on fitting a more powerful muscle in a limited space. A final curiosity: those cicadas that have evolved asynchronous timbal muscles retain ordinary synchronous flight muscles.

Clamming Up Forcefully

Flight takes lots of power, but it can pay off in quicker routes to food, in exploitation of more distant or otherwise unreachable food sources, in avoiding climatic extremes by migration, in rapid dispersal, and so forth. To us fellow frenetics it makes a certain sense. We spend lots of energy just staying warm and being active; a human-size alligator uses about a quarter as much energy as a person. Like the fliers, we live life in the fast lane.

The world, though, contains all manner of slow-lane animals. If you don't expend much energy, then you needn't eat much, the tactic of the sitting spider or alligator. But neither gets close to the limits of inactivity, certainly nowhere near as close as some of the invertebrates that just sit on the bottoms of bodies of water and quietly sift out passing edibles. Clams typify these creatures, phlegmatic in both mind and body. Consider a clam lying buried in mud or sand, with only a pair of tubes sticking up into the water. It draws water in one tube, passes it through its sifter, and pumps it out the other. A clam, I'm told, has to process a ton of water to get an ounce of food, and a typical clam pumps an amount of water equal to its own volume every few seconds. That effort leaves little energy for doing much else.

One thing a clam does well is clam up. A pair of muscles connects the two halves of its shell, as in Figure 4.3, and these generate a force respected by anyone who has tried to pry one open. What's most impressive, however, isn't the magnitude of the force but its persistence and its economy. In theory or in physics class, as noted in the first chapter, exerting a force need cost no energy or power; three chapters later the chandeliers still hang in their accustomed positions, their chains exerting enough force to balance their weight. Our muscles, though, require a steady input of energy to exert their forces—not as much energy as they need to shorten at the same time (recall Hill's work, in the last chapter) but a substantial energy input nonetheless. Clams can do better, as can other bivalve mollusks, such as mussels and scallops. These animals produce a muscle that can contract and then lock itself in the contracted state. Again, the trick may be unusual, but it's no radical bit of physics; it does no more than you do when you hoist a flag and then tie the rope to a cleat so that you needn't hold the rope until sundown. Nor is this some recent recognition. That some muscles of bivalve mollusks could "catch" in the contracted state has been known for more than a hundred years.[24]

A catch muscle when caught behaves more like a simple rubbery elastic than like a muscle in an ordinary sustained (tonic) contraction. Impulses from the nerve supplying it make it catch; impulses from the same nerve make it uncatch. The nerve, as it happens, contains two kinds of nerve cell running out to the muscle. The excitatory nerve cells work like those described earlier, instigating contraction by the usual triggering of calcium release around the myofibrils. But then the myosin and actin remain attached by the cross-bridges. They remain attached until stimuli arrive from the relaxing nerve cells. Like a light switch rather than a piano key, a catch muscle must be turned off as well as on.

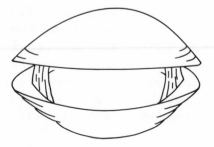

FIGURE 4.3. The location of the catch (shell adductor) muscles of a clam.

The action of either of the two kinds of neuron can be mimicked by bathing the muscle in the agent that kind of neuron releases where it contacts the muscle. That much is easy. Going further has been harder, and we still don't fully understand the mechanism. It seems to involve an unusual form of myosin.

While catch muscle sounds like a wonderful device, something that would ease the problem—and pain—of holding odd postures for long periods, it does have a downside. It goes into action very slowly, and it produces little force for its cross section. Were our locomotory and postural muscles of this kind, the massive investment in muscle mass might compromise a lot of other activities. We do use something a little like catch muscle to propel food through our digestive systems. The so-called smooth muscle (because it lacks nicely aligned cross-striations) of the walls of our stomachs and intestines shares both its slow speed and its dual excitatory-inhibitory stimulation system.

Trading Force against Speed

In our collective genetic wisdom we vertebrates take the 2.5-micrometer, ten-thousandth-inch sarcomere as sacrosanct. The rest of the animal kingdom rejects that prejudice, making sarcomeres that range from less than half that length in squid tentacles to more than ten times as long in some mussel muscles. The variation doesn't follow animal size but instead goes into refining the match of structure with function.

Muscles with short sarcomeres shorten rapidly but can't pull hard, while muscles with long sarcomeres generate lots of force but don't move loads all that fast. At one level, that rings true right away. Power should depend on the overall mass of muscle, not on sarcomere length. Since speed times force gives power, it's easy enough to see how a particular power could be achieved with different combinations of speed and force. More subtle is how changing sarcomere length adjusts the mix. The diagram in Figure 4.4 will help explain the trick.

Assume, for a start, that cross-bridges always push filaments of actin and myosin past each other at the same rate. If a muscle consisted of one long sarcomere, that rate of filament sliding would correspond to the overall rate of muscle shortening. The overall force would then be the sum of the forces of all the cross-bridges pulling at a given time. Now imagine that a sarcomere attached to the left tendon pulls on a Z-disc

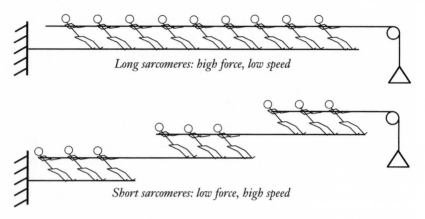

Long sarcomeres: high force, low speed

Short sarcomeres: low force, high speed

FIGURE 4.4. Pulling with long versus short sarcomeres.

partway along the muscle. That Z-disc will move left at the cross-bridges' sliding rate. From that Z-disc the next sarcomere extends to the right, and it should shorten at the same rate as the first, pulling on the next Z-disc. But that second Z-disc will move left twice as fast as the first, since the opposite end of the sarcomere moving it is already in motion. And so it will go along the muscle, with each successive sarcomere moving left faster than its immediate leftward neighbor. Meanwhile each Z-disc has fewer cross-bridges pulling on it than would pull on the ends of the one-sarcomere muscle; only first and last sarcomeres pull on the load itself. Result? More speed but less force.

A squid catches prey by shooting out its pair of tentacles, and shooting out depends on muscles inside the tentacles themselves. In particular, it depends on the rapid shortening of muscles that run across the tentacles, muscles that make it skinnier and therefore longer. Faster has to be better, and according to measurements by my good friend Bill Kier (who moves pretty rapidly himself), the strike takes only around a fortieth of a second. In one of the species of squid that he examined, the sarcomeres of the key extender muscle were only nine-tenths of a micrometer long, almost three times shorter than ours.[25]

At the other extreme are the muscles that pull on the threads with which mussels attach to rocks. Their sarcomeres can be over thirty micrometers long. Mussels being hit by storm waves may have to pull hard, but they're going nowhere fast. Nearly as long are the sarcomeres of the claw-closer muscles of some crabs, which approach twenty micrometers. These crabs crush mollusk shells, no mean task, and their muscles generate tensile stresses (forces adjusted for muscle cross section) four or five times higher than anything known among vertebrates.[26] While the mollusks have a few tricks of their own to make life more difficult for crabs, a rapid getaway isn't the usual thing. So the slow closing speed of a crab claw shouldn't hurt, or at least it's a fair trade for the extra force.

Not that we vertebrates can't trade one component of power for another. But we don't interchange force and speed by messing around with the proportions of our sarcomeres. I can only guess that at some point in our evolution we lost this particular element of genetic versatility, however useful it might be. Maybe we shouldn't be too smug about the large size of our genome.

CHAPTER 5

Knowing What We're Doing

A very simple reflex is the "knee-jerk," which the physician uses as a test of nervous health. It is a slight quick straightening of the knee to a light tap given below the knee-cap. Its variability is its main service to the physician. His tap on the tendon releases within the muscle a nerve-signal which on reaching the spinal cord fires one back again into the muscle, making the muscle twitch. That is all; but how delicately variable. . . . It tests the state of different parts of the nerve-system. The brain, the cerebellum, the centres for ear, the bulb, parts quite remote from the knee's own nerves and nerve-centre. The reflex has, as it were, its fingers on them all; or rather they on it.[1]

SIR CHARLES SHERRINGTON, neurophysiologist

CLOSE YOUR EYES, GIVE A GOOD WHOLE-BODY STRETCH, AND then make your forefingers approach each other, one coming inward from the left and the other from the right. Can you aim and steer the fingers accurately enough to make them touch? Most of us can, if barely. Now try to make them meet behind your back; curiously, the latter task presents little increase in difficulty. Question: what must your neuromuscular system know in order to do the trick?

Stand, slightly crouched, on both legs, and then lift one leg off the floor. Notice first that you don't fall over. Notice also that your head has shifted to a position above the supporting leg. Notice finally that you don't sag down toward the floor, even though the load on the remaining muscles holding you erect must have doubled. Questions: what adjustments have you made in the activity of your muscles, and which muscles matter in this simple action?

In power output for its weight and in the efficiency with which it uses fuel, muscle remains just another engine—albeit one that's soft, wet, and contractile rather than hard, dry, and rotating—that's no particular paragon of performance. The ways in which the neuromuscular machine controls itself, though, deserve the envy of any engineer. Its operation remains impressive whether we look at what it does in seconds and fractions of seconds, as in these examples; in seconds and minutes as it takes on some demanding task; and over much longer periods as it guesses future demands from past experience and reshapes itself in anticipation.

Reflexes and Our Position Sense

We speak of five senses—seeing, hearing, feeling, tasting, and smelling—and allude to a sixth sense when we mean to imply special insight or understanding. We know we have internal senses as well. We may feel hunger or some sense of urgency about a toilet; we sometimes feel dyspepsia; a few of us feel angina. Rarely do we realize that besides these monitors of our external and internal worlds we have yet another sensory system. It looks at the forces at play within our muscles and tendons and at the positions of our parts, and it works in concert with our sensors for the direction of gravity and for bodily acceleration. The tasks just asked of you make heavy use of this sensational (in both senses) proprioceptive system. The name comes from *proprius* for "one's own," as in "proprietor." One's own muscles, tendons, and joints contain the system's main receptors.

This distinction among three receptive systems—exteroceptive, interoceptive, and proprioceptive—originated with Sir Charles Sherrington (1857–1952). This greatest of neurobiologists gave us the basic picture of how nervous systems work as systems. Only the way nerves conduct impulses escaped his profound insight and experimental genius. A solution to that problem demanded tools beyond what he had available and awaited Alan Hodgkin, Andrew Huxley, and the late 1940s. In particular, our modern understanding of reflexes rests on Sherrington's insights, which were as sophisticated as the interrelationships among the reflexes themselves. His classic book of 1906, *The Integrative Action of the Nervous System*, still deserves to be read as it approaches its century mark. It transformed its domain much as Darwin's *Origin* transformed evolution, and it remains elegant, informative, and a bit sobering when we recall its age. Moreover, nowhere is the central importance of physiology better defended than in his 1940 *Man on His Nature*.

Consider the simplest of reflex activities. You touch something hot and yank your hand away. What's the inside story? For one thing, even though you feel the heat—some sensor in the skin has sent a message to the brain—your brain doesn't then decide to withdraw your hand and so instruct the muscles that perform the action. No, the hand pulls back without the intervention of your brain, and it will do so perfectly well in a brain-dead animal. The brain gets informed strictly post hoc by nerves ascending from your spinal cord. Perhaps the information might guide some further action, immediately or well in the future. What happens is reflexive. Impulses go from a sense organ in the skin, where such organs are numerous and diverse, to the spinal cord. There their nerves make connections (synapses) with other nerves. A group of these latter provides the link to the motor units of the withdrawing muscles (flexors). Another group goes upward, both anatomically and metaphorically, to inform the brain. Sherrington called this the flexion-reflex, as do we, doing no more than dropping his hyphen.

Were that the whole story, its role here would be trivial, merely pointing up the fact that every connection between sense organ and muscle—between sensor and effector—goes through the central nervous system. In fact, there's more to it. A third set of connections in the spinal cord affects another group of muscle-driving nerves, one linked to the muscles with the opposite action. You can't just stimulate flexor muscles; you have to make sure that the extensor muscles are turned down or turned off. Sherrington (as do we) called this the crossed extension reflex. He demonstrated the reality of both flexion and crossed extension reflexes with clever and unambiguous experiments done, of course, without electronics, computers, or molecular techniques. Figure 5.1 puts the two reflexes in diagrammatic form.

The combination of flexion and crossed extension reflexes illustrates two general phenomena. First, they're cases of negative feedback, in which a system looks at the results of its own activities—hence "feedback." Cessation of the hot stimulus stops flexion; the system "knows" when it has accomplished its goal. Goal directedness comes from taking action to offset or neutralize a disturbance—hence "negative" to convey the sense of taking an opposing rather than augmenting action. Devices based on negative feedback pervade our lives. The float in the tank

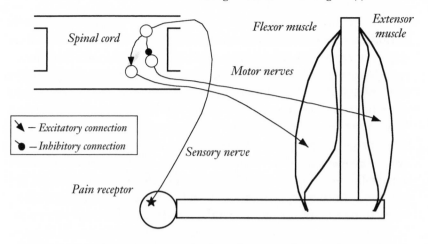

FIGURE 5.1. The circuitry involved in flexion and crossed extension reflexes and the location of their major components.

behind the toilet bowl senses water level; if the level drops, it opens a valve that admits more water. The water level thus gains independence from slight leakages or evaporation, and it restores itself after a flush. The thermostat in the oven watches its temperature; if the temperature is too low, it turns on the heater element. The power steering unit of the car is only a little fancier. It checks for any difference between the direction the car is going and the desired direction, as signaled by the steering wheel, and its power goes into reducing any difference it detects.

As for the second phenomenon, these reflexes require not only exciting some muscle-driving nerves (motor neurons) but inhibiting—turning off—others. Those extensor muscles might be doing something just when you burn your finger, and impulses going out to them must be turned down or off. Inhibition may be even more important than excitation. All kinds of biological systems, from genes to brains, remain inhibited most of the time. Since the number of wrong actions always exceeds the number of right actions, we might argue that keeping the lid on what we might do represents our greatest accomplishment. Failure to inhibit one or another action runs a gamut of unpleasantness from convulsions to cancer. To use Sherrington's analogy, when a person uses a telephone, more important than getting the right connection is not getting any of the vast number of possible wrong ones.

Muscles participate in another class of reflexes. These particular proprioceptive ones provide the elegant and efficient machinery we use to perform the kind of ordinary task asked of you a few pages back. A little stage setting will introduce their cast of components. Muscles contain stretch receptors of at least two types. The simpler sort, the Golgi tendon organs (named for Camillo Golgi, the Italian physician who discovered them in 1880), monitors tension in the tendons that connect muscle to bone. With them our central nervous systems keeps track of how hard we're pulling. Among other things, these receptors signal when tension rises to dangerous levels, to those that might tear muscles or tendons or even pull tendons off bones. We do tear muscles and tendons occasionally, but it usually happens in a sudden landing or with other impulsive loads, those applied so quickly that the muscle can't be turned off in time.

Tension receptors provide the input for so-called stretch reflexes. Stretch a muscle, and the muscle responds, as (yet again) Sherrington showed, by pulling harder. The tension receptors have signaled the central nervous system to increase the rate at which it stimulates the muscle. We make use of (or subvert) this reflex to check for trouble in the nervous system, as described in the quotation that begins this chapter. When a doctor or nurse taps a dangling leg just below the knee, the leg should swing forward. The tap stimulates only briefly and is over by the time the muscle responds, so the leg takes action too late to do any good. Functionally useless in this clinical version perhaps, but success means that a lot of personal machinery must be working properly.

The second kind of sensory receptor in muscle plays the central role in these stretch reflexes. A small number of fibers in our muscles do something besides pull bone toward bone. These skinny fibers run, tendon to tendon, in parallel with the bulk of a muscle. Even working maximally, they make next to no difference to the muscle's force. Each little group of fibers has, midway between its ends, a swelling that contains some stretch receptors, together called a muscle spindle. Odd—the scheme appears at first glance both needlessly complicated and redundant with stretch receptors of the Golgi type. But consider the limitation of a receptor in series with a muscle. It can't distinguish between an externally imposed stretch and the muscle's own contractile response—both pull on it—so it makes a lousy basis for a general-purpose negative feedback system. The muscle's response to a stretch would stimulate the receptor further, impelling the muscle to pull still harder. Instead of offsetting a change with negative feedback, the system would amplify a change with positive feedback. Any little stimulus, and the system would pull as hard as it could, not at all what we want.

A receptor in parallel to—running alongside—the main mass of muscle avoids the problem. Adding a load stretches both the main mass of muscle and the receptor system. The receptor system tells the central nervous system that the muscle ought to work harder and increase its tension. The central nervous system then makes the muscle contract more forcefully, relieving the stretch on the receptor system. So the muscle gets back to its original length, but now with the higher tension needed for the greater load. Neat—muscle, a tension-producing machine, can be told to maintain a constant length.

But it won't yet do the whole job. More specifically, the scheme as described will work for only one particular muscle length, the one that just triggers the stretch receptors. It would do the same job if stretch receptors in the muscle spindles were connected to the tendons of the muscle by some passive component, such as a pair of strings. To the rescue come the special puny fibers. Putting the receptors in these muscles permits us to adjust the length at which the system operates. It makes the system more complicated, but Figure 5.2 tries to sort out what goes on

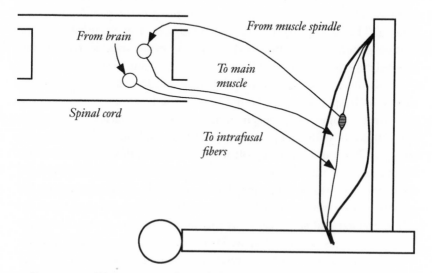

FIGURE 5.2. How we control muscle length in the face of varying loads— the circuitry and location of the components.

in diagrammatic form, and the next paragraph tries to do it in words. Bear in mind that these long, skinny fibers put muscle to a use that's different from anything we've yet met.

Consider how you might flex your forearm, bringing a glass closer to your lips by bending your elbow. Direct stimulation of the biceps muscle of your upper arm while inhibiting the antagonistic triceps muscle won't suffice. While these actions would bend the elbow, the amount of bending would depend on the load, on whether you lifted a large, full mug or a nearly empty piece of delicate crystal. But instead of (or, depending on circumstances, as well as) stimulating the main motor units of the biceps, your central nervous system stimulates those special puny parallel fibers. So they develop tension, but, too weak to move the load, they don't shorten. Their tension, though, sends nerve impulses back to the central nervous system. Those impulses in turn trigger impulses that return to the muscle, now to the main motor units. The main fibers develop tension and shorten, and shorten, and shorten . . . until they relieve the tension on the little parallel fibers, lowering the rate at which the stretch receptors send off impulses.

Strangely obtuse, with two stages of stimulation, but effective—the rate at which the little parallel fibers receive impulses sets the length to which the muscle will shorten. It sets that length whatever the load. The main muscle shortens until the stretch receptors in the parallel fibers say "enough" by ceasing to send extra impulses back. You can tell a muscle, a tension-generating engine, to develop not some specific tension but whatever the tension needed for it to shorten by any desired amount.[2]

With these little parallel muscle fibers and the stretch receptors in their muscle spindles, your body "knows" the location of its parts. You can direct a finger to approach another finger without looking because you can direct your various pieces to change their locations by specific amounts. You can lift one leg off the ground, thus doubling the tension on the extensor muscles of the other leg, without sagging down toward the ground as a result. The other leg simply works harder, increasing the activity of its extensor muscles just enough to offset the extra load. So self-contained and self-effacing is this proprioceptive system that conscious awareness needn't intrude. Worse than operating subtly, it hides; it works so well that its very presence remained unsuspected until the last century.

When teaching a general course in biology, I used to do a stunt that illustrated this basic length-setting stretch reflex. Before class I'd stick a pair of electrocardiogram electrodes to my upper arm, on the skin above my biceps muscle. Neither the electrodes nor their wires showed beneath my shirt—until, at the appropriate point, I rolled up my sleeves to show the electrodes, pulled the plug from between my shirt buttons, and connected myself to the public-address system. (For some reason, students found this last amusing.) With just a small additional amplifier, the electrical activity of my biceps could be heard as noise throughout the lecture hall. The position of my forearm made little difference to the noise level since with only its weight as load, a small number of motor units operated at a low level. But when I held out my arm and someone placed a brick on my hand, things got noisier. My arm twitched, dropping slightly, and then went back up; the drop increased the activity of my muscle spindles and the rest of the reflex readjustment. A second brick gave the same twitch and subsequent restoration of position, with a lot more noise. A third did the same, although as my brick-holding capacity was approached, the noise took on a slightly buzzier character. When near capacity, our motor units begin to synchronize, something we may notice as a slight tremor.

People can work their muscles with seriously impaired proprioceptive systems, but they have to learn to substitute visual for proprioceptive feedback. In fact, all of us supplement proprioceptive with visual and tactile feedback for delicate tasks; you'd hesitate to do much knife-wielding kitchen work with your eyes closed. You can bring together the index fingers of your hands with proprioception alone, but you can do it more easily and dependably if you watch what you're doing.

Even under normal circumstances, however, proprioceptive feedback often proves inadequate. For the most rapid actions it's simply too slow. Nerves don't conduct impulses at anywhere near the speed at which wires conduct electrical pulses. We can manage only about a hundred meters per second (around two hundred miles an hour), and even that's exceptionally fast by animal standards. A full trip of, say, two meters from sensor to spinal cord to muscle takes about a twentieth of a second. Transmission from sensor to nerve, from nerve to nerve in the spinal cord, and from nerve to muscle imposes further delays, as does activation of the muscle itself. So the responses of the system are none too rapid. Once you start swinging ax, hammer, or baseball bat, your muscles are for the most part on their own. Neither a praying mantis snagging a fly nor a squid shooting out its tentacles to catch a tiny shrimp can make much use of proprioception or any other neural feedback. You or mantis or squid can only take aim.

One additional part of the muscle control story needs mention, although we'll ignore the details. Besides sensors for position, you come equipped with sensors in the inner ear for the direction of gravity and for acceleration in any direction. These come into play in such movements as the adjustments you make when lifting a leg off the ground. Besides avoiding sagging, you change your orientation to relocate your center of gravity over the supporting leg. No strength or guile could keep you from toppling were the center of gravity anywhere else. You know up from down, although sitting in a banking aircraft, you can be fooled into thinking that rather than you and the plane, the earth has tilted. Since you lack specific receptors, your sense of velocity is poor. Similarly, beyond about ten feet your binocular vision loses effectiveness, so your sense of distance deteriorates. In both cases, you rely for clues on landmarks and familiar objects. But you can detect gravity, rotation, and acceleration quite well, and you use the information to aid your proprioceptive sensors as they tell your muscles what to do.

The righting reflex of small animals, such as dogs and cats, gives a dramatic example of the integration of these sensory systems with muscular engines. These creatures compensate for spins and misorientation with remarkable speed by doing nothing more than moving one body part with respect to another—with no aerodynamics at all. Try dropping a cat a few feet onto a bed, and you'll see a performance that leaves in the dust anything the robotics community can yet manage. The trick does require mobile appendages of decent size, though; physics can't be violated, and with nothing to push against, the rule about a force's generating an equal and opposite force (Newton's third law) doesn't help. That leaves only the principle of conservation of angular momentum, the explanation of why a skater spins faster with flexed than with extended arms. Testing that righting reflex, Kenneth Roeder, my undergraduate physiology teacher, discovered that his short-limbed dachshund performed poorly. In addition, out of either pique or embarrassment, the normally accommodating animal nipped him.

We use the machinery of the righting reflex in activities less dramatic but still important. In righting, an animal first levels its head, using information from its gravity detector, and then deals with the rest of the body. Even when not in free fall, mammals and birds can keep their heads on the level while their bodies thrash around. Watch the head while moving the body of a small cat or dog, young human, or, best, pet bird. Or watch the head of someone pitching or hitting at a baseball or swinging a golf club. Where are the proprioceptors with which we adjust the positions of our heads relative to the rest of our bodies? The muscles of the backs of our necks contain as dense a population of muscle spindles as those in the muscles that do the delicate work of our hands and fingers.[3]

Beyond the clever design of these controls, we should be impressed by their spectacular performance. We routinely handle eggs; a more apt version of the adage might be that we can't make omelets without not breaking eggs. That's no easy thing for a robot. In industry, we still use human operators to get particularly fine control, taking advantage of our great neuromuscular machinery to do tasks that need little or no human judgment. Automation often depends not on more delicate machines but on redesigning things so their manufacture doesn't require particular delicacy—so they can be assembled by ham-handed machines.

Controlling the Fuel Supply

Work demands fuel, and muscles do work. As their rate of working varies, their fuel consumption varies, and so must the rate at which fuel is supplied. A muscle amounts to a combustion engine, even if its enzymes manage to "burn" fuel at body temperature. As in every such engine, its fuel consists of two components, oxidizer and what's oxidized. Think of it as a rocket engine, in which both components need special provision for storage and supply, rather than as a jet engine, which takes atmospheric oxygen *au naturel*. Quite a lot of what's oxidized—fat and carbohydrate—can be stored in the muscle itself, and more comes without fanfare through the circulatory system. The oxidizer proves more troublesome, even with our ability to store oxygen in myoglobin. If it didn't have to transport oxygen as well as its by-product carbon dioxide, a circulatory system could be far smaller and simpler, as in insects, in which tracheal gas pipes relieve the circulatory system of oxygen delivery duty. So we need to examine how oxygen gets metered out and delivered. In doing so, we view another aspect of the control of muscular action, one that takes seconds rather than milliseconds.

Skeletal muscle accounts for 40 to 45 percent of the weight of an ordinary mammal like any one of us, and it accounts for 50 to 75 percent of the body's protein. But when we're at rest, only about 20 percent of the blood that leaves our left ventricles makes its way to our muscles. However, when we work hard for more than a few seconds, the picture is transformed.[4]

First, the heart pumps out more blood, mainly by beating more often but also by pushing out a little more blood in each beat. In sustained heavy work, the normal rate of five liters per minute, about three ounces per beat, goes up around fourfold, to twenty liters per minute. The time needed for a bit of blood to make its way around the whole circulatory circuit goes down by the same factor, so it takes only about fifteen seconds instead of a minute. Thus, when working hard, we use our blood, with its oxygen-carrying hemoglobin, about four times as often as when we're at rest.

Second, each unit of blood transports more oxygen. A small part of the increase comes from raising by about 5 percent the amount of oxygen in each bit of arterial blood that leaves the lungs. Some commercially stimulated current lore notwithstanding, our lungs load up our hemoglobin almost fully at normal atmospheric oxygen levels. The largest part of the increase in transport comes from a threefold reduction of the oxygen level in venous blood; when an organ idles, blood leaving it contains almost as much oxygen as does blood coming in. In all, our muscles extract about four times as much oxygen from each passing bit of blood when we work hard. Combining the increase in blood's flow rate with the increase in oxygen extraction, we get an increase in oxygen supply no less than sixteenfold.

Third, the blood gets routed differently, with a far higher fraction going to the muscles. The body can raise blood delivery to the muscles from the resting 20 percent to nearly 90 percent—another fourfold increase and then some. The last time I was tested on a treadmill, my whole-body metabolic rate (usually measured by following oxygen consumption) rose sixteen times over its resting level. So the metabolic rate of my muscle must have increased by about seventy times over its resting rate! And I'm certainly no young athlete. Since we don't work all our muscles equally hard, some muscles must increase their power consumption far more. (Curiously, the power consumed by the heart goes up far less in the process, only about doubling.)

A few data will illustrate the dramatic shift in blood supply that accompanies this fourfold increase in the output of the heart:[5]

Region	Rest		Maximum Exercise	
	%	Liters/min.	%	Liters/min.
Digestive system	24.1	1.20	1.1	0.22
Kidneys	19.0	0.95	1.0	0.20
Brain	12.9	0.64	3.0	0.60
Skin	8.6	0.43	2.4	0.48
Other	10.3	0.52	0.4	0.08
Heart	4.3	0.26	4.0	0.80
Skeletal muscle	20.7	1.04	88.0	17.60
In all:		5.04		19.98

Notice that the brain gets as much blood during exercise as at rest; the fourfold decrease in the fraction it gets matches the fourfold increase in the overall supply. The blood supply to the skin actually increases a little; exercise produces lots of heat, and we use the skin for cooling. How much we divert to the skin depends on, among other things, air temperature. But we kiss off our kidneys and digestive systems for the duration, part of the reason we risk stomach cramps when we do hard work after a meal.

How does the circulatory system manage this major reapportionment? A hint comes from something that doesn't happen or at least that happens less than you might guess. By the rules of fluid mechanics, pushing a liquid like blood through a set of small pipes four times as fast ought to take four times as much force. That ought to translate into a fourfold increase in blood pressure when a person works maximally. Blood pressure does go up, but it doesn't even double. A law-abiding animal that doesn't raise its blood pressure must decrease the resistance of the system to flow. That can be done in either of two ways. It might increase the bore of the pipes, or it might increase the number of pipes running in parallel.

We do both. The relevant physical rule turns out to be unexpectedly helpful. Double the diameter of a pipe, and sixteen times as much fluid will flow through it, all else (pumping pressure, for instance) being equal. So a little alteration of bore can do a lot of reapportionment.[6] How to change the bore? The greatest part of the resistance of a circulatory system comes from the small arteries rather than from the veins, capillaries, or big arteries. Muscular sleeves surround our small arteries, and chemicals rather than nerve impulses, as in skeletal muscles, control these particular muscles. In anticipation of hard work, our brains tell us to produce certain chemicals that instigate widening (dilating) the small arteries of heart and muscle while triggering constriction of a lot of other arteries.

Release of these chemicals also lets blood enter previously closed vessels. Once hard exercise begins, carbon dioxide and other by-products of muscle activity further stimulate vessel opening and increase flow. During exercise, between ten and a hundred times the normal number of capillaries in the muscles may be open for passage of blood. That not only increases the blood supply but also reduces the distance across which oxygen (helped by myoglobin, of course) must diffuse to get to the mitochondria. Warming up before strenuous exercise starts up all this flow shifting, so its value in reducing cramping and even cardiac insufficiency represents no mere ritual say-so. With the huge increase in metabolic rate of muscle when active, the term "warming up" can be taken literally, even if temperature increase isn't the main benefit.

Cranking up the circulatory system in strenuous activity makes a big difference; we might wish it made still more. We humans do better than many mammals, such as cats and rabbits, in increasing our oxygen consumption during sustained activity, but we don't do as well as dogs and horses. The bottom line is that the limitations of our circulatory and respiratory systems preclude running all our muscles at their capacities simultaneously. Were oxygen supply and removal of wastes and heat not limiting (and assuming some worthwhile activity that used all our muscles at once), we might do four or five times better. Imagine what that would do to athletic records that turn on aerobic performance!

Remodeling the Machine Itself

Several times each week I push, pull, lift, and otherwise displace weights on a series of machines that would have graced the Spanish Inquisition. I've been assured that the resulting increase in the sizes and effectiveness of my muscles will yield more than cosmetic benefits. Several years of distinctly nonfanatic activity have allayed my instinctive skepticism. I notice a difference, and once in a while the extra strength proves useful. Exercise, and your muscles get more massive; put a limb in a cast, lie for a week or so in a bed, spend time in a spacecraft, or even write books on a word processor, and your muscles atrophy. Somehow the system keeps track of how it's being used and uses that information to adjust itself in anticipation of future demands. Nor are muscles the only body component that routinely remodels; they're just the largest and most conspicuous. Blood vessels and bone do likewise, as do many of our other parts. An archaeologist guesses how a tool was used by the way its material has worn smaller; the same archaeologist guesses how an arm was used by the pattern of hypertrophy of its bones, the way its material has worn larger, I might say. In short, we exert control not only over times from milliseconds to minutes but also over weeks, months, even years.

We've known since the classic work of Rudolph Schoenheimer in the late 1930s that all the proteins within all our cells undergo continuous breakdown and resynthesis. What he described as "the dynamic state of body constituents" provides the context for this remodeling. In an adult human between 1 and 2 percent of all muscle protein breaks down each day, and the person's muscle fibers, directed by their nuclei, resynthesize almost precisely the same amount. So exercise and inactivity don't induce changes in an otherwise static system but instead shift the balance between breakdown and synthesis.[7]

For muscle, one other component enters the picture. To keep muscle intact, its nerve must supply both impulses and some chemical factors. A muscle deprived of its nerve will degenerate. Over about two months the contractile machinery disappears, leaving little besides tendonous tissue. Reinnervation can reverse the degeneration, but only if it happens within (roughly) that two-month period. Sherrington puts best the wondrous character of this machine that must function to maintain itself but that knows just how to restore its functionality:

> One brief glimpse must not let us suppose that when the embryonic phase of life is over, this power of the parts of the body to "become" reaches its goal and ceases. Suppose a wound severs a nerve of my arm. The fibres of the nerve die down for their whole length

between the point of severance and the muscles they go to. For eighty years and more these, my nerve-fibres, have given no sign of growth. Yet after this wound each fibre, whether motor or sensory, would start again to grow, stretching out toward its old goal. It would arrive finally, after weeks or months, at the wasted muscle-fibres which were its goal. These too it would, as it were, recognize forthwith. With them it would unite at once. . . . Nor would one nerve-fibre of all the thousands join a muscle-fibre which another nerve-fibre had already begun to repair. When all the repair was done the nerve's growth would cease. The wasted muscle would recover.[8]

What happens when a muscle enlarges? Its cells don't divide, since fully formed muscle has given up any strictly cellular organization; muscle fibers, we remind ourselves, are large, multinucleate entities. Two possibilities remain. New muscle might generate from unspecialized muscle precursor cells that have hung around—myoblasts they're called during embryonic development—or individual fibers might enlarge, with their nuclei signaling production of the various proteins needed to make more contractile, triggering, and energy-processing machinery. Enlargement, the second alternative, is what happens. The nuclei of a fiber stimulate production of new protein, but the fiber doesn't divide, so a muscle enlarged by, say, strength training has the same number of fibers—genetically or prenatally determined—as it had initially. At least that's the way things look by an indirect test, since one can't slice a muscle and then train it. A needle stuck into a muscle can be used to extract a sample for microscopic viewing and chemical analysis without causing a person particular injury or discomfort. Comparing samples taken before and after a period of training shows that fibers enlarge in close proportion to muscle enlargement as a whole.

A muscle can also be persuaded to change length aside from any generalized growth of an animal. Immobilization of the hind leg of a small mammal in a position in which a muscle is slack impels the muscle to shorten by removing sarcomeres from its ends. Similarly, an immobilization that exerts a pull on a muscle stimulates the muscle to add sarcomeres, again to the ends of its myofibrils.[9]

Not only can muscle enlarge when persuaded by increased use, but it can change character as well. We learn that by comparing equivalent

muscles in different people or by taking samples before and after a training regimen that differs from previous activity. Both our worries about
an aging and sedentary population and our athletic fanaticism promote
such investigations, so we've learned a great deal about muscle's various
transformations.

Thus how one trains makes a difference, something now recognized
by every serious coach. Also, muscles respond to the stimulus of training
with both long-term and short-term changes. Normally, slow (red) muscle receives sustained nerve impulses of low frequency, while fast (white)
muscle gets intermittent bursts of more rapid impulses. It's possible to
switch the nerve supply between members of a pair of muscles. Over
time the muscles gradually change their contractile characteristics, the
slow one getting twitchier and the fast one becoming slower and better
able to sustain activity. Implanting tiny stimulators in experimental animals for months at a time shows that impulses alone are sufficient to
induce the transformation. A particular pattern of impulses turns out to
be sufficient to activate particular genes to synthesize more of a particular set of proteins; one needn't invoke chemical control through the
nerves.[10]

These transformation experiments mimic what we normally do,
beginning shortly after birth. Initially, muscle is muscle, all much the
same. (This underlies the pale color of meats, such as veal, that come
from young mammals.) How we use a muscle, how the central nervous
system sends signals to it determine how it will develop, not just in size
but in structure and biochemical properties. We're perpetually in training, intentionally or incidentally. My efforts with the weight machines
just attempt to compensate for the degeneration that would otherwise
accompany the inactivity of my occupation and advancing age. We've
long recognized that the best training for a specific athletic task consists
of doing either the task itself or exercises similar to it. That's not just a
way to improve coordination or other aspects of neural skill; it enlists the
normal plasticity with which our muscles adapt themselves to changing
requirements.

Endurance training doesn't much affect the size of myofibrils or, for
that matter, the overall size of the muscles involved. That makes sense.
Larger myofibrils would increase the distances across which oxygen has
to diffuse and thus gain little in the long run. Furthermore, force isn't
what endurance is about. Nonetheless, the muscles of sedentary individ-

uals and trained long-distance runners turn out to be very different. Runners have both more and larger mitochondria as well as a lot of fat droplets associated with the mitochondria, as in Figure 5.3. The effect isn't trivial; mitochondria take up about 50 percent more of the volume of the leg muscles of trained runners than of sedentary humans of similar age. Also, peak oxygen-consuming ability parallels mitochondrial density.[11] Of course we might argue that people with better mitochondrial endowment tend to go in for long-distance running. But a study of identical twins closes that alternative: a period of vigorous endurance training of one of each pair increased both mitochondrial density and oxygen-consuming ability, each by a satisfyingly coincident average of 15 percent.[12]

Long-distance runners may be skinnier, on average, than the rest of us, but for all their training, they don't look radically different. Atrophy of fast white fibers about balances hypertrophy of slow red fibers. Endurance training, sustained activity against small loads, does induce increase in muscle mass—but the mass of the heart rather than that of

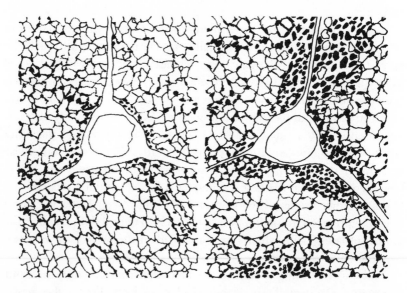

FIGURE 5.3. Equivalent muscles of sedentary (left) and endurance-trained (right) individuals. These are tracings of electron micrographs; the similarity in layout should not fool anyone into thinking that these are before and after pictures of the same material!

the skeleton. By contrast, resistance exercise, making muscles briefly exert near-maximal force, changes the mass of skeletal muscle but not of heart muscle. The appearance of bodybuilders and weight lifters announces their deviant behavior—whether or not you judge the difference attractive. Incidentally, using other animals to learn about such transformations, however wonderful in theory, can encounter special difficulties. As Alan McComas, an exercise physiologist, put it, "The ingenuity of the investigator in devising an animal model of exercise has often been exceeded by that of the rat (the usual choice) in minimizing effort."[13] Animals, sensible creatures, prefer sloth.

The story has yet another chapter. As implied by what's been said about fuel supply, increasing your power output takes more than cranking up your contractile machinery or multiplying your mitochondria. The blood supply of a white muscle can't manage the demands of a red muscle. Not unexpectedly, when a muscle transforms, its part of the circulatory system readjusts as well. Any bit of body that consumes oxygen at more than its normal rate releases chemicals that have come to be called angiogenesis factors, just a descriptive label for stuff that stimulates the growth of blood vessels. Angiogenesis factors have received recent attention because of the way growing tumors, which we abhor, produce them. So not only can an active muscle trigger a short-term increase in its blood supply by telling its capillaries to open, but it can instigate longer-term changes by stimulating the proliferation of capillaries and other vessels.[14]

That's only half the transformation of blood vessels. Capillaries may be standard items, each about one two-hundredth of a millimeter in diameter, but the rest of the vessels come in every size above that—in us up to about an inch in diameter. Moreover, the diameters of the rest of those vessels don't get fixed once and for all. Send blood more rapidly through a vessel, and the vessel lining detects the greater flow. Its cells begin to divide and in turn stimulate division of the rest of the cells that form the vessel wall. Over a period of weeks to months the blood vessel enlarges, and you should recall that even a slightly larger vessel will carry a lot more blood. The main rule for the resizing, a neat mix of mathematics and fluid mechanics put to physiological purpose, has been known for several decades. What matters here are (1) that all lining cells work as detectors, (2) that all, whether in large vessels or small, have exactly the same threshold speed that triggers division, and (3) that all these local decisions turn out to produce a hydrodynamically optimal branching system by a lucky turn of physics that evolution could not have invented.[15]

Brains and genes and embryonic development may be impressive, but equally impressive are the mechanisms by which local control and demand-driven changes continuously fine-tune the organism. We're not built once and for all according to some blueprint in our genes, but instead the genes provide general instructions that, like recipes, are interpreted and refined according to local circumstances.

Something other than the training of Olympic athletics provides the epitome of dramatic muscle remodeling and audacious intervention. It originates in an even more fanatic concern of contemporary humans. Heart disease dispatches more of us than any other category of affliction, at least in well-nourished and well-medicated (or perhaps overnourished and overmedicated) first world countries. So we explore therapies from life-style and dietary change to various kinds of medication, surgery, and mechanical prostheses. To circumvent the present limitations of transplantations and mechanical substitutes, a few surgeons have tried a different (and equally heroic) way to help a failing ventricle.

The particular bit of surgery specifically addresses congestive heart failure, a condition in which the walls of the main pair of pumps, the ventricles, enlarge, thin, and apply only weak squeezes to the blood inside. The body needs more and better muscle, and the procedure supplies that from the patient's body itself. Thus, like bypass surgery, it avoids awakening the immune system. But where can the muscle come from? Most muscles won't wrap around something shaped like half an egg, functional muscle requires both nerve and blood supply, and diversion of many muscles would leave a person awkwardly impaired. Current activity centers on a particular muscle, the latissimus dorsi, and a procedure called dynamic cardiomyoplasty.[16]

The latissimi are a pair of big, flat muscles of the back. The upper end of each is attached by a tendon, about three inches long, to the bone of the upper arm (the humerus) near its upper end—in other words, just beneath the shoulder. The muscle widens out below its tendon and runs under the lower parts of the shoulder blades (scapulae) toward the middle of the back. It attaches to the bony rearward extensions of the lower vertebrae of the chest and torso (thoracic and lumbar vertebrae) by way of a flat, tendonous sheet. The latissimi serve a variety of functions, but for none of them are they the sole actors. They mainly help swing your arms downward and rearward. If you're climbing and have your arms in front, they pull your body upward and forward.

Because a latissimus doesn't lie near the heart, it has to be moved. One of its attractions is that it can be moved without cutting its nerve and blood supplies, which enter near its upper end. But the procedure, seen diagrammatically in Figure 5.4, makes something of a mess. First the surgeon frees the lower, broad end of the muscle—either the right or left latissimus can be used—and inserts stimulating electrodes near its upper end. Then a window gets cut in the second or third rib, and the free end of the muscle is threaded over the shoulder and down through that window into the chest and pericardial cavity. The surgeon wraps the muscle around the pair of ventricles, secures it in place, and runs the wires from the stimulating electrodes and some sensing electrodes to an electrical unit beneath the skin of the abdomen. Stimulation of the latissimus makes it contract and squeeze the ailing heart, assisting the heart's own muscle. Some additional benefit may come just from wrapping an elastic material around a distended ventricle, making its own efforts more effective.

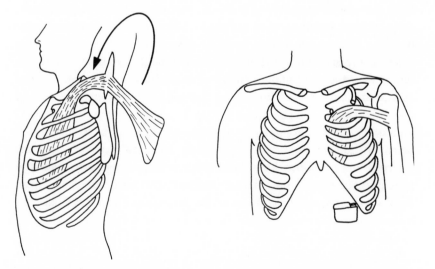

FIGURE 5.4. Cardiomyoplasty. The diagram on the left shows the unhooked latissimus muscle before and after being threaded down into the thorax; on the right is the finished job, with the implanted circuit box.

All that trauma, pain, and tricky rehabilitation bring little immediate benefit. Because the latissimus dorsi is a fast white muscle, inappropriate for uninterrupted pumping, it must be transformed. After a respite of about two weeks to allow the muscle to adhere to the heart and to readjust its own circulation, stimulation begins. Sensing electrodes detect the heart's normal rhythm, so the stimulating electrodes can make the latissimus join in at the right times. The muscle then begins transforming, and, after two or three months it has become a proper red muscle, entirely suitable for around-the-clock function.

Cardiomyoplasty remains far from routine. Since the first in 1985, nearly a thousand have been performed worldwide. The surgery may not require a cardiopulmonary bypass, needed for a lot of other heart surgery, but it's heroically invasive and lengthy, taking seven or eight hours. Nothing, I might quip, for the faint of heart. I suspect it will be used (to a limited extent) until some sufficiently effective mechanical prosthesis replaces it; experimental ventricle-assisting devices show considerable promise.

BEFORE WE move from how muscle works to how it's arranged in the body and how we make use of it, the message of this chapter needs reiteration. Muscle does more than any engine of our technology. Direct comparisons or even analogies mislead. Like almost every other body component, large and small, it has what one might call smarts. The system is dynamic, continuously sensitive to the results of its own actions and continuously responding to the demands we place upon it. Yes, we can and do build such goal-directed machines—the great American physiologist Walter Cannon (1871–1945) was wrong when he declared such behavior (homeostasis he called it) unique to living systems[17]—but the smarts of our machines, for all their fast feedback connections, pale into insignificance next to one that can redirect itself over time scales from fractions of a second to decades and does so with no overt fuss at all.

CHAPTER 6

Connecting Up Muscles

Who indeed would be stupid enough to look for a machine to move a very light weight with a great force, i.e. use a machine or a contrivance not to save forces but rather to spend forces? This is as if one who can move and carry a weight of one pound directly without any machine by exerting a force equal to one pound, with complete disregard for economy, was looking for levers, pulleys and other instruments requiring forces ten or a hundred times greater to raise this one pound. And, if this is rightly considered as stupid, how is it possible that wise Nature, everywhere looking for economy, simplicity and facility, builds with great industry in animals machines to move, not heavy weights with a small force but, on the contrary, light weights with almost boundless force. This seems strange and against common sense, I agree, but I can convincingly demonstrate that this is what happens. . . .[1]*
GIOVANNI ALFONSO BORELLI, Renaissance scientist*

MUSCLE REMAINS MUSCLE, ONE REMARKABLE ENGINE. THE same device powers a microorganism, thirty-thousandths of a millimeter long, that powers a whale, thirty meters (one hundred feet) or a million times longer. Muscles do differ among themselves—red muscle, white muscle, fibrillar muscle, catch muscle—but we mustn't lose sight of the underlying commonality of cross-striated skeletal muscle. In mechanism of action, in speed of action, in force production, in power output, muscles differ less among themselves than do the electric motors that power our home appliances.

We run marathons and assemble tiny watches; fleas jump and worms crawl; cicadas scream and hearts beat. The range of tasks animals do with muscles dwarfs the diversity of the muscles themselves. Animals use a wide range of coupling devices—transmissions—to deal with the problem of using one kind of motor to do all kinds of jobs. We've seen how muscle works, as far as is known; knowing how we work muscle takes some additional explanation and reveals further cleverness and further uncertainties. Our science remains a work in progress.

Unshortening Schemes

When stimulated, muscle fibers pull their ends together or at least try to. Since they pull rather than push, something has had to reextend every muscle that has ever contracted. By old habit, we call these reextenders antagonists despite the term's malevolent odor. Asking how nature achieves reextension requires a look at muscle antagonists, and these in turn provide an initial glimpse of nature's transmissions.

Another muscle provides the textbook antagonist. With the biceps of your upper arm you flex your arm and lift a weight in your hand. With the triceps on the opposite side of the arm you reextend, pushing downward instead of upward. The lower end of the biceps connects to the forearm in front of your elbow; the lower end of the triceps connects to the back of the elbow, behind the hinged joint of upper arm and forearm, as we saw back in Figure 5.1. But for their upward rather than downward forces, the two muscles play out their roles like a pair of children on a seesaw or the known and unknown weights on a beam balance. In fact, the picture oversimplifies; nothing requires that antagonists work as specific and unique opponents, and most often combinations of muscles antagonize other combinations. Nor must the forces (torques, strictly) of antagonistic muscles match; your biceps can overpower your triceps. The muscles with which you raise your lower jaw, close your mouth, and bite down generate a lot more force than the muscles that do the opposite; a general facial spasm is called lockjaw rather than lockgape. I've heard that holding together the jaws of an adult alligator takes no great muscle (but lots of nerve!), while no one tries holding them apart more than once.

A simpler device antagonizes the shell-closing muscles of clams and other bivalve mollusks—the muscles with the neat catch mechanism. Right near the hinge connecting the halves of the shell sits an elastic pad, shown in Figure 6.1. Contraction of the big central (and tasty) adductor muscle of a scallop or the pair of adductors in the clam, mussel, and oyster compresses the pad as well as closes the shell. When the muscles relax, the pad expands and reextends them. Cutting the muscles doesn't so much loosen the shell as permit that elastic pad to force the half shells apart.

A jellyfish, in the same figure, uses a similar elastic when it propels itself. Muscle running around the circumference of its hemispherical bell squeezes out water as it squeezes the soft but elastic sides and cap of the bell. Elastic recoil of the bell then reextends the muscle. A squid does a fancier version of the same trick. It contracts muscle that runs around its outer mantle, thereby putting the squeeze on water in the cavity beneath the mantle, which squirts out of the so-called siphon tube as a propulsive jet. The squid does have some muscle that reexpands the mantle—a layer of muscle the very short fibers of which run directly outward works as a mantle thinner—but much of the reexpansion of the mantle depends on simple elastic rebound. While we make some use of elastic rebound to reextend muscle, we do nothing as dramatic as do these various invertebrates. A hoofed mammal, such as a deer or camel, has a ligament that runs along the upper part of the neck (Figure 6.1); when the animal lowers its head to feed, the ligament stretches and thereby helps raise the

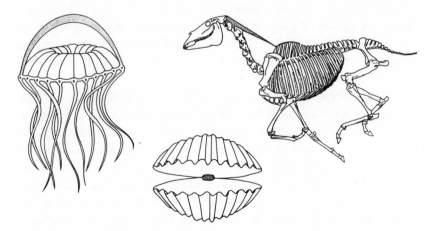

FIGURE 6.1. Three uses of passive elastic rebound to antagonize muscle: the bell of a jellyfish, the hinge ligament of a scallop, and the nuchal ligament of the neck of a horse.

head again, acting as an antagonist, although not the sole one, for the muscles that pull the head downward. Turkeys have a similar ligament. Sometime take a good look at a turkey neck before you drop it into the soup pot.[2]

Still another way to reextend muscle depends on hydraulics. When a liquid is squeezed, it exerts pressure in all directions, something we put to use in everything from car brakes to hoists for heavy machinery. Hydraulic connections can pull, but they do better for pushing.[3] Only the quality of the fittings and strength of the pipes limit their ability to push, but few can pull at pressures above one atmosphere without ruinous vacuum voids appearing in the pipes. Still, pushing counts most when you antagonize a puller, so there's no real problem.

No animals put the device to better use than spiders; hydraulics truly puts them a leg up. Spider legs have only flexor muscles, muscles that pull the legs in toward the thorax, and their lack of extensors long puzzled arachnoanatomists. In the late 1930s Herbert Ellis, then a graduate student in neurobiology, showed that extension occurred hydraulically. When a spider wants to extend a leg, it pumps the blood into the leg at pressures of as much as six-tenths of an atmosphere, nearly four times human blood pressure. The system requires both an inextensible outer wall for the leg and internal tissues through which blood can ooze. Spiders meet both requirements, but so do all the other arthropods, none of which extends it appendages hydraulically as it moves about. The nearest other case I know of happens in some insects. Just after an adult moth emerges from its pupal skin, it raises its blood pressure by detelescoping its abdominal segments. The increased pressure in its wing veins then expands its wings.[4]

(Some years ago I served with Herb Ellis on the board of our local Friends school. One night he asked the head teacher, also a biologist, and me if anyone paid attention to spider hydraulics. I told him that the phenomenon was now standard textbook stuff. He then disappeared briefly and returned with a yellowing reprint of the paper that had resulted from his thesis. He had never returned to the field and worked, when I knew him, in our local pharmaceutical industry. When something becomes sufficiently familiar, we often stop asking who alerted us to it in the first place.)

To refill the ventricles of our hearts, we have to reextend the ventricular muscle. This we do by using a combination of hydraulic pressure—from the atrial chambers—and elastic recoil. No vertebrate heart comes equipped with a layer of wall-thinning muscle comparable to that of the mantles of squids. So we're in the game, low-pressure players compared with spiders, if superior (when measured by pressure) to squid and jellyfish.

Muscle can reextend in yet another way, using what have come to be called hydrostatic skeletons. Figure 6.2 shows two versions. Like antagonism by hydraulic pressure, these take advantage of the incompressibility of liquids. The basic idea couldn't be simpler. If muscle covers a closed, flexible container, then a contraction of the muscle that reduces one dimension of the container will increase some other dimension. Make it shorter, and it will get fatter; make it skinnier, and it will get longer. By contrast with spider leg or butterfly wing hydraulics, hydrostatic skeletons need no hard outer shells. On the other hand, they're limited to cylindrical, ellipsoidal, and spherical shapes. While they look unfamiliar, they use muscle to antagonize muscle in a way not all that different from what we do with our biceps and triceps; the liquid in the container (the "hydro-" in the "hydrostat") just substitutes for a bony humerus. In one arrangement, on the top in Figure 6.2, contraction of lengthwise muscle running along one side of a cylindrical animal will make the other side

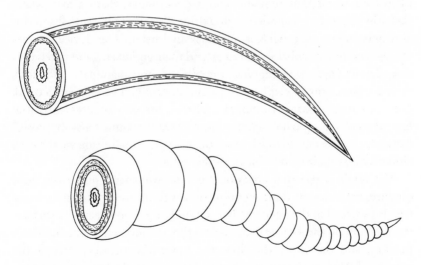

FIGURE 6.2. The location of muscle in a roundworm, which has only lengthwise muscle (dotted in cross section), and in an earthworm, with both lengthwise and circular muscle (the encircling dashes).

bow outward, extending its muscle. That's what small roundworms (nematodes) do, both those that live in the moisture between soil grains and sometimes bother the roots of plants and the ones that penetrate our tissues and cause such diseases as filariasis and trichinosis.

Earthworms and their ilk do something similar, but they use two sets of muscles in each of a series of separate segments, and they operate their hydrostats at far lower pressures than do roundworms. Contraction of lengthwise muscle in a segment makes it fatter as well as shorter, and this (on the bottom in Figure 6.2) stretches the muscles that run circumferentially around the worm. A worm alternately contracts the lengthwise and circumferential muscle of each segment, with the action of each taking place a fraction of a second behind the one in front of it. That plus some bristles that point backward, and, *voilà*, the worm burrows forcefully forward with the benefit of neither bone nor shell.[5]

The trick of using something incompressible but nonrigid has perhaps been pushed furthest in a variant of the hydrostatic skeleton called the muscular hydrostat. As Jan Swammerdam showed long ago, muscles are constant volume systems. So they resist being compressed just like water or blood. No core of blood, guts, and body fluid need be provided if muscle can be arranged to squeeze other muscle. Squeeze a water-filled balloon circumferentially, and it gets longer; that's a hydrostatic skeleton. Squeeze a cylinder of dough or clay circumferentially, and it also gets longer; it models a muscular hydrostat. The device enables squids (and other cephalopods) to extend their eight arms and two tentacles, lizards (and other reptiles, amphibians, and mammals) to extend their tongues, and an elephant to extend its trunk. The elephant's trunk does have nasal passages running lengthwise, but they're sufficiently well braced so changing trunk length doesn't give the animal a stuffy nose.[6] Muscular hydrostats do much more than simple antagonism; we'll have a closer look at their wonderful versatility later.

We retain a parental affection for our own ideas, no matter how obscure, so I cannot omit mention of a way to antagonize muscle on which I worked in the 1980s. When a liquid or gas flows across a surface, the pressure on the surface drops below the pressure where the fluid isn't flowing, and the faster the flow, the lower the pressure. That's the essence of Bernoulli's principle, derived in 1738 by the Swiss mathematician Daniel Bernoulli. When cars had carburetors, the reduced pressure of air being drawn into the engine sucked gasoline into that air. It still drops the pressure atop an airplane wing low enough to keep a plane aloft. But antagonize muscle? Consider a squid, jetting to escape a predatory fish. The squid has to refill its mantle cavity between each squirt, so it must expand the muscular mantle. It has the mantle-thinning muscle and elastic tissue mentioned earlier. But as in Figure 6.3, it also uses the flow itself to develop an outward force that helps it expand its mantle and draw in water. Something similar happens in scallops, as they repeatedly clap their half shells together during their brief episodes of swimming. Flow across the shell draws it open again, assisting the elastic pad near the hinge that was mentioned earlier. Finally, fin whales, which feed by raising the floors of their throats and squeezing

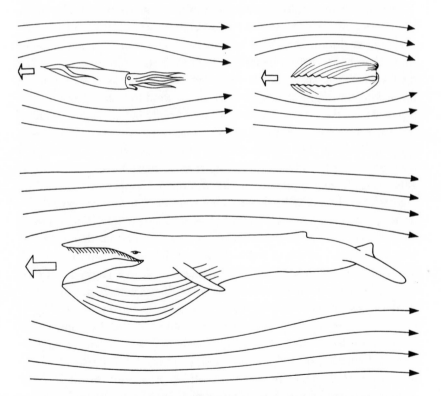

FIGURE 6.3. Three cases in which a flow-induced pressure decrease reextends muscle: reexpansion of a squid's mantle, reopening of a scallop's shell, and refilling of a fin whale's mouth. Because the animals move right to left, the flow goes left to right.

water through their baleen plates, use flow across their lower surfaces to help drop that floor again and thus draw in more water.7

A muscle may pull steadily, but once it has shortened, it generates only force and does no external work. To work (in the formal sense) for any appreciable length of time, it has to give a series of pulls, relaxing and lengthening between each. So it can only work (in both formal and vernacular senses) as a pulsating motor, and to do that takes what we might half-seriously term dynamic antagonism. Despite all her diversity and evolutionary ingenuity, nature oddly enough makes little use of two of our ways of dealing with pulsating motors. Dazzled by what nature does, we all too easily lose sight of what nature doesn't: her diversity and ingenuity are not unlimited.

An electric motor, for instance, doesn't push its rotor around steadily but instead gives the rotor a kick at several points (most often two) during each revolution. That's the main reason electric motors hum. Something then has to bring the next pole of the rotor into position for the next kick. What keeps the rotor going between kicks is its inertia; a massless rotor might be easy to start, but it wouldn't keep going. Like a top or yo-yo, the rotor stores energy very briefly by taking advantage of the way rotational momentum persists unless expended as work or dissipated by friction. The single-cylinder internal-combustion engine of a lawn mower uses a more extreme version of inertial energy storage. It gets its kick from a fire in its cylinder. But the fire happens not twice during each revolution of its flywheel but only once for every two revolutions. So the inertia of the flywheel plays a still larger role. The motor of a leaf blower or chain saw fires once per revolution rather than once per two revolutions, so it can make do with a lighter flywheel, but it still needs one. The trouble with the scheme for nature's engines must go back to her lack of wheels. Flywheels, after all, are wheels, not flies.

Besides such inertial storage, our devices occasionally use gravity to offset the intermittent pushes of their motors. In the development of both steam-driven and electrically driven motors, oscillating engines preceded rotating ones. Some of these used pendulums instead of flywheels to counteract the on-again-off-again actions of their basic movers, and pendulums depend on gravity. Two factors, though, reduce gravity's attractiveness as a counterforce. For one thing, the faster we run an engine, the more power we can get relative to its weight, so our modern engines have become speedy affairs. Even a short pendulum takes too much time to complete a swing for motors of any respectable power-to-weight ratio.[8] For another, gravity acts downward, so a pendulum must always keep the same end up. You can use one to make a clock, but a wristwatch much prefers a hairspring and brief elastic storage instead. Nature has a similarly equivocal attitude toward gravitational storage. She certainly uses it—as we'll see, it's crucial in walking—but not primarily for muscle antagonism. We also use it for tasks other than immediate engine action—we pump water uphill into reservoirs during periods of low electrical demand and then let the water run down through turbines when we need to recover the electricity—but we no longer make it complete the action of our pulsating engines.

Levers to Increase Distance and Speed

Archimedes (287?–212 b.c.e.) must be best known for crying, "Eureka!" ("I have found it") in the streets of ancient Syracuse when he hit upon his principle, the rule relating density and buoyancy. His second best-known quotation refers, with slight hyperbole, to levers: "Give me where to stand, and I will move the earth." Force multiplication: that was his point.[9] Since the application of muscles to tasks most often involves levers, we need a short digression on these marvelously versatile and ubiquitous devices. Bear in mind, though, that the levers used to apply muscle differ fundamentally from what Archimedes had in mind. To put it simply, they don't increase force; they diminish it.

The simplest lever consists of a rigid beam with a pivot point right in the middle, a seesaw. Pushing down on one end makes the other end push up. The lever reverses the direction of the force, but it does nothing else and needn't bother us further.

Things become interesting when the distance between the pivot and the place we apply force differs from the distance between the pivot and the place where force comes back out. Whether wrench or nutcracker, such an asymmetrical lever, as in Figure 6.4, engages in a bit of trading. What goes in consists of a force that pushes the lever a certain distance at a certain speed. What comes out consists of a force that pushes a load a certain distance at a certain speed. A lever such as a prying bar increases the force you apply. Since it neither uses nor supplies work (force times distance, remember), it therefore must decrease distance. So you push farther (if less force-fully) than the load gets pushed. Leverage follows one simple rule: the force

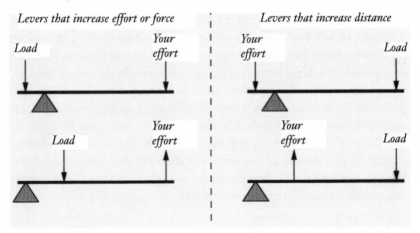

FIGURE 6.4. Basic levers that increase force (left) or distance (right). The former characterize most household devices; the latter include most muscular systems.

you exert times the distance through which you move that force just equals the force the lever puts out times the distance the lever moves the load. Force times distance going in equals force times distance coming out.

Since speed is distance divided by time, and since your input and the lever's output take the same time, a lever that decreases distance must also decrease speed; you push faster than the load gets pushed. So as a corollary to the rule, force times speed going in equals force times speed coming out.

Levers like wrenches or nutcrackers put out more force than you put in. By contrast, muscles produce lots of force but move things only short distances; recall that they usually shorten by 20 percent or less of their resting lengths. That of course was what puzzled Borelli's contemporaries. We can put the matter in more immediate terms. You swing your hand through an arc of nearly three feet by contracting your biceps muscle just inches. So nature must produce a longer, not a more forceful, motion from muscle—if we're to swing arms or legs far enough to throw or walk. Muscle-driven things like our appendages need to increase distance at the expense of force. Thus muscle-operated levers like arms and legs multiply distance instead of force.

At some point you may have run into the term "mechanical advantage" in reference to the force-increasing assistance provided by household or workshop levers. A lever's mechanical advantage is the number of times by which it increases force. But that's awkward for talking about how muscles work because the decrease in force when they're hooked up to bones and such gives them mechanical advantages below one.[10] Instead we'll use a different term, one of less ambiguity and prejudice. Let's say that a lever that increases distance rather than force has a *distance advantage*. If it increases distance threefold, we'll say that it has a distance advantage of three. Since speed change follows distance change, we just have to bear in mind that any lever that multiplies distance will multiply speed by the same amount. (Distance advantage, then, equals one divided by conventional mechanical advantage.)

Nature's distance advantages vary widely. After all, in some cases a lot of movement matters, and in others a fair bit of force must be exerted. Distance advantages for the muscles with which animals run and walk tend to be high, while distance advantages for the muscles with which we chew run much lower. Even a single muscle can be used at different advantages. For instance, an animal nips things with its front teeth, which, being a long way from the jaw-closing muscle, have a substantial distance advantage and move rapidly. It chews things with its back teeth, which, being closer to the muscle, bite more forcefully but don't move as far—a lower distance advantage. Also, equivalent muscles can be used at different advantages in animals that use them for different functions. The triceps of the upper arm of a mole operates at a much lower distance advantage than that of a human; burrowing, like chewing, takes more force and less speed than throwing. But keep in mind the underlying forcefulness of muscle; even our molars and the mole's arms operate with distance advantages greater than one.

A few numbers should add a note of reality:

- We use our forearm-lifting biceps at a distance advantage of 5.5 and the antagonistic triceps at about 22. That means that if we assume similar muscle cross sections and muscle type, the hand can be lifted or flexed more forcefully but lowered or extended more speedily. We're not well arranged to lift weights while upside down! On the other hand (I might say), a pitcher can throw a baseball at nearly a hundred miles an hour.
- We close our mouths with paired temporalis muscles, located just in front of the points where the lower jaw pivots against the skull. Front

teeth bite with a distance advantage of 3.5; molars chew with distance advantages of around 1.7. That low value gives great force, useful, if hazardous to the teeth.

- Kicking a leg forward ought to be done rapidly to be useful in locomotion, and the main kicking muscle, the quadriceps femoris of the front of the upper leg, operates with a distance advantage of about 17.5. But such a high distance advantage precludes pushing hard on a door just by pressing the lower leg and foot against it. You can kick a door effectively enough, but that's using the leg's mass and accumulated speed rather than its immediate muscular force. By contrast, pushing the toes downward to stand higher and perhaps peer over a fence requires lifting your weight. That takes lots of force but elevates you only a little. The main muscle involved, the gastrocnemius of the back of the lower leg, operates at a much lower 2.3.

Thus 5.5 versus 22; 1.7 versus 3.5; 2.3 versus 17.5—different enough to transform the output of the basic engine.[11] While values for advantage don't often get cited, the notions here are nothing new. Nor are the arrangements of muscles and bones that produce these differences. Borelli certainly understood how the details of their attachments set the leverage of muscles.

Within a stable structure, all forces must balance. Tension—pulls—must be balanced by compression—pushes. Muscles produce tensile force, so a stable structure must contain something that withstands compression. For us, the main compression resistor is bone. We surround our bones with muscles, connecting the two with tendons and other variously

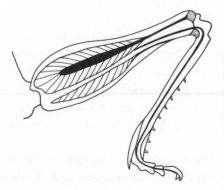

FIGURE 6.5. How the main jumping muscles in the upper hind leg of a grasshopper connect with its outer segments. Notice that the muscle that extends the leg is much bigger than the one that flexes it.

named straps, belts, and sheets. Arthropods do things the other way around, with the tension-producing muscles on the inside and the compression-resisting stiff stuff on the outside, as in Figure 6.5. In the jargon of the biologist, we have endoskeletons while they have exoskeletons. Leverage presents peculiar problems for a system in which the muscles are inside. Muscles, or at least their passive extensions (in arthropods, apodemes rather than tendons, but no matter), must extend across the joints. But since they're on the inside, they can't be above or below the places where the tubular skeletal components contact. An arthropod has no kneecap to permit a tendon to pull at least a little obliquely to the axis of the shinbone. So it has to operate its muscles at very high distance advantages; an almost imperceptible contraction of a muscle must give a wide swing of an appendage. To make matters still more extreme, many arthropods have the skinniest of legs; think of crane flies, spider crabs, or harvestmen (daddy longlegs) with their muscles inside their legs.

The extreme of the extreme occurs in the wing-beating systems of flying insects, appearing here yet again. As explained earlier, small size demands high rates of beating, and tiny insects beat their wings hundreds of times each second. Problem: A. V. Hill long ago showed that faster operation of muscle means less force produced and that above some optimum speed, faster means lower power output. Flight costs as much as anything any animal does. So the small insect must contract its flight muscles frequently but without at the same time contracting them at high intrinsic speeds. Solution: contract them only by a small fraction of their resting lengths. Most muscle works well at a shortening of 10 percent; flight muscle in flies, bees, and beetles typically operates at shortenings of only about 4 percent.[12] Consider what that implies for a fly muscle that powers a downstroke and must contract in one four-hundredth of a second. If it contracts by only 4 percent, its intrinsic speed stays down at sixteen lengths per second (400 x 0.04), a respectably and effectively low value.

Low shortening distances require high distance advantages; somehow flight muscle must make insect wings swing through an arc of at least 120 degrees. From anatomical data, I estimate the distance advantage as about 100.[13] Most insects achieve such high distance advantages with a radical arrangement of their main muscles. As shown in Figure 6.6, the muscles make no direct attachment to the bases of the wings. Instead they distort the thoracic container as a whole, with one set mak-

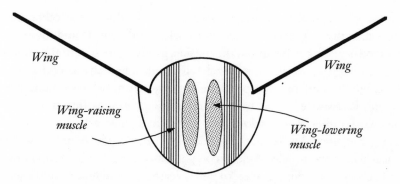

FIGURE 6.6. How the arrangement of flight muscles in the thoraxes of most insects allows them to raise and lower their wings—without the muscles' attaching to the wings.

ing it bulge fore and aft and the other set making it bulge top to bottom. Lots of muscle power can thus be brought to bear on the wing hinges with only a little change in the shape of the thorax.

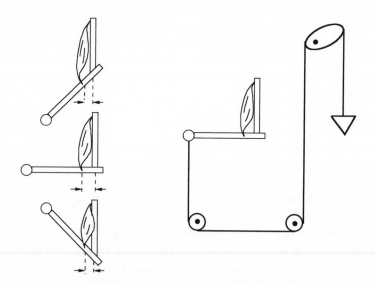

FIGURE 6.7. At left, how the lever arm changes as a forearm is flexed; the lever "arm" takes on a particularly literal meaning! At right, the way an exercise machine uses a noncircular pulley to vary the load it imposes, compensating for the changes in the lever arm.

These living levers may be even more sophisticated than the discussion so far has implied. Two tricks permit the distance advantage of a particular anatomical hookup to vary over the length of an appendage's swing. First, while a muscle doesn't shift its point of junction with a bone, its leverage depends on more than the location of that junction. To get maximum effect from a push or pull on a lever, you have to exert your force at a right angle to that lever. Any angle higher or lower than ninety degrees works less well, as Figure 6.7 illustrates. So the trade-off of force versus distance and speed varies with where in its swing an appendage happens to be at any instant. You find it harder to lift a weight with your forearm fully extended than you do once it has begun to flex a little. But the change in leverage doesn't account for the whole change in force as you swing your forearm. Interestingly, the biceps muscle, which does most of the work, passes through its resting length, the length at which it can exert greatest force, at that same ninety degrees.

A second trick depends on changing the trajectory of a muscle and its tendons in the course of shortening. The cable that lifts the arm of a small crane passes over a circular pulley where the crane's arm meets the boom. But a pulley need not be circular, and it might even take the form of some protruding strut; either arrangement will make leverage change with position. A lot of odd protrusions on bones as well as on such accessory structures as kneecaps tune the lever actions of our skeletomuscular systems in this way. The fancy machines we use for doing resistance exercises at fitness centers achieve analogous position-dependent changes in leverage. Lift weights with one machine or another, and most often you'll turn some strange, noncircular pulleys, as on the right in Figure 6.7. The radius of the pulley, measured at the point where the strap leaves its surface, varies with the angle to which the pulley has turned. As a result, the force needed varies with that angle. Such a machine thus fights your efforts to a degree that varies with where in its cycle you happen to be at any instant. Resistance training undoubtedly works; I'm less sure of the additional efficacy of all that mechanical complexity.

Making Even More Force

The fibers of some muscles are arranged in a peculiar way. They head not from one end toward the other but instead lie obliquely to the muscle's long axis. These muscles produce especially great force and exert it over particularly short distances. How do they do it? A short, fat muscle ought to produce more force with less shortening than a long, skinny one; force, after all, depends on muscle cross section, while shortening distance depends on muscle length. Their force amplification depends on making a long, skinny muscle behave like a short, fat one—what they accomplish by running the fibers obliquely rather than lengthwise. The oblique arrangement gives them, on the one hand, a lot more fibers but, on the other, much shorter ones. More fibers mean more force; shorter fibers mean less shortening. Such muscles, as in Figure 6.8, are called pennate or pinnate.

We've already seen how arthropods have to stuff their muscles inside their skeletons and how that must give the muscles high distance advantages. High distance advantages in turn demand more force but need less shortening distance. No surprise, then, that these superforceful pennate muscles find especially wide use in arthropods, so wide that we notice their absence more than their presence. Thus we're struck by the nonpennate flight muscles of insects, the consequence, it seems, of the way their location in the thorax allows a short, wide shape without any chicanery.

A bit more surprising is the fact that we use pennate muscles too. Some of our muscles can't take advantage of handy elbows and kneecaps or useful protrusions on bones, some of our muscles and their tendons

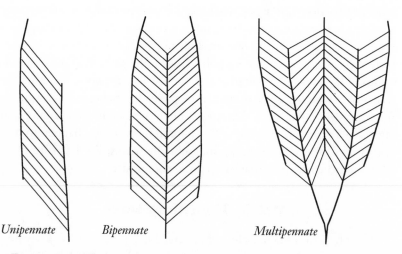

Unipennate   *Bipennate*   *Multipennate*

FIGURE 6.8. Three ways to realize the extra forcefulness of pennate muscles.

run close and parallel to the bones they move, and some get called on to do particularly force-demanding tasks. In such cases, anatomical or functional necessity drives high distance advantages, to which we respond with these offbeat or at least off-angle muscles.

Where do we use pennate muscles? The obvious case, the jaw-closing masseter, only starts the list. We flex a thumb, bending it toward the palm, with a unipennate muscle, the flexor pollicis longus, or long flexor of the thumb. Its belly lies well up in the forearm, connected to the thumb by a tendon that extends four or five inches handward from the muscle. That same tendon runs four or five inches back into the belly of the muscle to receive the pennate fibers. As a deep muscle, located close to the bone, this flexor suffers from low leverage; it needs a large distance advantage and therefore has to exert a lot of force.[14] Incidentally, most of the mass of muscle that runs the hand and fingers lies well up in the forearm; mobility and dexterity depend on keeping bulky muscles out of the hand itself. We do of course have small muscles in the hand, such as the flexor pollicis brevis, the pad of the palm next to the thumb, which works in concert with the flexor pollicis longus.

We raise a knee with another pennate muscle, this one bipennate, the rectus femoris. It runs downward from the outer part of the pelvis to the kneecap, again not far from the bone. We also have a pair of multi-pennate muscles, the deltoids. These thick, triangular muscles taper down and out from the clavicles, or collarbones, and the scapulae, or shoulder blades, to a tendon that attaches to a bump on the humerus, or upper armbone. With the deltoids we raise our arms.[15]

When two people drag something, each pulling on a separate rope, they normally pull in the same direction, with their ropes parallel to each other. We all know that if one pulls a bit to the left and the other to the right, they'll be less effective. By the same token, putting muscle fibers at an angle to the desired direction of pull sounds like a terrible idea, wasting force and reducing efficiency. Yes, it does these bad things, but no, it doesn't do them much. A muscle fiber angled at twenty degrees to a line between a muscle's ends exerts about 94 percent of its force in that overall direction of the muscle. Pennateness exacts only a minor tax.

You can see some bipennate arthropod muscles if you poke around in the briefly boiled large claw of a Maine lobster. Breaking the claw off the next segment usually exposes two thin but strong membranous structures. Diagonal muscle fibers attach to these central apodemes. Pulling on the larger should close the claw; pulling on the smaller should reopen it—and you pull forcefully but not far. Notice that the two big claws of a lobster differ in size and shape. The smaller and more sharply tapered one cuts, while the more robust claw crushes. Crushing takes more force, and the cutter not only has more muscle but uses it at a lower distance advantage, 3.0 rather than 6.25.[16] Both advantages are especially low. So beware, both go in for big-time force. Still, that figure of 3.0 exceeds the distance advantage of 1.7 of the combination of our lower jaws, molars, and temporalis muscles. One might guess that the pennate muscle of the lobster boosts the force so well that the distance advantage of its associated lever system needn't be quite so low.*

Pennate muscles have another advantage for arthropods. As a constant-volume system, a normal muscle must become fatter when it shortens. Fatter would spell trouble if the muscle contracted inside an insect's leg. Either the wall of the leg would hinder shortening by limiting the muscle's radial expansion, or the wall would bow outward and put the whole tube at risk of buckling. Not only don't pennate muscles change volume when they contract, but (perhaps counterintuitively) they also don't change their widths. They just don't become fatter. They do change shape, but in a different way, as Figure 6.9 shows. They take advantage of the rule that the area of a parallelogram equals the length of its base times its height. Sliding its top edge with respect to the bottom edge changes neither base length nor height, so it has no effect on area or, in three dimensions, on volume. At least that's the case if the top and bottom edges stay the same distance apart. For an example, consider a ream of paper resting on a table. Pushing on one end converts the stack's cross section from a rectangle into a parallelogram. Same height, same length of paper, same volume.

Borelli understood most of this business about pennate muscles, so we're not talking about any new revelation. Once again we have to bear in mind the excessive value the popular press puts on contemporaneity. Every day the soup de jour comes with the tacit promise of its superiority to all previous soups de jour. The journalists collude rather than collide with the publicity departments of every aspiring institution anxious

---

*Of course the arthropods play another force-increasing trick, one mentioned earlier and one that we vertebrates don't. They play with sarcomere length, increasing it at the expense of the number of sarcomeres end to end. Recall those crab claws, with sarcomeres approaching 20 micrometers instead of our standard 2.5.

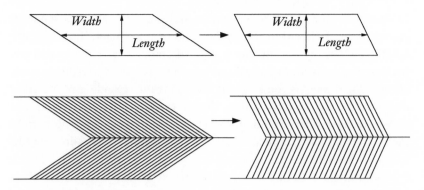

FIGURE 6.9. How pennate muscle works as a constant-volume system. Contraction leaves width and length—hence area and volume—unchanged.

to spread its name and fame. But most new work in science adds only small flourishes to old work, and it has always been so. We make progress, but we lurch rather than march forward. Our institutional memory plays tricks on us, so we forget what we once knew; we inadvertently spend time chasing will-o'-the-wisps into cul-de-sacs; we compromise, dissemble, and promise leaps forward in order to extract sufficient support to keep going; and meanwhile most of our ideas remain remarkably stable.

Hydraulic Leverage

As we saw earlier, water could provide a compression-resisting substrate against which tension-producing muscle could work. Water differs little from most solid materials in its ability to resist compression, the ultimate bottom line. Of course water has to be properly contained to do so since as a liquid it has the awkward habits of flowing, leaking, and converting a force pushing in one direction into a force that pushes in all directions. Put the squeeze on a liquid by pushing on a piston in a cylinder (as in a hypodermic syringe), and the liquid will come out through a hole pointing in any direction. That awkwardness, though, conceals opportunities, of which we humans make a little use and other animals make much more.

Consider a water-filled cylindrical balloon. The rubber withstands tension, the water takes compression, and the thing remains stable. If muscle encircles it and squeezes, the balloon gets longer; if muscle runs lengthwise and tries to shorten it, the balloon gets fatter. How can you increase the pressure in the balloon? Either circular or lengthwise muscle will do, as long as you keep the balloon from expanding in the other direction. If circular muscle does the contraction, then something, like nonstretchable fibers running lengthwise, must keep its length from increasing. Similarly, nonstretchable circumferential fibers allow lengthwise muscle to do the pressurization.

Nature makes an important modification to these schemes. A thin-walled cylinder with all its wall reinforcement running lengthwise and circumferentially has several bad habits. It has little resistance to twisting, and when bent, it tends to develop infoldings, or kinks. Things go better in both respects if the fibers run helically around the cylinder, with some fibers running clockwise and others counterclockwise, as in Figure 6.10. If the fibers run nearly circumferentially and the muscles run lengthwise, then contracting the muscles still increases the internal pressure. That's what roundworms (nematodes) do, generating enormous internal pressures and sufficient overall stiffness to penetrate flesh (ugh); in consequence we've been told to cook pork properly. If, conversely, the fibers run almost lengthwise and the muscle runs circumferentially, the pressure again will rise when the muscle contracts. That's what a squid does in order to raise the pressure in its mantle cavity so

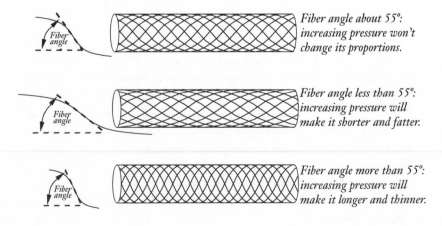

Fiber angle about 55°: increasing pressure won't change its proportions.

Fiber angle less than 55°: increasing pressure will make it shorter and fatter.

Fiber angle more than 55°: increasing pressure will make it longer and thinner.

FIGURE 6.10. Fiber-wound cylinders with different fiber angles.

that water squirts out the siphon tube. If the fibers run in helices with a pitch of about fifty-five degrees to the long axis of the cylinder, contraction of either kind of muscle will raise the pressure; that's the angle at which (if we assume the fibers can't stretch much and a few other things) the system has its greatest volume.

Thus these systems use the angle of the fibers in their walls to set the relationship among change in length, change in girth, and internal pressure. Muscle, running along or around a water-filled cylinder, can be operated at what amounts to different distance advantages. So these systems gain leverage, but from an anatomical setup different from anything we'd normally recognize as a lever. That various internal organs substitute for much of the water makes no difference at all; they're mostly water and about as incompressible.

This basic picture emerged during the 1950s and was described in a remarkable book, *Dynamics in Metazoan Evolution* ("metazoan" corresponds roughly to "multicellular animal"). The author, R. B. Clark, took the then-unusual step of integrating studies of how animals functioned with how they evolved, two separate traditions. He focused on worms, a group more diverse than ordinarily appreciated and one that occupies a central location in almost every scheme for the evolutionary history of animals. Diverse though they be, most worms remain spineless creeps. They invest heavily in these hydroskeletons, but they use them in ways varied enough to give a good sense of what's possible and of the practical outcomes of different structural arrangements.

Clark got people thinking about hydroskeletons, and as a result, we now know about lots of others—for instance, the tiny feet of starfish, sea urchins, and the like, and the bodies of even large sharks. We ourselves use some hydroskeletal support when we lift something heavy, hold our breaths, and tense our abdominal muscles, thus increasing our internal pressure. The behavior may take some force off the backbone, but it can't be recommended since it may prevent an adequate amount of blood from returning to the heart. We—in particular we human males—use another hydrostatic device, an erected penis. Oddly, the mammalian penis turns out to be the only hydrostat known to have its reinforcing fibers running lengthwise and circumferentially. But it gets its pressure from a combination of heart-driven hydraulics and muscle contraction, and it differs in other functional respects as well.[17]

Hydraulic leverage needn't operate only beneath some squeezing muscles. We use remote hydraulics to operate the brakes on automobiles, and we take advantage of leverage in the process. Two simple rules set the game, both depending on an unchanging volume of hydraulic fluid in the system. First, the area of the driving piston (in the master cylinder) times the distance it moves must be the same as the combined areas of all driven pistons (at the wheels) times the distances each of them moves. If a small driving piston moves a large driven one, the driven one will move less far than the driver. Second, force times distance will be the same for driver and driven, so within any such system the piston that moves less far will move more forcefully.

Nature makes only limited use of remote hydraulics and its leverage. We do, as noted, inflate penises, although their stiffness isn't entirely attributable to blood pressure. We use a heart to push blood through remote capillaries and to power the filtration stage of our kidneys. Probably spiders, extending their legs with pressure generated outside the legs, come closest to using the hydraulic devices common in present human technology.

But nature does use a remarkable version of hydraulic leverage based entirely on muscles, something distant from anything in our technical armamentarium. These are the muscular hydrostats mentioned earlier when I talked about reextending muscle. With such a simple principle, I wonder why we didn't notice until the 1980s how nature used the device. Perhaps the absence of an everyday example from the engineers deprived us of our usual clue. Biomechanics, I once claimed, most often studies how nature does what human engineers have shown to be possible. Anyway, the basic idea that saw the light of day in Bill Kier's Ph.D. thesis turns on the relationship between the length and the circumference of a cylinder of unchanging volume. Increase in either length or girth always requires a decrease in the other. But the relationship between increase and decrease depends on the shape of the cylinder: from short and fat to long and skinny.

At this point an example ought to help. Consider a cylinder whose circumference or diameter drops by 10 percent as a result, perhaps, of muscle contraction, as in Figure 6.11. A simple calculation shows that its length will increase by 23 percent, whatever its length and diameter. Here, though, comes the rub. For a short, fat cylinder, one with length and diameter initially equal, that shape change will translate into a distance advantage of 2.3 (from 0.23 divided by 0.10). If, by contrast, that

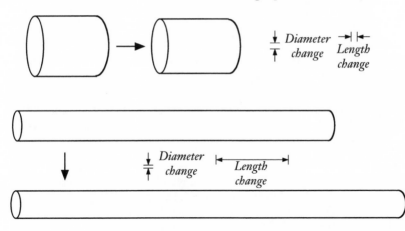

FIGURE 6.11. Diameter versus length changes in two constant-volume cylinders, one short and fat, the other long and skinny.

cylinder is long and skinny, ten times as long as it is wide, the shape change will give a distance advantage of 23 (from 0.23 times 10 divided by 0.10). Cylinders of different shapes thus give different amounts of leverage.

Several other factors increase the versatility of these systems. Muscle need not run just lengthwise, crosswise, and circumferentially but can wrap around a cylinder in helices of every possible pitch. That means muscle shortening can be divided between length and girth changes in any proportion. These systems also use muscles of surprisingly varied contractile behavior, including some that can shorten to unusually great fractions of their resting lengths.

Oh, yes, what creatures are we talking about? Kier's original work involved the way squid use muscular hydrostats to extend their arms and tentacles (Figure 6.12). A squid has eight arms, like its cousin the octopus; unlike the latter, it has two tentacles as well. Its arms are manipulative organs, capable of little change in length, in sharp contrast with the way tentacles shoot out toward its prey. Its tentacles are long and thin, giving it a high distance advantage and, of equal importance here, a high speed advantage. They extend by 70 percent of their initial length while shrinking by only 23 percent of their initial (and much lower) diameter. The crosswise muscle that causes extension contracts rapidly

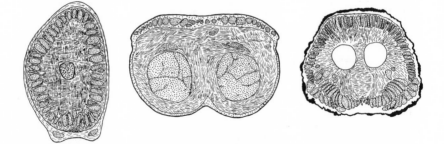

FIGURE 6.12. The arrangement of muscle fibers in cross sections of a squid tentacle, a lizard tongue, and an elephant trunk. These are tracings (with some simplification) from micrographs. Dots are lengthwise muscle; dashes are crosswise, helical, and circular muscles.

but not so rapidly as to compromise power. But its 6.6 lengths per second gets leveraged up to no less than 45 lengths per second for extension of the tentacle as a whole. Furthermore, as mentioned back in Chapter 4, squid compound that distance and speed advantage with especially short sarcomeres to achieve peak power at especially high shortening speeds. A squid takes aim and then shoots its tentacles at passing prey in about a fortieth of a second, extending them at twenty-five times the acceleration of gravity—the original predatory leveraged takeover.

At some point after his initial work on squid, Kier made contact with Kathleen Smith, a vertebrate functional morphologist and paleontologist who worked (among other things) on feeding in lizards. Besides marrying each other, they found that lizard tongues protrude the same way as do squid tentacles. They then found that elephant trunks do so as well. The internal structure of these systems differ little, but they show just the variations expected for their different amounts of leverage. Elephant trunks, for instance, lengthen and shorten slowly and for only a relatively short distance, but they do so forcefully—behavior opposite that of squid tentacles.

(In connection with a related project, I was once asked to build a weight rack that an elephant at our state zoo could grab and lift. I borrowed some weights from my son and hooked up some pipes and fittings to connect them to a loop of heavy rope. The elephant then put its trunk in the loop and lifted the weights. The rig worked exactly as intended but completely failed for its purpose. As shown on a videotape of markers painted on its trunk, at the capacity of the rig, the weights barely strained the elephant. An elephant trunk can shorten *very* forcefully.)

We now know of other muscular hydrostats that produce rapid elongation. Some frogs use the device to extend their tongues flyward, although the fastest tongue-flicking frogs use their tongues' inertia much the way we throw lassos.[18] Among reptiles, the chameleon currently holds several records. The peak acceleration of its tongue has been measured at fifty times gravity, twice that of a squid's tentacle. At the same time, a chameleon can shoot out its tongue a distance equal to the length of its own body![19] Once we recognize the first instance of some natural device, other instances, cases often independently evolved and of different appearance and function, become recognizable.

Stretching with Elasticity

Animals almost always move unsteadily. That's certainly how a muscle would view locomotion. Legs, wings, fins, and tails go back and forth; tongues and tentacles go in and out; whole animals burst forth, jump up, or duck for cover to chase prey or to escape predation. Even when we add steadily rotating wheels, we work our muscles unsteadily, pushing on the pedals of a bicycle first with one leg and then with the other. Sets of intermittent pullers, muscles, thus drive accelerating and reciprocating actions. On the one hand, the combination cries out for complicated transmissions; on the other, it presents unlimited scope for clever contraptions.[20]

In particular, intermittent motion opens the door for what we might call temporal levers. Levers, again, change the relationship between the force that an engine exerts and the force that acts on the load. In the process, work remains unchanged, except for a little leakage down the heat sink. But no rule requires that power, the rate at which work is done, or force times speed, stay the same. That's where temporal leverage comes in. In the long run the power coming out of anything that's not an engine can't be greater than the power going in. But in the short term no such conservation rule imposes itself. Where motion is episodic, the relationship between power going in and power coming out can be altered by short-term energy storage. What happens resembles the way an ordinary lever alters force. The basic rule restates the principle of energy conservation: power times time going in equals power times time coming out. Or, to put it another way, force times speed times time going in equals force times speed times time coming out.

Think of what spring-loaded devices can do. Winding up a clock, once a household ritual, takes (relatively) high power since you put energy in quickly (lots of force and speed, little time); the clock's spring then doles out the same energy, working at a minuscule power level for a long period of time (little force or speed, much time). Drawing a bow, you do the opposite, with a slow, low-power input and a fast, high-power output. I've seen lawn mower engines that were started by slowly cranking against a spring (a low-power input); a triggering mechanism then made the spring spin the engine (a high-power output). Reduction of the necessary power input allowed (ideally, at least) a small or elderly person to get the mower going. Short-term energy storage permits power amplification (bow, lawn mower) or attenuation (clock).

The intermittent motion of most appendages poses a peculiar problem. Accelerating an appendage uses force and does work, work in addition to any useful push against substrate or passing fluid. That work wastes energy just as if you made your car roar away from every stop sign and traffic light. Worse, the loss happens twice with every full wingbeat or complete stride. Worse yet, decelerating an appendage takes force and does work as well, even if our muscles happen to be more efficient as brakes than as motors. But the problem can easily be solved, at least in part. The system need only bank the energy of deceleration, making itself charge up some brief battery instead of being slowed by muscle power alone. It can then reinvest that energy to reaccelerate the appendage at the start of the next cycle of movement.

Nature uses two schemes for short-term energy storage. When we walk, we swing our legs back and forth with gravity as intermediary. Our motion resembles that of a pendulum, which trades energy back and forth between two forms, energy of motion (kinetic energy) and energy of height (gravitational energy). With two legs swinging as conventional pendulums and a body that goes up and down as an inverted pendulum, things get a little complicated, but one way or another, gravity provides the rechargeable battery. The earth's gravity sets the value of the acceleration of descending bodies—in this case, parts of our own bodies. Your size sets how far you can swing your legs upward at least without abusing your anatomy. Together, gravity and size set the top speed for efficient walking—or at least the speed above which we can cover ground more cheaply if we break into a run.[21]

Walking proves an unusual activity when it comes to energy saving through brief storage. Few other movements put gravity to such positive use. Much more diverse are the activities that use another scheme, brief elastic storage. Elastic elements may act as muscle antagonists, but they do a lot more than just act as simple antagonists. On Planet Earth, people of ordinary size shift from walking to gently jogging when doing a mile in about twelve minutes. The transition marks a shift from gravitational to elastic energy storage. Walking and jogging are completely distinct gaits, with no efficient intermediates, whereas jogging, trotting, running, sprinting, and so on represent a smooth continuum. You walk as an inverted pendulum with your head highest in midstride, when a foot is directly beneath your torso. You run as a system of springs with your head lowest in midstride as your torso bounces downward.

If running stores elastic energy from stride to stride, where's the energy housed? Bone, muscle, or tendon might serve as a battery. Choosing among them requires that we measure three things for each: how much force gives how much change in form, how much change in form occurs during activity (or how much force is exerted during activity), and how much of the work put in can be recovered elastically. These daunting measurements have been made on a variety of running and jumping mammals by R. McNeill Alexander and his various collaborators, mostly at Leeds University, in England. Bone turns out to do little or nothing. Muscle looks more promising as a place to store energy. But while it does get stretched during locomotion, relatively little of the work of stretch returns elastically when a muscle shortens again. Alexander's work leaves no doubt that the main player is tendon.[22]

At first glance, tendon looks like a poor spring. We've been talking about it here as an inextensible cable, and a spring must stretch to be of any use. In reality, though, tendon can be stretched, just not very much; it breaks when elongated beyond about 8 percent. To do its main job, tendon can't be too stretchy, or when a muscle shortened, it would just stretch its tendons rather than pull bones together. But to the extent that it does stretch, tendon makes a splendid elastic material, giving back about 93 percent of the work put in, about the same as rubber. We just can't use it the way we might use a strap of rubber, which extends 200 percent or more rather than a mere 8 percent. Think, though, about how we fashion steel springs. We form steel into coils so that no bit of steel deforms more than about 1 percent when we stretch the spring as a whole—even when the whole spring doubles or halves its length. Incidentally, in terms of energy stored per unit weight, tendon beats steel at least tenfold. Muscle still matters, at least a little; cross-bridges do store energy elastically. For instance, the gastrocnemius muscle of a wallaby stores a small but significant eighth as much energy as does its tendon. Relative to weight, of course, tendon comes out even further ahead of muscle than of steel.

Where are the tendons that matter? For a jumping dog, the tendons running down to the ankle may be the most important; Figure 6.13 shows the hookup. We run and jump in similar fashion, stretching our Achilles tendons by about 5 percent as we stress them with forces around seven times our weight. We also store energy in the tendons of our feet themselves. Feet normally arch upward, but the feet of a barefoot runner flatten just before each leaves the ground. With admirable fortitude, Alexander and his collaborators tested freshly amputated human feet[23] and found that a foot stores about half as much energy as an Achilles tendon. Camels and horses go to extremes; some of the muscles near the far

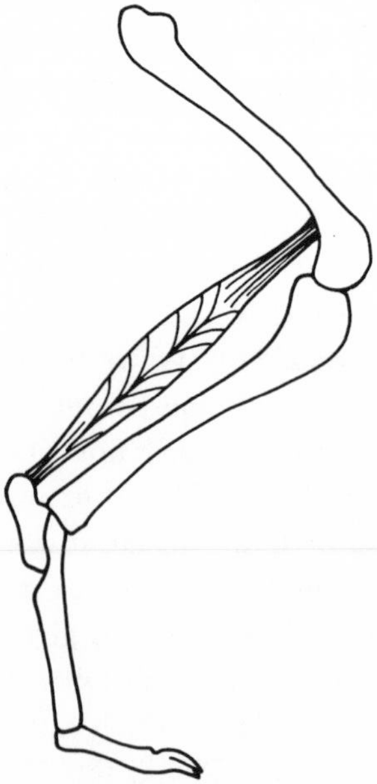

FIGURE 6.13. The way the main jumping muscle and its tendons are arranged in the hind leg of a dog.

end of each leg have extremely short fibers and very long tendons. So in any gait more of the change in muscle plus tendon length when a leg is in contact with the ground comes from tendon elasticity than from muscle contraction.[24]

How much energy does all this stride-to-stride or bounce-to-bounce storage save? Figures calculated for people running, for kangaroos hopping, for horses galloping, et cetera run around 50 percent. Thus, for large mammals, temporary energy storage in their slightly extensible tendons roughly halves the cost of locomotion.

Nor does running exhaust the ways nature uses elasticity for something fancier than muscle antagonism. A few other examples will emphasize the diversity of her devices and I hope extend your appreciation of what elasticity can do.

As an extreme performance insect flight has much to tell us about what muscle-powered systems can do, the reason, of course, for why it keeps appearing here. It involves specializations in every component of the body that plays any role in flight. Not unexpectedly, flying insects have taken elastic energy storage to extremes. Their wings may weigh little, even in relation to their light bodies, but they move rapidly and change directions many times each second. So a great deal might be saved by using the kinetic energy of wings moving in one direction to reaccelerate the wings in the other direction. The flight muscles have no tendons, so they can't help. The muscles themselves store a little more energy than ours do, and their flexible exoskeleton stores some as well, but they're bit players.

Major storage happens in pads of a rubberlike protein, called resilin, in the wing hinge. After deformation, resilin returns about 97 percent of the energy put into it, even better than the 93 percent of tendon or its main protein, collagen. No other rubbery material, natural or synthetic, can match this resilience. Why must resilin be so good? One idea is that high resilience solves a thermal rather than a mechanical problem. After all, 97 percent is only 4 percent better than 93 percent, just a slight gain. But any loss between deformation and recovery appears as heat. Thus 100 percent minus 97 percent gives a 3 percent loss to heat; 100 percent minus 93 percent gives a 7 percent loss. The 7 percent figure means that over twice as much heat leaks out as at 3 percent. Heat raises temperature, and systems built of protein don't do well when their temperatures get high.[25] Deforming a pad a few hundred times each second might well risk excessive temperature increase. Unfortunately, only arthropods seem to know (in the genetic sense) how to make this marvelous material; it's unknown in any other lineage in the animal kingdom.

Another insect story. Long ago Galileo calculated that all animals, whatever their size, should be able to jump to the same absolute height, at least if air resistance can be ignored. If we reverse Galileo's logic, equal height means equal takeoff speeds. The smaller animal, though, must reach that takeoff speed in a shorter distance and a shorter time; once its legs leave the ground, an animal can go no faster. The smaller the animal, then, the greater the necessary acceleration.[26] Consider the most extreme case, a jumping flea. The power phase of jumping, when it extends its hind legs, lasts less than a thousandth of a second. Doing that directly would defy what we know about how muscle operates. But the flea uses temporal leverage. By pulling on a leg segment, one muscle strains a pad of resilin. To initiate the actual jump, the pull of a second muscle acts as a trigger, and the resilin pad extends the hind leg. A flea is so small that it moves only about half a millimeter before it leaves the ground. It needs to accelerate at well over a hundred times gravity to reach its takeoff speed of about two miles per hour (a meter per second). But because it can put the work in more slowly than the work comes out, that absolute speed makes no direct demand on the power output of its muscles. The muscle works at an intrinsic speed of less than a length per second, a speed that lets it reach almost its peak force. The system makes nice use of conventional leverage as well; as the resilin pad unloads, the distance advantage rises steadily, so the increasing speed of the accelerating flea doesn't reduce the pad's ability to supply force.[27]

Locusts and click beetles also amplify power output during their jumps by storing up energy beforehand. A tenfold amplification in locusts sounds impressive until you learn that click beetles amplify power a thousandfold. Hence, of course, the palpable and audible click when one jumps from your palm.[28] As you might expect, mammals that jump store energy in tendons, but being a lot bigger, they don't need the same extreme accelerations and have little use for the same kind of preloading and triggering.

I can continue with such stories. For instance, the inextensible outer membranes of hydroskeletons do stretch a bit, and the small stretch stores energy for the unsteady locomotion of everything from small worms to large sharks. Moreover, whale blubber isn't just fat but contains a lot of collagen fibers that may act as a kind of whole-body energy-storing support stocking.[29] More important is a general point. All the stories here turn on unsteady locomotory movements and on materials—muscle, tendon, elastic pads—that can change shape. With our predilection for steady motion and stiff materials, we have to adjust our frame of reference to guess what nature might be up to. Guessing proves critical; hypotheses are no more than guesses with their clothes on. We can't simply gather data or construct models; we must have in mind some testable idea.

CHAPTER 7

Using Hand Tools

. . . it was bipedalism which started man on his separate evolutionary career. But tool use was nearly as early. Biological changes in the hand, brain, and face follow the use of tools, and are due to the new selection pressures which tools created. Tools changed the whole pattern of life, bringing in hunting, cooperation, and the necessity for communication and language. Memory, foresight, and originality were favored as never before, and the complex social system made possible by tools could only be realized by domesticated individuals. In a very real sense, tools created Homo sapiens.[1]

SHERWOOD L. WASHBURN, anthropologist

TOOLS: HOW WE HUMANS LOVE THEM! WE MIGHT BETTER CALL ourselves *Homo faber*, the hominid that makes things, than our pompous and self-congratulatory normal Linnaean binomial. Other animals do use tools, breaching another hoary line between the human and the beast, but the great apes use tools far more casually than we do. Moreover, while the Galapagos finch may depend on a thorn to play woodpecker, the thorn isn't fashioned for the task at beak. I suspect the occasional octopus uses some tool as it arranges its personal stone castle, but if so, it makes no big deal of the matter. Vulcan, the tool-maker, deserves a better place in our divine pantheon.

With typical arrogance, we explain our love affair with tools as consequences of intelligence, of planning, of consciousness. Less glamorous factors, though, bear as much relevance. Being bipedal permits us to carry tools without compromising locomotion, as Darwin pointed out,[2] and precious few mammals are bipedal. Having a digestive system that likes—indeed has come to require—some diversity encourages us to acquire and prepare novel foods. That puts a premium on dexterity, and the latter additionally preadapts us to tool use. Being only intermittently migratory and lately quite sedentary, we easily accumulate artifacts. Even our relatively large size might encourage tool use. With its longer or heavier swing, an arm with a club or stone can have more impact on the world. If we were much smaller, no such assistance would suffice; a mouse with a club would do no better than a clubless mouse, nor could a mouse throw stones of enough weight with enough speed to have much effect on its world. Tool use might even be as much a cause as a consequence of our smarts, as the late Sherwood Washburn pointed out. Improved reproductive success from an ability to imagine and craft tools would make intelligence a virtue, making braininess advantageous.

Hand tools, the present concern, work as handy extensions of our bodies. They apply the force and power of our muscles to the tasks we do. More than that, they readjust the direction of our efforts; they change the mix of force, distance, and speed; they let us accumulate work and then apply it with sudden, powerful effect. Pliers grab more forcefully than fingers; a hammer hits harder than a fist.

Amplifying Force

We return, then, to the matter of leverage, looking at it now from the tool's point of view instead of from the muscle's. We use a simple tool most often to increase the force we can exert. Nor does "simple" overstate, as in "some simple assembly required." No joints or parts that move upon each other are required; Archimedes' lever has neither lost its utility nor acquired high-tech appendages. Levers of the ordinary kind increase force at the expense of distance and speed, as explained earlier. Before the advent of the pull tab lid, no picnic could proceed without a can-puncturing tool—a church key so called—to access the beer. Its sharp point bore down with about four times the force with which thirst drove the hand upward. The wrecking bar with which I deconstruct my earlier creative efforts deals with sturdier stuff. Full arm and some torso provide far more force than forearm and hand, force that the bar then amplifies ten times further. As the bar rotates, that amplification drops and the bar no longer pries upward. So if the load hasn't loosened, I put a small block under its fulcrum to restore full amplification. Figure 7.1 puts the matter diagrammatically.

Irony (beyond the ferrous metal of the tools) underlies that force amplification. Our muscles produce force to a fault, our bones then act as distance amplifiers to reduce the force so our appendages can swing far and fast, and finally we invent all manner of devices to restore the force again. I imply nothing of principle, of goodness, of technology misapplied by the way the cultural evolution of tools undoes the biological evolution of bones and muscles. Using a long lever to roll over a rock makes as much functional sense as swinging an arm through a wide arc. These opposite imperatives, however, have led not just to the biomechanically awkward terminology noted in the last chapter but to our historical difficulty in appreciating how muscles must act. Were it not for the misleading analogies provided by our hand tools, the forcefulness of muscle would have been clear to the anatomists of antiquity. Instead Borelli had to begin by making the point, quoted at the start of the last chapter, that using a forceful engine to move light weights shouldn't be regarded as an inefficient design.

Simple levers may have provided our first force amplifiers, but we've added to them another scheme of equal utility. Putting the fulcrum entirely within the device frees us from what worried Archimedes: where to rest the lever with which he meant to move the earth. We add only a single part, a second piece pivoting about the first. With that we get shears, tongs, pliers, wire cutters, metal snips, and so on—the things in Figure 7.2. A few just provide remote operation or a safely distant cutting edge, without amplifying force and sometimes even diminishing

FIGURE 7.1. Wrecking bar and church key—both devices with middle pivots that greatly multiply the force you can exert.

FIGURE 7.2. The simple two-piece devices of the table on page 130.

it—our wooden tongs for freeing the toast from the toaster, our long-bladed paper or grass shears, our tweezers—but most hinged tools amplify force.

These force-amplifying tools come in two variants, depending on whether the fulcrum, now better called the pivot point, lies in the middle or at one end (the basis of the old distinction between "first-class" and "second-class" levers, respectively). Quick forays to the family kitchen and workbench produced the following examples of tools whose operation depends on grasping with good force.

The exact amplification of most of these hand tools depends on how you use them. To crack an especially hard nut, you grasp the handles of the nutcracker as far out as you can and put the nut as close to the pivot as it will go. To loosen an especially tight nut from its screw, you position it as far down between the jaws of the pliers as you can.

Device	Location of Pivot	Force Amplification
Garlic press	end	6.0 X
Nutcracker	end	5.3 X
Side-lever corkscrew	end	4.0 X
Jar opener	end	3.8 X
Wire cutters	center	6.7 X
Channellock pliers	center	4.6 X
Ordinary pliers	center	3.0 X
Needlenose pliers	center	1.6 X

The force with which any of these tools actually grasps, however, depends on much more than its amplification factor. How strongly you grab depends on the size of what you grab, with a maximum force somewhere between extended fingers and tight fist. Big, strong hands obviously do better. Cunning design of handles and choice of amplification factor, though, can go far to make a tool either specifically useful for a particular size and strength of hand or especially indifferent to such factors.

Neither these tools nor the wrecking bar mentioned earlier tests the limits of single-stage force amplification. I can amplify my effort by over twenty by applying a long pipe wrench (Stillson wrench) to a fitting with threads an inch in diameter, and I once used a wrench about six feet long to screw the top against the lead (metal, not initial) gasket of something called a bomb calorimeter. With this rig I could amplify a whole-body push thirty or forty times. Incidentally, screwing down the top yielded perhaps another hundredfold amplification, so I gave that gasket a harder squeeze than if it had been trod upon by a truck or an elephant.

Occasionally we want a tool's leverage to decrease distance, tolerating the increased force as an awkward consequence. An aspiring biologist of my generation had to be warned not to crank the objective lens of the microscope down into the glass slide on its stage. These lenses would attempt to penetrate slides, to the detriment of both, without increasing the resistance of the focusing knob enough to alert us. (Also, when cranking a meat grinder, you can put a pretty good squeeze on any fingers you may have pushed down its maw.)

Using and Designing Tools

The parent subject, neither unimportant nor unsophisticated, goes by the name of ergonomics; I wish its results received more attention than do guesses about consumer habits and prejudices or the economics of material and fabrication. That experienced tool users inevitably have their well-worn favorites reflects real variation in the quality of the underlying designs. I have too many tools that try my temper: screwdrivers with handles too narrow and too deeply fluted; an eggbeater and a grater neither of which can be held in position without abusing my hand; a wine cork extractor that can't be pulled up without cutting into my palm. I've also seen much worse, ergonomic horrors leering at me from store shelves and glossy catalog pages.

Having favorite tools reflects as well the differences among us in size, in strength, and even in handedness. We're a variable lot, and no amount of training can eliminate our individual differences. For one thing, men on the average grow significantly larger than women, and children are significantly smaller than either. We ought no more assume that tools of one size will suit all users than try to wear shoes of a single size. I started my son with a ten-ounce hammer and fourpenny nails and then bought him a thirteen-ounce hammer and a stock of six- and eightpenny nails before I encouraged him to use my trusty sixteen-ounce hammer. Were I as large as he is now, I'd supplement my sixteen with a twenty-ounce hammer, at least for sixteenpenny nails. But you'll find a ten-ounce carpenter's hammer of decent design and construction no common item.[3]

Not only do we differ in size, but we grasp and manipulate different tools in different ways and for different purposes. Sometimes we even use a given tool in several modes. Starting a wood screw requires more axial force than twisting force—you push lengthwise on the screwdriver. Once you start, the twisting force dominates, and you shift your hand on the handle. The well-designed screwdriver accommodates both motions. When you apply a power grip to a heavy hammer, both the thumb and the fingers curl around the handle, with your forearm extending to one side of the handle. To tap more carefully than forcefully with a precision grip, you must straighten the thumb and point it toward the hammer's head, angle the fingers obliquely away from the head, and align the forearm itself nearly as an extension of the handle. For maximizing power, a fat handle works best, with an optimal grip diameter of around an inch and a half. Handles of decreasing diameters permit increasing precision for both screwdrivers and hammers, down to the shaft size of about a quarter of an inch of a jeweler's screwdriver— not coincidentally about that of a pencil.[4]

A third grip, the hook grip, comes into play when the fingers pull on a bar. We grasp quite a few tools this way, including all those in the table on page 130 as well as wrenches, many saw handles, and the pulling handles used to start small engines. For these hook-held handles, the best diameter runs around three-fourths of an inch. My bow saw, carpenter's saw, hacksaw, socket wrench, and pipe wrench come close to that diameter, but most of the others are narrower. If they had fatter handles, I'm certain that some of the offenders, simple wrenches in particular, would be more comfortable for me to use, and perhaps they'd be more effective as well.

In this spectrum from maximum power to maximum precision we shift the muscles and joints we employ. Fewer joints, smaller muscles, and a shorter length of unsupported appendage mean more precision. We swing an ax, a maul, or a baseball bat with the entire torso as well as with the arms; where and how it strikes vary a lot even for a skilled operator. Once it's launched, time and available force put a limit on what steering the feedback machinery can do. We swing a heavy hammer toward a nail with a full arm and shoulder but with little trunk muscle, achieving better accuracy but applying less force. A tack hammer held in a precision grip depends on the movement of the wrist and elbow but not the shoulder. When using a sharp pencil, most of us brace the wrist, limiting movement mainly to the fingers. We can then position lines with an uncertainty of less than a millimeter. With only a little effort, most people can draw two parallel lines that can't be distinguished by eye at a normal reading distance.

In fact, our muscular control systems can do better than our proprioceptive feedback systems normally permit. I was taught that to dissect small insects under a microscope, I should brace both wrists and palms, leaving only the joints outward from the knuckles free to move. With a microscope to improve visual feedback prodigious feats of dexterity become almost routine. You do need lots of practice, and biologists can tolerate failure rates that would make any surgeon blanch. I have a friend who can casually glue the sharp end of a dog hair to a microscopic crustacean. People working in pairs have transplanted glands between houseflies. One person handles the donor, the other the recipient, passing across a table an isolated gland scarcely visible to the naked eye. But for still finer manipulation, as when one of my colleagues tweaks the chromosomes of individual cells, muscles need help from special distance-reducing manipulators.

The use of shorter muscles to get greater precision squares with what we know about how muscles operate. Very simply, a 1 percent shortening of a short muscle produces a smaller movement than the same relative shortening of a long muscle. Bracing unused joints also makes sense in that any unbraced appendage must be actively stabilized by one's proprioceptive equipment. It takes some perturbation to trigger a feedback response, and the minimum perturbation of, say, biceps or triceps would far exceed the minimal motions that your fingers can manage. Just try to write legibly on a piece of paper taped to a podium while standing erect and touching nothing! I recall a vain attempt to teach me to write in cursive script with shoulder motion and a locked wrist. Elementary schools saw some virtue in this so-called Palmer method. I assume a few people have the talent to produce results better than my nearly illegible scratchings.

Of course sometimes a given device must accommodate all users. How big should a door handle be, and how high should it be mounted? I think, unless I'm up to my old self-deceptive tricks, that very old houses have lower handles, commensurate with the smaller average size of humans a few generations back. I'd guess that doorknobs ought to be more nearly spherical than disklike to provide lots of area for hand contact, and perhaps they should be slightly fluted so we don't have to grab with extra force to prevent slipping. Still, I perceive (and applaud) a trend away from knobs and toward levers that extend sideways. These latter ought to have greater intrinsic versatility since we have more choice in how to grasp a lever than a knob—an elbow and a foot suffice for a door that opens outward, a boon for the bearer of bundles. Similarly, faucet handles in the form of lateral levers of reasonable length or knobs with protruding spokes should be easier than smooth, round knobs for those of us of small size, low strength, or soapy fingers.

Opening Doors and Turning Screws

Doors present instructive ergonomic problems. Even cheap hinges have little friction, and a normal door does no work against gravity.

What, then, makes some doors so hard to open? Not weight, but a variable inseparable on the surface of the earth from weight—namely, mass. Consider what happens, as in Figure 7.3, when you open a door. You grab the handle and pull. The bigger person has an easier time of it, but a bit counterintuitively, strength enters only indirectly. Pulling just draws you and the door into closer proximity. How can you make the door move a lot and yourself only a little? In part, you lean backward, so your center of gravity lies behind your legs; you pull on the door by leaning farther backward and thus moving downward. Or, as Newton showed, you can equally well look at the equal and opposite force; the swinging door tries to pull you (and your center of gravity) forward and upward. In either view, the key is your vertical motion. The bigger person, being weightier and taller, doesn't have to lean backward at as great an angle. So it's the door's mass against your weight. On the moon, masses stay the same, but since gravity pulls less strongly, weights drop to about a sixth of their terrestrial values. So the door would move less and you'd have to move more by leaning farther backward to get it open. In a spacecraft or when you're scuba diving, where you're weightless, a door would win hands down. I once lost 30 percent of my weight in a fairly short period; once-friendly doors got less cooperative for a while.

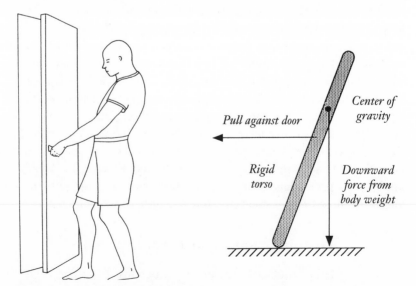

FIGURE 7.3. The forces at play when you open a door.

Until I readjusted my habits and leaned farther backward, I tended to move more while the doors moved less. After a few months the awkwardness receded. Men, on average taller and heavier than women, should indeed be our default door openers.

Another door matter. We sensibly mount a door's handle as far as we can from its hinges. The longer the distance we have to pull or push the handle, the less force we have to exert, as with any other lever. Pushing against a width-wise bar will do the job, and such bars provide a safety factor for crowds in public places. But pushing a bar takes less effort when we know which way the door will swing and can push on the more effective end. Perhaps the more effective end, the one farthest from the hinges, should have some universally recognizable marking.

Screws joined our armamentarium of fastening devices much more recently than nails and rivets, which go back almost to the beginning of metallurgy. Not all their uses were apparent early on. The ancient Greeks used screws to amplify force, the way we still use them in vises and clamps. Archimedes invented an inefficient water pump that worked like a hand-cranked meat grinder, with its operator turning a screw within an inclined, ascending pipe. The Romans used screw presses to squeeze juice out of grapes and olives. But screws were hard to make by carving or filing, and nuts, with internal grooves, gave still more trouble. Modern screw technology began with the development of mechanical clocks and of screw-cutting lathes—need and means, respectively. The first-known mechanical screw cutter was a German clockmaker's tool of 1480, shown in Figure 7.4.[5] A hand crank turned a screw through a

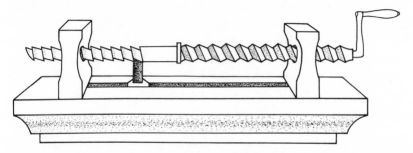

FIGURE 7.4. The arrangement of the earliest-known screw-cutting lathe.

headstock; the advance of that screw turned and pushed a cylindrical blank past a cutter and into a tailstock. Thus the cutter inscribed a helical groove in the blank. In the modern version of this lathe the blank turns around without moving lengthwise while the cutter advances along its length, but the principle remains the same.

Screws can be made to advance when turned either clockwise or counterclockwise, and one kind works no less effectively than its mirror-image twin. Yet almost all our screws move forward when turned clockwise and backward when turned counterclockwise. We call them right-hand screws. The terminology smells judgmental: "right" for correct, just as "dexterous" alludes to both right-handedness and agility and as "sinister" suggests both left-handedness and malevolence. Was the choice accidental, with the same benefit of standardization whichever direction won the primal coin toss? Or was it a rational choice? Or, a third possibility, was it a consequence of some other bias? The screws of classical juice presses came in both handednesses; I've encountered no claim that one or the other predominated. But pictures of early screw-cutting lathes in operation show right-hand screws. I think a likely culprit is the crank of the screw-cutting lathe—the other bias.

Think about any cranks you've turned. Two points. First, turning a crank in a vertical plane (one with a horizontal turning axis) seems more natural if the handle moves away from you near the top of its circle and toward you near the bottom. Not only does it feel better, but you can crank more forcefully. Second, a machine built with such an overhand crank and designed to be operated by a right-hander will have its crank on the right end. That allows the person either to watch its operation while facing it or to stabilize or manipulate the machine with the left hand. For an overhand crank on the right to advance the screw, the screw must have a right-hand thread. All hand-operated meat grinders (as in Figure 7.5—so rare have they become that a picture is needed) use such right-sided overhand cranks, and all have clockwise-advancing right-hand screws. Windup appliances use clockwise cranks. Crank-started automobiles used clockwise cranks, and automobile engines still turn in the direction once set for the cranking convenience of right-handed people.

The left-handed among us have good reason to get cranky about this biomechanical bias. Few of our devices use hand-operated vertical cranks any longer, but the discrimination has persisted and even rami-

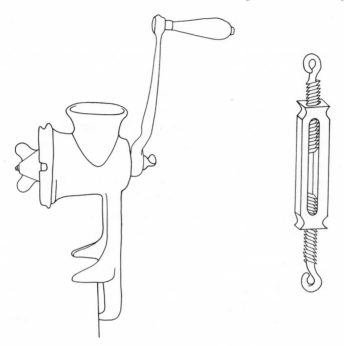

FIGURE 7.5. Meat grinder and turnbuckle.

fied. The muscles with which you turn an arm along its length develop unequal force. You can turn an arm outward—that is, turn your right arm clockwise or your left arm counterclockwise more forcefully than you can turn the same arm inward. To put it technically, you supinate your arms (turn the palms up) with more force than you pronate (turn the palms down). You turn outward with your biceps muscle, which happens to be a supinator as well as a flexor—it's sometimes called the corkscrew muscle for its combined action—but you lack a pronator muscle of equal effectiveness. The difference in force (or torque) seems to be about 20 or 25 percent, based on my poorly controlled tests using a crude contraption on people entering my office. I asked each to turn a large screwdriver, which rotated a lever, which pulled on a spring scale. Interestingly, the difference, as well as overall force, diminished when subjects were asked to turn the screwdriver with an extended arm; the variable leverage of the biceps has relevance here as well.

What's more, the muscles of your dominant hand are as a whole stronger than those of the other hand. Thus the right-hander can, all else being equal, drive a screw or tighten a jar lid with a right-hand thread, as well as crank a grinder, more forcefully than a left-hander can. Right-hand screws work best for right-handed people. Left-handed scissors can be purchased, and a better world would include left-handed meat grinders, but I shudder at the thought of a technology compelled to maintain some politically correct balance between screws of alternative handedness.

In this society you learn at an early age the convention that clockwise tightens and counterclockwise loosens. I wanted to watch the one case we had at home, but I first paid attention when my son was about three years old and already knew the rule. He reports that my number one grandson lost his innocence before the age of two—before talk or toilet. Meanwhile I'm watching how people loosen lids. Unless your nondominant side is especially weak, it makes most sense to tighten lids with the right hand and loosen them with the left. Some of us do that; others (such as me) persist in using the dominant hand for both actions.

Incidentally, left-hand screws may be uncommon, but they aren't mere novelties like counterclockwise clocks. Turnbuckles (see Figure 7.5) pull by drawing together the threaded ends of a pair of rods. To make both ends pull inward when you turn the turnbuckle, one rod has to have a left-hand thread. Moreover, turnbuckles come in a wide size range, from the tiny ones we sometimes use to tighten or square up a screen door to giant industrial fittings. Before the mid-1960s the Chrysler Corporation used left-hand screws and lug nuts to fasten the tires onto the hubs on the left sides of all their cars. It intended, I was told, that a lug nut accidentally left loose would spontaneously tighten as the car was driven. I don't know if the scheme worked. I can, though, attest to the nuisance of having two noninterchangeable but visually identical sets of lug nuts and of having to think about which side of the car I was on before turning the wrench. Many of us find lug nuts hard enough to remove without initially overtightening them! Our neighborhood garage owner tells me that in the seventies he was afflicted with young mechanics who, working on those decreasingly common cars, persisted in trying to turn the nuts the wrong way and in the process breaking off their screws.

Making an Impact

Levers, wrenches, and the like operate in real time. When you work, they work, and either force or work measures what you put in and what you get out. With a different group of tools, you work for a while, and then they work ever so briefly, the way a flea manages its jump. Swing a hammer, and you've invested kinetic energy to endow it with momentum; it collides with a nail, transferring a lot of that momentum. The nail then moves forward, tearing and compressing the wood in front of it. The good hammer weighs enough to provide adequate mass but not so much that the operator can't accelerate it to a decent speed, mass and velocity being the ingredients of both momentum and kinetic energy.[6] It has to be hard enough so impact doesn't deform its face, both wasting energy and damaging the tool. It also has to be tough enough so impact doesn't crack it.

Nails may be only a few millennia old, but percussive tools such as hammers and axes antedate our species. The material of choice must first have been stone. Hominid remains routinely come with stones of appropriate sizes and shapes for chopping, smashing, and hammering, so we can assume stone tools dominated, even allowing for the greater persistence of stone artifacts than those of wood and bone. Stone, harder and denser, serves better for impact tools, and stones of handy shapes and sizes can be selected for use without tricky tool-requiring preparation. A huge literature deals with the stone tools of antiquity—how they were made, how they were used, how they changed over time and varied among cultures—and museums store vast quantities of them.[7]

Cultures based on stone tools can be complex and sophisticated. Don't use a few technologically primitive tribes from Amazonia or New Guinea as benchmarks. Think instead of the Inca and Aztec civilizations at the time of the Spanish conquest. These civilizations used metal for little except decoration, yet they built huge structures out of stones shaped so accurately that you can't slip a piece of paper into their junctions. Times change. Beyond a few abrasives I have no stony material on my bench, so completely have metals displaced it. Metals exceed stone in density—copper is 3.4 and iron 3.0 times as dense as flint—but stone does better in hardness. Flint, the main stone of ancient choice, will take and hold an excellent edge, and it was extensively mined and traded over long distances to supply prehistoric people with axes. Obsidian, rarer, holds a better edge yet: comparable to our best metals in sharpness and superior to them in resistance to wear; good enough for shaving or surgery.

What makes metal better than stone is its toughness, its resistance to crack propagation. In general, an increase in hardness comes with a decrease in toughness; hard materials fracture more easily than soft ones. The trade-off, though, follows no rigid formulas like the rules for leverage. In particular, metals combine hardness and toughness remarkably well, as on the graph in Figure 7.6, compared with either simple stuff like stone or complex biological composite materials. This advantage of metal, of even early copper or bronze, transformed both the shapes of tools and how they were used. When percussive tools of metal become available, people switch over to them. Their fundamental superiority can

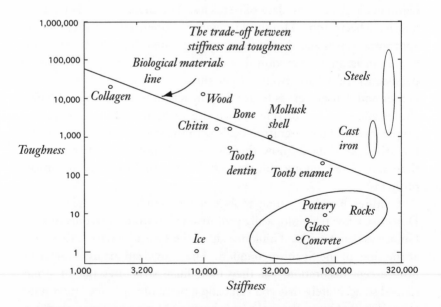

FIGURE 7.6. The trade-off between stiffness and toughness. Being high on one axis usually means being low on the other. (The units are joules per square meter for toughness and meganewtons per square meter for stiffness.)

be resisted only by the strongest tradition, the worst poverty, or the most extreme isolation.

Not to put too fine a point on it, stone is a fragile material. Consequently, stone tools must be delicately applied, counterintuitive as that sounds. A hand ax—that is, one with no handle—makes reasonable use of stone. The operator has good control of its motion, and it can't be moved too fast. For felling trees, a handle will help, but for all the difficulty of attaching it, it can't help as much as you might expect. You can't risk a hard, full-body swing; the swing can't be directed with consistent enough accuracy. Some part of the stone head will at some instant get loaded in tension. That will make any preexisting crack open and propagate, cleaving or shattering the tool. So stone axes, as in Figure 7.7, however well shaped and sharpened, remain arm tools, not torso tools. You must hit rapidly instead of swinging powerfully but less often, and a three-foot handle like that of a modern ax would risk damaging the tool.

If a modern test means anything, though, stone axes must have been effective tools. A few years ago some Danish paleoanthropologists fitted handles to ax heads more than four thousand years old, handles modeled after one of the rare kinds that have been preserved rather than the longer, modern version. After some practice, trees could be felled at rates not radically inferior to what can be done with modern axes with steel heads and long handles.[8] The experiment proved popular; lots of analogous tests have produced a large literature (and, I suppose, blisters) of axing and felling comparisons. Metal wins, whether in terms of penetration depth per time or energy consumed relative to penetration depth. But . . . steel and bronze work equally well. Metals gain much of

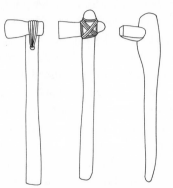

Figure 7.7. Several kinds of stone ax. Securing a stone to one end of a stick challenges human ingenuity!

their advantage by virtue of the more acutely angled tapers permitted by their superior toughness. For trees under about ten inches in diameter, any ax head and any length of handle work about as well as any others.[9]

So land could certainly have been cleared with stone axes. They may even have had an energetic advantage over modern axes. Felling with modern axes provides the fine cardiovascular exercise expected from an activity that works a large mass of muscle hard and often. By contrast, felling with stone axes, using the wrist and forearm, could not have been particularly aerobic. How could they work as well as they did? Perhaps the short penetration of each stroke has a peculiar advantage. In a well-struck, deep bite, an ax pushes a lot of wood sideways, and compressing that substantial volume takes a lot of work. A deeper bite means a larger area of slice, which is good, but a larger volume compressed, which is bad. Double the depth of the bite, and you increase the sliced area fourfold, but you increase the compressed volume eightfold; by this crude analysis, cutting with fewer, deeper bites should cost more, even if it's faster. I know from youthful experience that one can fell a fair-size tree with lots of short strokes from a hatchet; on that score at least, we can't dismiss the legend about young George Washington's dropping a cherry tree with his hatchet.

So there's rationality as well as tradition in our use of metallic percussion tools. Furthermore, rationality (and admittedly a little tradition) underlie their design and the ways we use them. Those admirable tools, hammer and ax with steel head and wooden handle, represent highly evolved forms—in how they're tuned to human operator as well as to cultural task. The garden-variety hammer with a one-pound head and a handle about a foot long gets swung with the entire arm, using the muscles that cross the shoulder joint as well as those of the upper arm. The ax, with a four-pound head and a three-foot handle, gets swung with the torso as well as with the arms. With more force available, a greater mass can be swung in a wider circle, and far more energy can be transferred to whatever is being percussed. As noted already, chopping steadily with an ax taxes the cardiovascular system—more so than hammering home a row of preset nails, which just tires the arm.

Whether or not head weights and handle lengths approach ergonomic optimality, they clearly reflect the preferences of their users. Hammers come in a range of weights, and handles can be grasped at a range of distances from the head—or a longer handle can easily be fitted by the end user. Some years ago a forestry graduate student here at Duke surveyed tool use among southern loggers. He found that when the workers were given a wide choice by their employers, most selected axes with ordinary 3.5- to 4.5-pound heads and 35- or 36-inch handles.[10]

The next time you're in a hardware store note the wooden handles of most hammers and axes. The use of wood has persisted since the first tools were hafted, perhaps a hundred thousand years ago. Tools of North American manufacture usually have hickory handles, whereas elsewhere (hickory being native here and rare elsewhere) ash is the wood of choice. I cheerfully admit a fondness for my personal sixteen-ounce hammer, which has a rubber-sheathed tubular steel handle, but all my other tools use wooden handles. I've used hammers with solid steel shanks and with fiberglass handles, and I find them far less agreeable. Why wood? After all, modern manufacturing technology ought to prefer a material that can be cast from some standard monomer and fixed more simply and permanently to the head. For one thing, we find repeated vibrations running from tool to arm unpleasant. Wood's low resilience means that it transmits vibrations poorly; a cylinder of metal, when hit, rings loudly, while a cylinder of wood gives a dull thud. For another, the particular woods used give about the right strength and stiffness vis-à-vis to handle diameter.

As important as either low resilience or sufficient strength and stiffness is wood's low density. We sometimes hear talk of a "well-balanced" tool. For a hammer or an ax, good balance requires that the impact occurs somewhere along a line that crosses the axis of the handle at a particular point, the sweet spot, in the jargon of the impact tools of sports. As in Figure 7.8, that spot, the center of percussion, lies slightly outward from the center of gravity. (The same factors, length and weight, determine the centers of percussion and gravity, but they combine in slightly different fashion.) Hitting at the center of percussion imparts least bending force to the handle, which means better use of the tool's energy. More important, it passes back the least energy to bother the user at the other end. The lighter the handle, the closer to the center of the head the center of percussion of the tool will be. So stick with wooden handles. But do make sure the handles adhere to the heads, bearing in mind that only friction mates wood and metal. The expression "fly off the handle" alludes to just this hazard, and it's still both dangerous and common, as it was in 1825, when the expression was first used.[11]

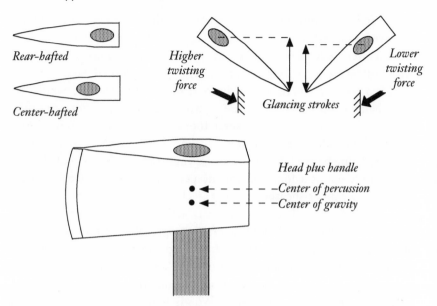

FIGURE 7.8. Left: The haft position of two axes, seen down from the top. Right: The sideways forces when these axes hit the tree with glancing blows. Bottom: The centers of percussion and gravity of the combination of ax head and handle.

Not one but two axes matter in the design of an ax. (The third axis of course would be relevant only if axes came, as we tell gullible city folk, in left- and right-hand versions.) Besides the location of the head along the length of the handle, we have to worry about the position of its mass in front of or behind the handle. Where in the head should the handle be inserted, or for old stone tools, where should the handle wrap around the head? Many stone heads, to judge from their shapes, were hafted in the middle. Metal heads, for most of their history, were hafted nearly or fully opposite their heads. Getting the handle as far from the work as possible would seem right and proper, especially if it's a damage-prone wooden handle. Not so—it turns out that the makers of stone axes (at least some) got things right, if only through the virtue of necessity.

Consider what happens when an ax head hits a tree, as in Figure 7.8. Ideally, the ax person has swung the ax so the head travels exactly along its length from sharp edge to back end. "Exactly," though, implies more than anyone can count on delivering. If the handle's long axis enters the head either in front of or behind the center of gravity of the head, then even a slightly misaligned stroke will cause the head to twist one way or the other. That twist in turn will cause a sudden lengthwise twist of the handle that will be felt by the operator. So energy gets wasted, energy the ax wielder feels as pain.

In practice, you wouldn't dare swing such a tool as hard as a centrally hafted one. On that difference hangs a small item of technological history. European axes were traditionally rear-hafted, as were the first North American axes. Rear hafting not only keeps the handle away from the biting bit but also simplifies its manufacture. A head needn't be made with a hole running through it, but it can just have a rear slot that, after forging, can be closed over a wooden handle. Or a head can be made out of a long slab of metal, one about twice the final length of the head. That slab gets bent across its middle around a rear handle hole. Where its ends come together, it's then fitted with a strip of harder metal to act as cutting edge.

For whatever reason, perhaps less rigid social or occupational structures or just the actions of amateur forgers, eighteenth-century colonial North Americans started making axes with polls—rear extensions of the blade behind now-central handles.[12] Maybe the initial impetus came from the utility of a rear that could drive wedges, so a single tool combined ax and sledgehammer. Not that the center-hafted ax was truly novel; it had been invented and then lost. The Minoans of Crete used double-bitted, center-hafted bronze axes and adzes almost four millennia ago.[13] New or old, this polled and thus center-hafted ax chopped with alacrity. In trials, three times as many trees could be felled with a New World ax as with the traditional design.[14] The rapid increase in the population of eastern North America during the eighteenth century may have been made possible by this new efficiency in clearing land and preparing timber. A visitor as late as the 1850s could still remark on the American ax: "The erection of a log cabin proceeds with unbelievable speed; seldom are more than three or four days required. . . . With the exception of a few nails, [building material] is furnished by the locality itself in the form of trees which the practically constructed American ax fells and properly smooths in about as many hours as the German ax would require days."[15]

The biologist derives some amusement from the similarity between the evolution and geographical expansion of the American polled ax and scenarios for the origin and spread of evolutionary innovations in organisms. In some part of an organism's range, a feature found useful even in slight and rudimentary form becomes exaggerated as more extreme versions prove better still. The earlier form and early changed forms persist for a time but gradually give way. The mature changed form then reinvades the rest of the range and displaces its predecessors except in some isolated, relict populations. Still, we're talking about similar, not identical, processes. Natural selection driven by function advantage, yes, but reproductive success as the measure of advantage, not likely; in no frontier legend does the girl seek the logger with the longer poll on his ax.

The design and use of percussive tools have to factor in yet another variable: gravity. We all know that hammering downward takes less effort than hammering sideways. In one case the weight of the hammer adds to your force as you accelerate its mass, while in the other the weight gives you no help. With a sufficiently heavy hammer and a light enough task, you can use weight alone to bring the hammer downward, investing your entire effort in raising the hammer for the next blow. We ordinarily use the same axes and hammers for downward, sideways, and even upward blows, but we adjust the way we use them. See, for instance, if you grasp its handle in the same place when swinging a tool in different directions.

We find hammering upward particularly awkward. The weight of both the tool and the arm opposes our efforts, so we have to work harder. Worse, perhaps, is a mismatch between simple tool and skeletomuscular system. In swinging downward, you extend your forearm, turning it through a wide angle around your elbow. Reversing the motion to swing upward requires flexure at the elbow, with the tool increasing its speed as it approaches you. Moreover, flexure uses the biceps rather than the triceps. The resulting reduction in distance advantage to 5.5 from 22 reduces the speed with which you can swing. At the same time, it buys no great force; you can push vertically upward only about half as hard as vertically downward.[16] Also, arm flexure requires that the work be in an odd position both physically and visually, so you can't just reverse the downward swing. Nor, for anatomical reasons, can you start with your elbow bent the other way, so you've no choice but to swing upward from the shoulder. That's troublesome, both because accuracy is lost and because muscles are poorly positioned to do it with decent power. Nor can you use a counterweighted hammer; that would put the center of percussion bang up (literally) where you hold the tool. So you either contort yourself or lie on some scaffolding when gravity works the wrong way.

Gravity works wonders when we split wood with a sledgehammer and wedges or with a splitting maul, a blunt, fat, heavy ax. A long swing starting above the head gives gravity lots of distance and time to do its acceleration. Some mauls get heavy enough—16 pounds or so—that we have only to guide rather than to force them downward. For felling trees or cutting thick trunks, a maul has another use. Keeping open the groove (or kerf) made by a saw, whether by hand or powered, ensures that the saw can run freely. A vertical tree tends to fall toward the side that the cut leaves unsupported, and the trunk of a felled tree often sags as it's cut; either way the width of the saw's groove is reduced. To maintain a sufficient groove as the saw advances across the trunk of a tree, a wedge often has to be driven into the cut behind the blade. The maul preferred by loggers for driving these wedges weighs 6 to 8 pounds and has a 32-inch handle.[17] That's shorter than mauls chosen mainly for splitting and is also lighter than most of them. Even I, relatively small and light, split logs with an 8-pound, 36-inch maul. The difference is that for splitting logs, we can work downward; during felling, wedges must be driven sideways.

Cranking

Much of the best machinery uses some kind of rotating wheel and axle. Never mind wheeled vehicles; just think of all the gears and pulleys and shafts and cams and capstans and windlasses. To them, we add obscure things you've probably never encountered, such as Geneva mechanisms and rolamites. Muscle only pulls, and no macroscopic engine in nature rotates like our electric motors and turbines. Muscle-based human technology thus bumped into a curious problem a few thousand years ago, when (at least in the Old World) rotational machinery came into use. How can muscles work a potter's wheel or a chain of buckets lifting water from a well?

Without hesitation the modern mechanic would prescribe some kind of pushrod and crankshaft, perhaps with a flywheel to smooth the motion, something of the kind shown in Figure 7.9. We use these for lots of tasks, if perhaps a little less frequently than a generation ago. I

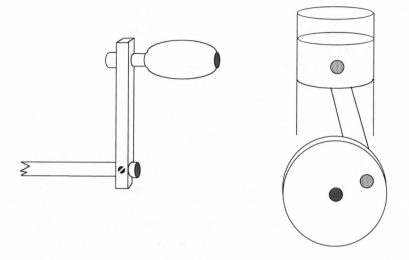

FIGURE 7.9. Two devices for converting reciprocating to rotary motion: a hand crank and a cylinder with a piston, a piston rod, and a crank wheel.

recall the treadle on my mother's first sewing machine; she rocked a footrest to and fro, and a rod from that treadle cranked a large wheel halfway up to the machine itself. I also recall standing on the platform of a train station as steam locomotives, big, noisy things that challenged a small boy not to back away, came and left. Right down at my level, a pushrod from a cylinder cranked the wheels. Treadle sewing machines and steam locomotives have gone, but down in the bowels of every automobile a pushrod from each cylinder turns the crankshaft.

Replace the pushrod with a forearm and the crankshaft with a hand crank, and you get the simplest version of the device, again in wide but recently reduced use. Hand-cranked telephone generators, centrifuges, and butter churns have gone, but a few meat grinders, hand-turned drills, eggbeaters, pepper mills, and flour sifters remain. With a tiny crank I rewind the film in my nonautomated camera.

Curiously, this seemingly self-evident solution to the problem of getting rotation from reciprocation was a late addition to our toolbox. Our ancestors used a variety of other ways to make muscle rotate things, several of which are shown in Figure 7.10. With a greasy enough bearing and an occasional kick, a heavy potter's wheel would keep going. An ani-

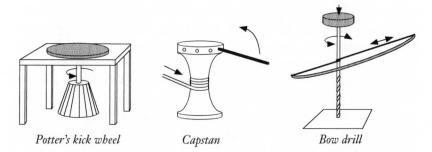

Potter's kick wheel   Capstan   Bow drill

FIGURE 7.10. Three muscle-driven devices that produce rotary motion without using a crank.

mal could work a bucket chain by walking in endless circles while attached to a radial bar. A wheel could be equipped with protruding radial spokes that a person grabbed and pulled, one after another. A rod could be inserted into a sequence of holes in a vertical drum or capstan and used to push it around. Or a bow drill could be turned with a wrapped cord first a few turns one way and then a few turns in the other direction so no net rotation occurred at all.

Classical civilization thus did not turn on cranks. A few years ago I ran across a description of a kind of crank that might have been used for drilling by the ancient Egyptians.[18] The device looked at once unfamiliar and plausible, so I built one. I moved a handle on the top in a small, horizontal circle; that motion made a weight on a string swing around a wider circle, which turned the shaft of the drill. With only a little practice, I could drill shallow holes with ease; without a top support, the thing was too prone to wobble to be useful for anything deep.[19] (But it could be operated by one hand, leaving the other to fix the work; a modern brace and bit takes two hands.) By no means, though, could this drill provide a general-purpose motor for hand-cranked machinery—even if it hadn't wobbled. In particular, dependence on gravity limited it to rotation in a horizontal plane and thus to driving a drill downward. More recognizable cranks appeared in China in the first century C.E. The earliest European illustration of a proper crank, according to one source, goes back only to 850 C.E. It depicts a crank that turns an abrasive wheel.[20]

We might have expected that the crank, once introduced, would have spread rapidly throughout any muscle-powered technology. But not so. For a millennium at least, grain was ground between the upper and lower rotating stone disks of querns, rotating, but usually with radial pegs or something other than cranks. The late Lynn White, the great technologically savvy medievalist, noted that in neither China nor the West did cranking take hold. As he put it, "The mechanical crank is extraordinary not only for its late invention, or arrival from China, but also for the almost unbelievable delay, once it was known, in its assimilation to technological thinking."[21] Making a general point of the matter, he commented, "The historical record is replete with inventions which have remained dormant in a society until at last—usually for reasons that remain mysterious—they 'awaken' and become active elements in the shaping of a culture to which they are not entirely novel."[22] In the West cranks caught on in the fifteenth century; in China they remained rare for several centuries more.

No question that cranks use human muscle effectively; bicycles, foot-cranked, provide efficient transportation worldwide. We have in fact additional data that come from something other than pedaling. A recent study looked at some humans who might be especially adept at power-demanding arm work, paraplegics.[23] Subjects were asked to crank against different loads until moderately fatigued. As expected, work against lower loads could be sustained for longer. The asymptote, the level that could be maintained almost indefinitely, turned out to be about thirty watts of power. As expected from relative muscle mass, the sustainable power for a leg-based activity, such as pedaling, beats that value several-fold. But thirty watts, efficiently invested, can produce a fine return, and it's impressive for an arm-powered activity.

Thus, much as I try to tie history to physiology, I have to admit that something else has to explain the late arrival of widespread cranking. Our wheel-based technology derives mainly from the medieval explosion of ingenious ways to use complex rotary machinery. The classical world cared less about such things. We're exposed to classicists at an early age, and we absorb a sense of an advanced culture from all the great classical literature and art. But technologically the classical world stagnated, while the medieval world, innovated, as Lynn White in particular persuaded us. Still, I hesitate to be dogmatic. Heron (or Hero) of Alexandria (first century C.E.) recognized five simple machines—lever, winch, pulley, wedge, and wheel—three of which rotate.

Finally, which way should a crank be cranked? We noted earlier that for a vertical crank—that is, for one turning a horizontal shaft—we do better with an overarm push and underarm pull. Underhand, as in another context, is wrong. As we saw earlier, that means a right hand or clockwise motion for a shaft if the shaft extends to the left of the crank, as it will if the operator cranks with the right hand and braces either appliance or self with the left. For a horizontal crank, direction probably matters less, but a right-handed operator ought to prefer counterclockwise motion. That means pushing directly outward, which we can do about 30 percent more forcefully than pulling directly inward.[24] It also means pulling inward nearer the center of the body and across the chest, a reasonably good direction for bending an elbow. But that counterclockwise motion reverses the clockwise motion with which we drive a screw downward. Different muscles do the two jobs, so we've not contradicted comments about the ease of clockwise screwing. Do we in fact turn horizontal cranks counterclockwise? We're inconsistent. We often adopt the less effective clockwise cranking convention that works better for the more common and powerful vertical cranks. We learn the right-hand screw convention early and rarely err with screws or vertical cranks. But both my wife and I have made trouble for ourselves on occasion by turning the horizontal crank of a small ice-cream freezer in the wrong direction. It wants to go clockwise; we (both right-handed) find counterclockwise cranking more natural.

CHAPTER 8

Working Hard

To me there seems nothing wanting to make a man able to fly, but what may easily enough supply'd from the Mechanicks hitherto known, save only the want of strength, which the Muscles of a man seem utterly uncapable of, by reason of their smallness and texture, but how even strength also may be mechanically made, an artificial Muscle so contriv'd, that thereby a man shall be able to exert what strength he pleases, and to regulate it also to his own mind, I may elsewhere endeavor to manifest.[1]

ROBERT HOOKE, seventeenth-century scientist

WE AGING ADULTS ARE URGED TO GET MORE AEROBIC exercise. The current prescription suggests at least three episodes per week, each lasting at least thirty minutes and keeping the heart rate at 80 percent of its maximum. How to do it? The cardiovascular system cares little, so we can choose among fast walking, running, swimming, or using one of the many energy-wasting machines designed for corporeal as well as corporate profit.

For a time I used a machine in which I skied against invisible nonslip snow while swinging my arms in a parody of what one does with ski poles. This so-called Nordic Track came not from Scandinavia (nor from Toledo, Spain) but from Minnesota—perhaps, to be fair, made by ex-Norwegian ex-farmers. The motion seemed easy enough, but it left me breathless in short order. What had happened confirmed the claims of its perpetrators. Since I distributed my efforts over a large mass of muscle, I worked with minimal impact (metabolic or physical) on any particular piece of my anatomy. No normal sensory feedback from muscles working harder than they'd prefer therefore made a conscious impression.

With hand tools, force, agility, precision mattered. Only for a good ax used steadily and skillfully or for an unusually hard bout of cranking did we need to consider the other criterion for muscle output, power. We humans, however, can act as engines of decent power output relative to either weight or fuel consumption—at least by mammalian standards, if no longer by those of human technology.

Hitting the Limit: Expending Energy

First a reminder about energy and power. Energy relates to power just as distance does to speed. Distance measures how far you go; speed measures how far you go in relation to the time the trip takes you. Energy (given as calories or joules) measures something you consume or expend; power (as watts or horsepower) measures your consumption or expenditure relevant to the time you spend doing so. As we define their units, one watt of power equals one joule of energy per second; unfortunately, joules have yet to oust calories as everyday measures.[2]

How powerful are humans? The top rate at which we can put out energy—power—should perhaps reflect how hard a muscle can work and how much muscle we have. A datum for muscle output appeared earlier; we took as a round number 200 watts per kilogram or 90 watts per pound. From that we figured that a human of about 40 percent muscle ought to be able to put out 5,600 watts or 7.5 horsepower. Again, we're not unusual. Muscles from a wide range of animals—insects and reptiles as well as mammals and birds—yield about the same maximum as do muscles removed from their animals. So do human muscles contracting on verbal command of an investigator. One of A. V. Hill's collaborators, Douglas Wilkie, did the latter, using a simple task, elbow flexion, with the setup shown in Figure 8.1, the source of the data given earlier in Figure 2.8.[3] A high school physics lab could check his numbers without difficulty. Just think, 5,600 watts, 7.5 horsepower—wow!

But muscle doesn't limit how hard we can work in any sustained fashion. As far as muscle goes, each of us has plenty of unused capacity. It's also unusable capacity, so much water over the dam, so to speak. To get real, we have to incorporate three more factors. First, we have to put the muscle in the machine, weightwise. An automobile or airplane repre-

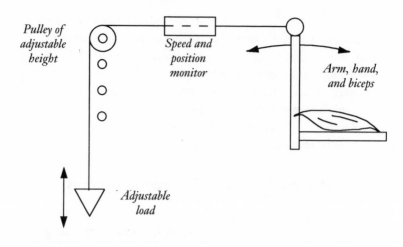

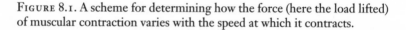

FIGURE 8.1. A scheme for determining how the force (here the load lifted) of muscular contraction varies with the speed at which it contracts.

sents a better analogue of person than does the disembodied engine of either. Second is something not so obvious. The limit on the whole-body power of our muscles comes not from the muscles themselves but from our ability to supply them with oxygen from our respiratory and cardiovascular systems. For any sustained activity, the rate at which blood takes up oxygen as it passes through the lungs sets that limit, and that oxygen consumption rate provides a near-perfect indicator of how hard the muscles are working. Third, we have to account for sustainability and motivation. We can work anaerobically by producing lactic acid only for a minute or two. After that the heart's output runs the show, and it sustains a much lower work rate. Even beyond that brief period, though, our ability to work continues to drop with how long we're asked to keep it up; other factors, both physiological and psychological, increasingly sap our performance.

One more preliminary: At first contact the published record on human power output presents a daunting mess. Data may be expressed in horsepower, foot-pounds per second, liters of oxygen per minute, or (all too rarely) watts. Worse, the data contain odd inconsistencies since some sources talk about power input, or metabolic rate, whereas others talk about power output, or work done against a load. We need a few equivalents and benchmarks to keep things straight. We power our engine with an oxidant, oxygen, and a fuel, either fat or carbohydrate. The relative proportions of oxidant and fuel can't be varied, so measuring either one can allow us to track power input. In practice, oxygen turns out to be easier than fuel to monitor, in part because we store little and use it as we go. Whichever fuel we burn, consuming a liter of oxygen per minute produces 330 watts of power. That energy comes out as some combination of heat and work. If we do nothing at all beyond rest peacefully, we use about a quarter of a liter of oxygen each minute and produce about 80 watts. That's what we call basal metabolic rate, and we release that much heat into the room. These figures set the input: oxygen consumption and the corresponding metabolic rate.

Input data—metabolic rates—provide a yardstick for comparisons among animals. Our basal rate of eighty watts makes us ordinary for mammals of our size. Big mammals have higher rates, and small mammals lower rates, as you would expect. Less obviously, big mammals have somewhat lower rates relative to their weights than do small mammals. While our rate may be ordinary by mammalian (and, incidentally, avian) standards, mammals (and birds) deviate dramatically from the rest of the vertebrates. Both are high-output specialists, self-heated warm-blooded creatures. Since heat is energy, staying warm takes power—unless we're perfectly insulated or our surroundings are sufficiently warm. In other words, if we're hotter than our surroundings, we lose heat to our surroundings, and that heat has to be replaced if we're not to cool down. But that easy explanation cannot tell the whole story. A mammal has about five times the power consumption of a reptile of the same weight at the same body temperature—in essence, power while the engine idles. We have to consume lots more food than reptiles of our weight, whatever the temperature. We dump heat even when we're not keeping warmer than our surroundings; we find rooms at body temperature unpleasantly hot. Given the inexorable calculus of evolution, staying hot and revved up must somehow pay well enough to be worth the price.

So in resting metabolism, humans are ordinary members of an extraordinary class. We're out of the ordinary, though, when we get active. The garden-variety rabbit or household cat can run itself at ten times its resting metabolic rate when push comes to shove or dog comes to call. These more typical mammals do fine at producing energy anaerobically, so real rabbits get off the starting blocks like their proverbial conspecifics, and a cat shifts from lurk to dash with the very best. But they can't keep the pace. In a large, open area with neither hole nor tree, a fit human can run down any cat or rabbit—or squirrel or goat. Just how manyfold a person can raise the resting metabolic rate depends on conditioning, motivation, and length of activity. Loosely, someone in good aerobic condition can put out more than twenty times the resting rate for minutes and more than ten times the resting rate for hours.

I hasten to add that we're by no means unique in this capacity for aerobic—that is, sustained—power output. Among domesticated animals, dogs and horses do better than we, as do lots of nondomesticated animals. Some of the best data come from the work of an old friend, the late Dick Taylor, and his various collaborators, some obtained at his lab near Boston and some at Nairobi, Kenya. One has to recognize that in this business data do not come easily. Determining condition is tricky, and providing motivation is worse; animals don't have the investigator's sense of the importance of working as hard as possible. On one occasion, a cheetah expressed its antipathy toward task and taskmaster by jumping off the treadmill and doing Dick quite a lot of damage.

Maximum metabolic rate can be given as the number of times by which an animal's resting metabolic rate increases. We call that multiplier metabolic scope. The table below gives a few values, arranged in order of increasing size of the animal.[4]

Animal or Group	Metabolic Scope
Small rodents	6 to 8
Birds and bats (flying)	15
Genet cat	10
Grant's gazelle	9
Dog,* wolf, coyote	30
Goat*	12
Calf*	12
Human*	20
Horse (pony*)	21
Eland (large gazelle)	12
Zebu cattle*	10

*Domestic; the others caught in the wild.

For the most part mammals of low metabolic scope catch prey or escape predators by lurking and lunging, while mammals of high scope catch or escape by sustained running: behind for predator, ahead for prey. The two extremes differ profoundly: in the sizes of their hearts and lungs, in the proportions of red and white muscle, in the amount of carbohydrate and number of mitochondria present in muscle, even in the leverage of muscles and bones.

That we're built for sustained power output must say volumes about human origins. Agriculture imposes no great aerobic demands, however long the hours of planting, weeding, or harvesting. So our penchant—or, better, our suitability—for marathon running must go back way farther than our ten millennia of farming. We've most likely been bipedal runners for at least three and a half million years.[5] As runners we win no prizes for efficiency, perhaps the price of being bipedal instead of properly (for mammals) quadrupedal. But we do have both endurance and versatility. We can run with about the same cost (per distance, not per time) at a wide range of speeds and over wide ranges of temperature and terrain. All animal locomotion produces lots of heat, and we're well arranged to get rid of the heat. We sweat copiously, more per unit area of skin than any other species. We're nearly hairless, so any little breeze, even that caused by running itself, improves our evaporative cooling. Moreover, sweating, as opposed to panting like a dog, decouples heat loss from breathing, most likely a good thing if one is running hard. As David Carrier, a biomechanic at the University of Utah, points out, the evolution of hominids must have involved strong natural selection for endurance running. The better endurance runner must have brought home the bacon.

The idea that well before weapons arrived, we chased down our prey as endurance predators finds support in the hunting habits of extant humans of a number of cultures. Bushmen in Africa, Tarahumara Indians of Mexico, and Aborigines of northwestern Australia do just that. Nor are they hunting small or slow stuff; gazelles, deer, and kangaroos don't dawdle. Neither are these short runs; some have been reported to last as long as two days. Carrier notes that several things offset any inferiority in efficiency: We can pick running speeds that force prey to run at less than their most efficient speeds, and we do better at regulating our body temperatures while running. Marathon runners may be a masochistic lot, but they don't do anything physiologically perverse or historically novel.

Our unusual ability to run with almost unchanging efficiency over a range of speeds has a peculiar basis, one shown by David Carrier's graduate adviser, Dennis Bramble, also at the University of Utah. Running steadily taxes an animal's respiratory capabilities, and fast quadrupedal runners, such as dogs (Bramble's main subjects), assist their lungs in a novel way. Their guts sling back and forth as they run, repeatedly pushing like pistons on the lungs. But the pendulumlike scheme works best at specific speeds, giving them speed-dependent running efficiencies. We erect humans, by contrast, don't go in for gut slinging.[6]

Hitting the Limit: Working Hard

Enough for now about the input side, what about the numbers for power output? The rate at which we do work at some energy-requiring task rarely exceeds a quarter of our metabolic rate; as muscular machines we work at efficiencies no higher than about 25 percent. Most of the time we don't come close to that figure. So even though a liter of oxygen per minute represents a power input of 330 watts, the same liter of oxygen per minute, even if we assume effective coupling to some hard task, produces no more than 75 watts of mechanical output.

With these caveats, then, what power output can we sustain for more than a few minutes? That depends on the nature of the task and on whom we consider. The most highly trained and naturally endowed male athletes, working on inclined treadmills or bicycle ergometers (stationary bikes with calibrated loadings), can reach a maximum oxygen consumption of about six liters per minute. That's an input of almost 2,000 watts, and it corresponds in practice to an output of about 450 watts.[7] Although it's ten times less than what our muscles might do, it's still far above what any ordinary human does at any ordinary task with any ordinary sense of urgency. A typical well-conditioned young adult male consumes oxygen at about four liters per minute, sustaining an output of about 300 watts. Doing sustained exercise on a treadmill or rowing machine, I manage 600 kilocalories per hour, which translate into 167 calories per second or 700 watts—input. That's an output of 175 watts, for which I figure I consume a little over two liters of oxygen each minute.

Serious interest in the sustained power output of human as engine goes back a long way. In 1734 John Desaguliers, in England, estimated that a man with a good pump could raise water at a rate that corresponds to 120 watts. (Of course he used other units; James Watt wasn't born until 1736.) Half a century later John Smeaton put that rate at 90 watts for a good laborer working all day. Smeaton, the founder of British civil (originally as opposed to military) engineering, didn't lack credentials for quantitative work. Besides specific feats of engineering, such as the third Eddystone lighthouse, near Plymouth, he did careful work on the design and performance of windmills and waterwheels. His 90 watts (perhaps rounded up to 100 to emphasize the imprecision of the datum) provides a good benchmark still.[8] The number may be good, but the impetus for the estimate may reflect the systematic dehumanization that came with the Industrial Revolution: person as just another item in some cost accounting. Should the eighteenth-century industrialist buy one of Thomas Newcomen's steam-powered pumps or should an equivalent number of people be employed as pumpers?

A rowing crew provided the material for the earliest properly scientific study that attempted to determine maximal power.[9] Not just any rowing crew but the Yale University crew that won the Olympic championship in 1924. Furthermore, not just a simple set of measurements but approaches from several directions, approaches that gave a satisfyingly consistent answer. Two remarkable physiologists, Yandell Henderson and Howard Haggard, towed a weighted racing shell and estimated what power each oarsman would have to put out to propel it at the record speed. They also built a rowing ergometer on which each subject pumped water against a calibrated resistance at the stroke rate of a race. Finally, they measured the oxygen consumption of each of the oarsmen while rowing on the ergometer. They found that a 180-pound oarsman used about 4.6 liters of oxygen per minute, corresponding to an input of 1,500 watts. (That's what our toaster oven uses and puts out. Hot work, rowing!) The oarsman put out 370 watts of useful power, an efficiency just short of 25 percent, for periods of around ten minutes.

That figure of 370 watts for sustainable male human output remains reliable. More recent, higher figures, such as the 450 watts mentioned earlier, result from improved circumstances and different experimental conditions, not from earlier error. We both know more and have become much more fanatical about training serious athletes. With more people trying to break into big-time aerobic sports, we must be selecting individuals of greater natural endowment. Also, with the larger data set afforded by easier (although not necessarily more accurate) techniques, we can select more extreme values. Still, the successors to Henderson and Haggard face the same underlying problem: "We recognized that the most willing and cooperative oarsman could not in cold blood and between two college lectures drop in at our room and in five minutes develop as great a power as when rowing on a crew and striving to win the Olympic championship, or to beat Harvard." At least the oarsmen don't try to bite the investigators.

Henderson and Haggard picked rowing as an activity that used an exceptionally large mass of muscle, guessing that such an activity would provide the best indicator of our highest possible power output. While they had the right idea, we now know that a well-conditioned athlete can reach peak output using less muscle. Anything that makes good use of the muscles of the lower torso and upper legs can absorb everything that our respiratory and cardiovascular systems can put out. Consequently, similar numbers come from measurements using inclined treadmills, stationary bicycle ergometers, Nordic Tracks, bicycling, and cross-country skiing, as well as more exotic ways to make us max out.

However the data are obtained, the same famous curve emerges: the

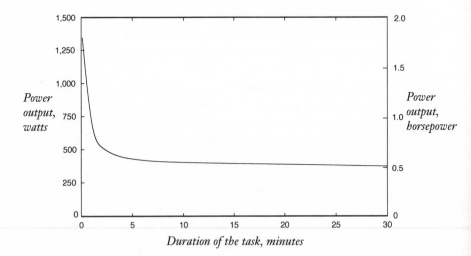

FIGURE 8.2. Douglas Wilkie's graph of power output against duration for humans. Note, again, that duration refers to how long the best of us can work steadily at tasks that impose different demands, not how output diminishes with time as a person does a single task.

one first published by Douglas Wilkie in 1960.[10] I've given the curve in Figure 8.2, but I hasten to add a few words about its interpretation. The curve doesn't track one's output as activity continues, starting intensely and then diminishing as the person tires. Instead it assumes steady output and indicates how long any particular level of output can be sustained. Its descending line says that the longer the activity has to go on, the less intense it must be. One can't do much better than the terse words of the abstract of Wilkie's paper, just updated by converting horsepower to watts:

> . . . the usable external power output of the body is limited in the following manner for the reasons stated:
>
> (1) In single movements (of duration less than one second) to less than 4500 watts; by the intrinsic power production of muscle. . . .
>
> (2) In brief bouts of exercise (0.1–5 minutes) to between 370 and 1500 watts; by the availability in the muscles of stores of chemical substances that can yield energy [anaerobically].
>
> (3) In steady-state work (5 minutes to 150 minutes or more) to between 300 and 370 watts; by the ability of the body to absorb and transport oxygen.
>
> (4) In long-term work, lasting all day, to perhaps 150 watts; by wear and tear of muscles, the need to eat, etc.
>
> All these figures refer to championship athletes; ordinary healthy individuals produce less than 70–80 per cent as much power.

In short, we're not like gasoline engines or electric motors but more like a combination of such a motor and a storage device like a battery or flywheel. Its undersize lungs and heart make no difference to the pouncing cat, and oversize lungs and heart give a person no advantage in the hundred-meter dash. We start anaerobically and then gradually phase out anaerobic work and phase in aerobic work. The maximum anaerobic work that a well-trained human can do equals the aerobic work that would be done by his or her consuming eighteen liters of oxygen. Thus, after intense exercise, a person "owes" the system that much oxygen—equivalent to about four minutes of maximum oxygen consumption. No wonder we pant after stopping some vigorous activity. Not that a person need take as much as four minutes to incur the debt. Bear in mind Wilkie's curve—output in a brief episode of exercise can be more intense than anything one can sustain.

We shift, although not abruptly, to aerobic power after incurring an "oxygen debt" of eighteen liters. Doing a hundred meters, the runner never gets that indebted, so the entire run is anaerobic. Lungs and heart just don't matter. A two-hundred-meter run reaches that maximum debt, but just barely; it's about 82 percent anaerobic. The fifteen hundred uses about half and half; a ten thousand is 90 percent aerobically powered.[11] Each increase in distance reduces the speed that wins, as it must from Wilkie's curve. The curve in fact predicts remarkably well the relative times for world records at different distances, whether for swimming, running, speed skating, or cycling.[12]

Running provides historical perspective on the maximum power a human can exert. We're runners by nature, as David Carrier and others have so persuasively argued, and the basic motions and equipment for running, unlike those for pole vaulting or high jumping, don't change very much. In Figure 8.3, I've plotted the winning times for Olympic competition in the fifteen-hundred-meter men's run, the longest race for

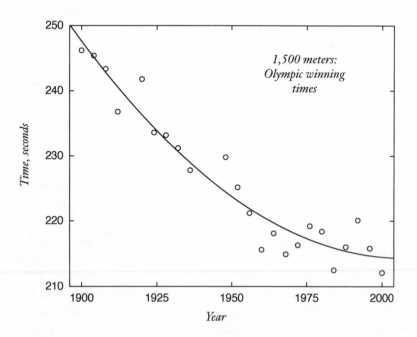

FIGURE 8.3. Record times for the Olympic fifteen-hundred-meter men's run.

which we have a hundred-year record.[13] Notice how little the times have changed over the past thirty or forty years, and that's despite the way we select competitors from an ever-wider population; the way training techniques must reflect increasing experience, advances in science, and ever-larger financial input; and ostensible (at least highly touted) improvements in shoes.

One can take this graph (or others like it; the fifteen hundred by no means stands alone) as evidence that we're bumping into some underlying physiological limit. In practical terms, two figures may define the cardiovascular capabilities of humans: an oxygen consumption rate of four liters per minute for a well-conditioned adult female and a rate of six liters per minute for an especially well-endowed and then well-conditioned male. We know a lot less about the aerobic capabilities of human females. Since females are on average slightly smaller than males, we expect lower maximum outputs; for good reason do the sexes compete separately in most athletic events. The politically incorrect biologist wonders whether ancestral sex role differences, cultural and biological, influence present physiological limits. Evolution has a lot to answer for, and it may matter here.

Curiously, we get the same descending and then leveling curve as for the fifteen hundred from data for horse races.[14] Here, again, animals have been persuaded to work at exactly the same tasks as hard as possible over many years—some races have been run for well over a century—but the factors behind changes in performance may differ. All the commotion about breeding lines and bloodstock means that one has to add genetic change. I must add as well the most intense and single-minded training; these creatures have no personal lives to speak of. On the other hand, racehorses represent an unusually homogeneous lineage rather than a small group selected from a huge population. The fragility of present-day racehorses, especially their susceptibility to injury, comes in for frequent comment. Much is made of the dysgenic effects of inbreeding. Jim Jones, at the veterinary school at Davis, California (and a former graduate student in my department), makes a strong case that the problems have a physiological rather than a genetic origin. Muscles, bones, heart, lungs, and so on change with training. Jim suggests that ever more intense training for races has produced animals that are ever less competent to meet any other demands placed on them—whatever their genes say.

That ought to worry us, at least a little. We don't deliberately breed Olympic athletes, a prospect that makes us first shudder and then thank our unusually great intergenerational period. But we train them for competition with ever-increasing efficiency and fanaticism. Do our highly trained athletes suffer long-term compensatory disabilities? Looking into the matter goes against the grain of the contemporary culture, but we ought to pay attention to the possibility. The 1999 edition of *The Merck Manual*, the standard quick reference for physician and nurse, points out that ten million sports injuries receive treatment in the United States each year. It details injuries and therapies in ten pages of fine print but without saying a word about residual and chronic factors. Nonetheless, it recognizes how much endurance training changes a person: "Resting sinus bradycardia, third and fourth heart sounds, systolic murmurs, a variety of ECG abnormalities, and cardiac enlargement on chest x-ray are characteristic. This syndrome, which would be considered abnormal in an untrained person, is a successful adaptation to endurance exercise and should not be misdiagnosed as heart disease."[15]

Paddling and Rowing

No other animal propels itself or its craft across bodies of water with paddles the way we do, even if we include animals that grow and use their own living paddles. Our stroke seems so obvious, alternately dipping into the water, pushing, withdrawing from the water, and moving paddle or oar back again. Obvious, yes, but as a biological device thoroughly unnatural. Still, we began using this system of propulsion before we became fully human and proper members of *Homo sapiens*. We continue to do it recreationally and competitively, and for short distances we keep using it for practical purposes. I taught for several summers at a marine laboratory in the Pacific Northwest. Rowing across the harbor rather than walking around it got me into town faster and facilitated my return with bags of beer and groceries.

Paddling and rowing prove as biomechanically effective as they are biologically strange. In essence, they depend on pushing rearward on water with big, flat surfaces oriented broadside to the water so they have the greatest possible drag. That surface then comes out of the water and moves forward (relative at least to the boat) in the less resistant (by about eight hundred times) air. Many animals paddle: turtles, muskrats, ducks, and tiny crustaceans, to mention a few. But they keep their paddles underwater for both power (rearward) and recovery (forward) strokes. Instead of depending on the different densities of water and air, they produce a high-drag power stroke and a low-drag recovery by changing the shape and orientation of their paddles. I know of only one animal that paddles with an aerial recovery, a human, swimming with a crawl, butterfly, or backstroke. These, oddly, rank among our fastest strokes.

Since nature's paddlers use an underwater recovery stroke, none could have shown us the particular scheme we use to propel boats. Yet the scheme is both old and widespread, so in some way it must represent an easy innovation. Perhaps our ancestors first learned to swim—whether with aerial or submerged recovery I hesitate to guess. A swimmer sitting on a floating log could have paddled quite casually. But although the basic act may be self-evident, the action, based on interaction of blade and water, involves subtleties that even the nineteenth century could deal with only empirically. However, let's defer the hydrodynamics.

First, which works better, paddling or rowing? Paddlers kneel and face forward. During the power stroke they bend forward at their hips and lower their backs while extending their inboard arms, high on the paddles, still farther forward. At the same time, their outboard arms, lower on the paddles, move backward, and they twist their torsos. By contrast most, but not all, rowers sit and face backward. They bend backward (toward the front of the boat) and flex at the shoulders and elbows of both arms when applying power to oar or oars. The paddler can transfer the force generated by the blade's drag to the hull of the boat only through the body. That requires tensing a lot of trunk and leg muscle to keep the torso from bending, aside from the work of the arm and upper-body muscles as they supply power to the paddle. The rower, though, can make use of the oarlock or its equivalent to transfer force. For efficiency, rowing wins; more power goes into the task in relation to the power consumed by the person. Try paddling a rowboat, and you'll get an object lesson in the advantage of rowing. Or if you think that's in some way unfair, try rowing a canoe. An Adirondack guide boat, more or less a canoe with oarlocks, goes wonderfully fast with remarkably little effort.[16]

Still, for both aesthetic and practical reasons, we retain a preference for facing forward. How much nicer and handier to paddle and see where you're going than to row and view only where you've been! Paddling came earlier and has been used by more cultures with more kinds of boats. Those from the cultures of Asia, Oceania, and the Americas preferred to use paddles or oars while facing forward; oars stroked while facing backward predominated among the seafarers of the Mediterranean and North Atlantic. Paddling works better in tight quarters inside or out—whether going through a rocky passage or a narrow stream or when cargo pushes you up against the gunwale of a boat. Neither paddled nor rowed boats store elastic energy from one stroke to the next, the way we store it between strides when running or an insect stores it between wing strokes. So all the effort expended accelerating and decelerating either a paddle or an oar goes to waste. Paddles, though, weigh less than oars; that should lower the cost of back-and-forth motion and in part offset their poorer use of the muscles.

How good is rowing? We should put aside modern racing shells, with their sliding seats. These permit effective use of our big leg muscles and drive humans to their aerobic limits—recall Henderson and Haggard's Yale team—but they carry nothing but their crews and can't tolerate anything but the most placid water. Among practical rowing craft, most attention is paid to ancient oar-driven ships, in particular those of Mediterranean civilizations, where rowing (as opposed to paddling) goes back at least three thousand years.[17] They've been analyzed on the basis of information gleaned from old accounts, Egyptian carvings, pictures on Greek vases, some surviving models, and an occasional underwater salvage. A full-size replica, *Olympias*, now in the Greek Navy, tests modern notions of how these ships were built, were operated, and performed.

Most shipping in the Mediterranean consisted of commercial transports. These vessels, some of great size, had a few oarsmen but mainly relied on sail. Voyages were timed to take best advantage of seasonal winds since the ships couldn't beat upwind and did poorly in crosswinds. Speed? Two miles per hour counted as decent progress. By contrast, while warships could sail, they took rowing far more seriously; for battles close to port they left behind all sailing gear, even the mast.[18]

The classical galley warship, the Greek trireme (Figure 8.4), used 170 15-foot long oars in three banks (hence "trireme"). One person rowed each oar; using several oarsmen on each came later, something perhaps begun in Syracuse (a Greek city on Sicily) and then picked up by the

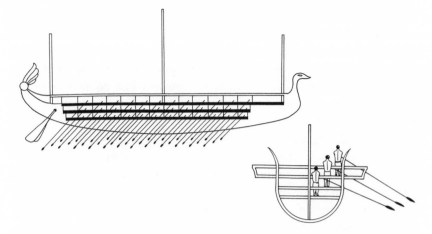

FIGURE 8.4. Side and cross-sectional views of a trireme of the size and arrangement of the *Olympias*.

Romans.[19] Experience with the *Olympias* suggests a maximum stroke rate a little less than 50 per minute, not all that different from a modern racing shell. It corresponds to an impressive output of 330 watts per oarsman. But of the 330 watts, fully 130, or 40 percent, go into manipulating the oar: accelerating and decelerating, raising and lowering. The remaining 200 watts drive the ship at a top speed of a little over 10 miles per hour. That agrees with calculations from classical accounts that a trireme could maintain 8 miles per hour for a considerable period. For surface ships of this size and speed, a lot more power translates into only a little more speed. Doing 8 miles per hour takes less than half as much power as hitting 10.[20] Either number should impress us.

We should be particularly impressed because galley warships on the Mediterranean couldn't reach those figures two thousand years later. Those of sixteenth-century France might have done brief bursts of up to seven miles per hour. Why the deterioration? Using several people on each oar reduces efficiency in a number of ways. The range of motion of the oarsman at the end of an oar limits its motion. One obvious solution, using successively larger oarsmen inward from the hull, transcends practicality, so oarsmen closer to the hull work with decreasing effectiveness. Bigger and thus heavier oars also take their toll, with an increasing fraction of power output going into accelerating mass and lifting weight, and the oars could be massive—in one account a length of forty-seven feet and a weight of 350 pounds. Stroke rates go down, so the muscles of the oarsmen, especially those closest to the hull, will contract too slowly for the best power output; recall the curves of Figure 2.8. Finally, some diminution in performance may come from using as oarsmen convicts in poor physical condition rather than highly motivated fighting men.

In general, larger means faster for ships. The larger ship has less surface to drag it back relative to its volume, and it can get up to a higher speed before it suffers serious resistance from its own surface waves. But the galley didn't scale up particularly well beyond the size of a trireme, and even the trireme probably exceeded the optimum. The problem arises because rowers can be increased in numbers but not in size, unlike the engines of our technology. For an analogue, imagine increasing the number of masts and sails rather than their sizes as you scale up a sailing ship. Eventually you have a ship propelled by hundreds of tiny sails, none far from the deck, and all inefficiently interacting. (A solution to the scaling problem, used for riverboats in China, never caught on in the West. You just use paddling side wheels, as was done later with steamboats, and connect them to treadwheels within the hull. More on treadwheels and such in a later chapter.)

To put it in more formal terms, the drag-causing surface area of its hull increases with the square of a ship's length, while the number of oars that can be accommodated increases only with the first power of length. Double the length, and drag goes up roughly fourfold; thrust goes up only about twofold. As we've noted, using larger oars with more oarsmen on each provides at best a partial solution. A trireme weighed (or displaced) about 50 tons; galleys two millennia later weighed 200 to 400 tons.[21] (By comparison, seagoing sailing ships of the sixteenth century ranged from roughly 100 to 1,000 tons; Magellan and Drake circumnavigated in 100-ton ships; in the nineteenth century, Darwin's *Beagle*, considered small, weighed 235 tons.)[22] Although galleys persisted into the age of transoceanic sail, they made only brief forays at times of low wind or places of shallow water or when their maneuverability gave them some advantage.[23]

We musn't forget how often oars shared the task of propulsion with sails, from the earliest Egyptian boats that went to Lebanon to procure wood on through the feluccas in use by North African pirates a century ago. The galley, whether of ancient Athens or medieval Venice, could be sailed or rowed, depending on wind and circumstances. The combination gave maneuverability in close quarters and reduced dependence on wind direction. But the combination created severe design problems. Sailing ships must be ballasted to sail crosswind or (occasionally) still closer to the wind, and that extra weight limits the practicality of rowing. Sailing other than downwind makes a vessel heel over, and a ship that can heel must have high sides, complicating the positioning of oars. Oared ships can have neither much depth nor much freeboard. Insisting that oared propulsion be retained must have precluded designs less dependent on wind direction and thus less dependent on oars. Still, that Catch-22 aside, oar plus sail didn't ensure disaster, at least for small vessels. We shouldn't forget that sail-and-oar Viking ships brought the Norse across the stormy North Atlantic to Iceland, Greenland, and Newfoundland.[24]

Even if better than pushing with paddles, propulsion with lots of oars has some drawbacks that are rarely mentioned. Oars arranged in a long series interact with one another. Only the foremost few oars encounter water without the swirling turbulence caused by other oars on previous strokes. Any such disturbance reduces an oar's effectiveness. Since the oar's task is to accelerate water rearward, the worst disturbance is water already moving rearward, as it will from the action of oars farther forward. If the water already moves rearward, the oar can't give it as great an increase in speed. Few airplanes have one propeller behind another; when we do build an airplane that way, we minimize the problem by making the propellers turn in opposite directions. The closer the spacing between oars, the worse the interaction, and both triremes and galley warships had oars as close as the actions of the oarsmen permitted. Modern racing shells move a long way between strokes, and that helps. They also have more widely spaced oars; they use a single bank in which one oar extends leftward and the next rightward.

Every animal that propels itself with a series of paddling appendages minimizes the interaction problem with a sequential stroke. The rearmost paddle pushes, the one next forward pushes, then the next, and so forth. Who are these creatures? We're talking about a lot of cilia-covered protozoa, some odd jellyfishlike creatures called ctenophores (the *c* is silent) that have plates of bundled cilia, some marine worms, and many little crustaceans. Having never heard of sequential rowing by humans, I asked a local coach about the possibility. He said that the rhythm would be very hard to maintain, especially when rowers were asked to change pace. While the gain for racing shells would be less than that for galleys, I think the scheme merits a fair trial. Incidentally, insects that fly with two independently beating pairs of wings do very much the same thing. Locusts, dragonflies, and the like keep hind wings ahead (in time) of forewings, so the hind wings (except when the wings pass each other twice in each full stroke) encounter undisturbed air.

None of the paddling animals just mentioned achieves great speed, even by animal standards, suggesting some serious drawback of both oars and paddles. Excluding jet propulsion, water (or air) can be pushed rearward by either of two fundamentally different mechanisms, as in Figure 8.5. We've been talking so far about the obvious one. Push back on the water with a high-drag object, and the water will pick up momentum from the object. Conservation of momentum demands that the rearward momentum of the water be balanced by the forward momentum of the ship. We then have to get the high-drag object, oar or other, forward again. The recovery half stroke may be in the water—it's just necessary to reorient the object so it faces lower drag when moved for-

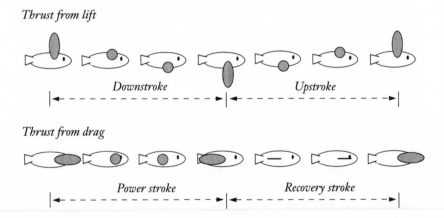

FIGURE 8.5. Two ways of getting thrust in water (or in air). In the upper series, thrust comes from the rearward lift of a hydrofoil (or airfoil) that goes up and down—crosswise to the direction of the craft. In the lower series, thrust comes from the difference between a high-drag rearward power stroke and a low-drag forward recovery stroke.

ward—or the recovery half stroke may be in the air. For a decently designed appendage, not a human hand, submerged recovery costs only a few percent. So recovery in air gains less than we might expect. Of more consequence, the necessity for a recovery stroke isn't what limits this drag-based propulsion.

All of nature's impressive swimmers (and aircraft) use the other mechanism. Here the appendages—wings, flukes, flippers, and so on—move not back and forth in the direction of the craft's movement but crosswise to that movement. They go up and down, like wings or a whale's flukes, or side to side, like fish tails, rather than fore and aft. They thereby avoid having any recovery half stroke at all, but again, recovery strokes cost little. More important, they produce forward thrust by an entirely distinct mechanism. Just like an airplane wing, they generate lift. Lift consists of a force at right angles to the flow across something. For a fixed airplane wing with air going rearward across it, the force is upward. But if the wing, or its equivalent, goes sideways by spinning or beating, the force can be directed forward, and however produced, a forward force constitutes thrust. Paddles use the difference between the drags of power and recovery strokes to make thrust; beating tails, flukes, and flippers use cross-flow movement and lift to make thrust.

Why the overwhelming preference for this devious way of getting thrust? In a word, efficiency. That's why paddling airplanes, proposed long ago, proved a dead end. Of more consequence, that's why propeller- (or screw-) driven ships displaced paddle wheels in the middle of the nineteenth century, even before anyone understood the subtle requirements of a good propeller. But the strangeness of thrust from lift, even for a world that knew whales and windmills, long kept it from our grasp; an understanding of how it worked came only in the twentieth century. Under just two circumstances does the direct backward push beat the thrust-from-lift scheme. One happens when sudden acceleration determines quality: Swing a big paddlelike appendage rearward, and you'll spring forward. The other occurs when an appendage must do more than swim. Playing thrust-from-lift depends on the details of airfoil or hydrofoil design, and appendages that do the task well can do little else. The thrust-from-lift flippers of seals or sea turtles don't work well as walkers. The legs of other turtles walk better, but they work as drag-based paddles, and their owners swim poorly.[25]

Nothing stops human-powered ships from using lift-based propulsion; Figure 8.6 shows a few possibilities.[26] Just equip a pedal boat with a propeller either in the air or underwater. For that matter, we use the scheme, if crudely, when we scull a boat, swinging a rearward paddle side to side. You can even use the mechanism to sail directly upwind; just connect a topside windmill to a propeller down in the water. But we've moved well beyond the present subject, which needs one more comment.

What an innovation was the oarless oceangoing sailing ship! Giving up the possibility of muscular assistance led to a radical redesign, to a broad, high, and heavy seaworthy ship capable of long voyages with a smaller crew and much more cargo space. Nothing better illustrates Lynn White's point about the remarkable technological innovations of medieval Europe. Nothing better also illustrates the point made by the late Joseph Needham about the technological sophistication of China than the appearance of an equivalent seagoing sailing ship, the ancestor of the Chinese junk, about the same time. One may have gotten the idea

FIGURE 8.6. Three radical human-powered boats. The paddle wheel makes good use of our muscles and uses a large enough wheel so little energy goes into pushing water up or down, but it suffers the usual limitation of drag-based propulsion. Pedaled underwater propellers work well, and variants of this design are commercially available. A propeller for use in air has to be a lot bigger. The boat shown here, Mark Drela's *Decavitator*, set a record for human-powered watercraft of 21.3 miles per hour over one hundred meters in 1991.

from the other, but differences in construction suggest independent development further along.[27]

Cranking Up Bicycles

Pedaling resembles aerial recovery paddling in being both biomechanically effective and biologically strange. In its history, though, it couldn't differ more. Paddling appeared early and often. Pedaling, which uses a crank to get rotary motion, came late and with difficulty—and that despite the fact that pushing downward and rearward on something solid differs in no way from what we normally do with our legs.

In 1839 a Scottish blacksmith, Kirkpatrick Macmillan, built the first self-propelled bicycle—that is to say, one in which the cyclist didn't push directly on the ground. In Macmillan's sensible design, the cyclist alternately pushed each of a pair of pedals forward, and pushrods (or, more descriptively, pullrods) from the pedals then cranked the rear wheel. His scheme, shown in Figure 8.7, works the same way pushrods from the cylinders drive the wheels of a steam engine (with which he may have been familiar) and the way piston rods turn an automobile's crankshaft (of which he was of course unaware). What could be more natural for a human than alternate extension of the legs? Indeed, one contemporary machine, a Nu Step recumbent exerciser, asks a person to make exactly the same movements.

Even though modern bicycles again apply power to their rear wheels, rather than the front-wheel drive popular in the 1870s and 1880s, we no longer crank the wheels with Macmillan's arrangement. From about 1874 on rear-drive bicycles have instead driven the wheels with a chain extending back from the pedaling shaft.[28] Pedal cranking, at or near an optimal cranking rate, turns out to be a highly efficient way to couple person to machine. But such an unnatural motion! Moving the feet in a

FIGURE 8.7. The way the Macmillan bicycle drove its wheels.

pair of circles, with the two feet at opposite points in their respective circles, bears little resemblance to the motions of any gait of any bipedal or quadrupedal animal. Neither gravitational nor elastic energy storage, as we use in walking and running, can play much role. About the only "normal" aspects of cranking pedals are the opposite movements of the two legs and the use of the powerful extensor muscles.

Yet so good is cycling that every successful human-powered aircraft has coupled person to propeller with bicycle pedals. A casual cyclist maintains a speed about four times that of a pedestrian, roughly 12 rather than 3 miles per hour. Metabolic cost per unit time may be greater in cycling than in walking, but as speed goes up, cost doesn't increase anywhere near as much. If measured by the cost of going a given distance, cycling leaves walking in the dust. To put things more specifically, cycling at 4.5 miles per hour takes 2.2 times less energy than walking. Cycling at 9 miles per hour takes 3.7 times less energy than running. Or to use a different comparison, an output of 7 watts per pound of body weight propels you at 19 miles per hour on a bicycle, but only 9 miles per hour if you run.[29] Cycling, using muscles in its strange fashion, incurs a much lower cost per unit distance traveled. You reduce the cost of moving yourself from one place to another by attaching 25 or 30 pounds of unpowered machinery to your body! Who would have guessed?

Not only does cycling beat walking for casual transport, but competitive cycling beats running by a similar factor. Consider—a 10,000-meter cycling race takes about the same time as a 5,000-meter run, 12 and 13 minutes respectively. So they put their participants at the same place on Wilkie's curve—that is to say, the winners of each will put out about the same power. But at that power level the cyclist goes twice as far, thus covering any given distance at about half the cost. In particular, record-level competitive cycling costs 60 kilocalories per mile, while running costs 115 kilocalories per mile. Incidentally, speed skating almost matches cycling, at 65 kilocalories per mile, while swimming pales beside cycling, skating, and running as a way to get oneself from place to place, costing no less than 525 kilocalories per mile.[30]

However unnatural, pedaling must still take its cues from anatomy and physiology. The length of the cranks usually defies easy meddling (although it matters), but the seat heights of all bicycles (and bicycle ergometers) permit adjustment. Each leg should approach full extension at the bottom of the cranking cycle for you to use your muscles at as near rest length as practical. Then the speed of cranking must be picked to make reasonable use of the force versus speed relationship for the main leg muscles. As explained back in Chapter 2, at zero speed you exert the most force but put out no power, which, again, is force times speed. At top contraction speed you put out no force and again put out no power. Maximum power output has to come somewhere in between.

Picking pedaling rates gets complicated, however. Fast rates, up to 100 full turns per minute, against low loads turn out to be more efficient than slower pedaling against higher loads; muscles hit both top efficiency and power when contracting rapidly against relatively low resistance. But the advantage of rapid spinning appears only at high-power outputs, which can't be maintained for very long. The casual cyclist does best at no more than 60 revolutions per minute and should shift gears accordingly. The competitive cyclist does best at higher rates, up to 90 rpm or so. Bicycles with a huge choice of gear ratios have become popular in recent years, but the casual cyclist who puts out about 40 watts will get indistinguishable results anywhere between 30 and 60 rpm.[31]

A few words should be added about the downside of cycling relative to perambulation. Running cares only a little about terrain; efficient cycling requires hard, smooth surfaces. Hikers can look with disdain at even the most adept trail bikes. Also, any modification of a bicycle to decrease its fastidiousness, such as fatter, lower-pressure, bumpier tires, extracts a price in efficiency.

Flight, If Low and Slow

No activity tests our power output as does human-powered flight. Twenty years ago, just after the first successful human-powered flights, we imagined that it would become a routine high-level athletic activity, a mix of soaring and bicycle racing. It has not become ordinary for the same reason that it took such enormous effort in the first place: it bangs up against our physiological limits.

Interest in human-powered flight goes back to antiquity, since birds fly so easily and engagingly and since humans provided the only engine in the toolbox of the ancients. We may be clumsy swimmers by animal standards, but we can't fly at all. Still, we dream of Apollo's horse-drawn chariot, Santa Claus's sleigh and reindeer, and various fanciful flying carriages drawn by fettered fowl. By no means are angel equivalents exclusively Christian, and Daedalus and da Vinci lead a long cast of inventors of varying degrees of real existence and realistic prospect.[32] The dawn of scientific aerodynamics in the nineteenth century brought sobriety, but the arrival of airplanes at the start of the twentieth revived the dream. Still, measurements on ergometers—Henderson and Haggard weren't the first to do them—gave numbers for what the engine could do. At the same time, work on wing performance gave a good idea of what the engine had to do.

Between the two world wars, the French and then the Italians and Germans attempted various forms of human-powered flight. A French aircraft succeeded first, in its fashion. In 1912 Robert Peugeot, builder and namesake of automobiles, offered a prize for a winged bicycle that could get up to speed, take off, glide ten meters, then repeat the performance from the other direction. The prize was won in 1921 by Gabriel Poulain, a pilot and bicycle racer. It sounds like fun, and I wonder why bicycle-assisted gliding hasn't retained popularity in an era of streamlined bicycles, hang gliders, and general athletic fanaticism. Italian and German human-powered aircraft were successful enough, both with catapult launchings and in short flights with unassisted takeoffs, to provide encouragement to their successors before World War II put a stop to such impractical activity.

Interest picked up again in the 1950s, and the path that led to the first successful flights has been repeatedly recounted. A British industrialist, Henry Kremer, was persuaded to put up what became a series of prizes for ever more difficult self-powered flights, the two most notable of which were won in 1977 and 1979 by an American, Paul MacCready. *Gossamer Condor*, MacCready's first craft, flew a carefully prescribed figure-eight course around a row of three pylons spaced a quarter mile apart—in all, a course slightly over a mile. His second craft, *Gossamer Albatross*, crossed the English Channel.[33]

Most accounts mention muscle physiology only in connection with the role of bicycle ergometry in picking and training pilot cum engines. In fact, it played a major role in the saga. The committee of the Royal Aeronautical Society that drew up the rules for the Kremer competitions included (and paid attention to) none other than Douglas Wilkie, probably the best choice they could have made. The committee tried to set rules that put the task within but close to the limits of the possible, and it wanted to exclude physiological as well as physical evasions. Wilkie envisioned only a brief involvement, but he ended up writing the definitive papers on the possibility of human-powered flight.[34] Indeed setting the Kremer parameters stimulated him to gather the data for the Wilkie curve (Figure 8.2). As he later commented, the committee got it about right—the prize provided a sufficiently strong goad, with many groups trying, but also a sufficiently distant goal, in that it took twenty years to be won.[35]

From data on the size of animals and the scaling of their muscular capabilities, Wilkie argued that human-powered flight ought to be possible—but just. Birds and bats do it, so one might suppose that we simply have our muscles in the wrong places. That's a solvable problem, technologically, if not surgically. Still, the heaviest fliers in nature—the trumpeter swan, wandering albatross, and condor—weigh only a little over twenty pounds. That three separate lineages of birds, birds with divergent styles of flight, reach the same maximum size has inauspicious implications for the flight of still-larger creatures. Metabolic scope may increase slightly with body size. But the underlying metabolic rates, both resting and maximal, decrease, relative to weight, as animals get larger. The larger the animal, the lower its maximum power vis-à-vis its weight. Wilkie figured that a dog, properly employed, could fly. A human stood at the intersection of the lines defining power needed and power available, so we might do so—but just. The best racehorse could not.

Inauspicious physics doesn't help either. To keep power down to a level that permits sustained flight requires, for reasons that I won't dwell on here, low speed.[36] That has several evil, indeed mutually vexatious consequences. Low speed puts the flier at the mercy of even minor winds, gusts, and air turbulence. It also demands huge wings. Dropping speed from forty miles per hour, that of a particularly low-speed airplane, to ten miles per hour, a speed typical of MacCready's prizewinning *Gossamer*s, requires a sixteenfold increase in wing area. But things get worse yet since the need for lots of lift but little drag precludes short wings. Wings not only must have great area but must be long and thin. The *Gossamer Condor* now hangs in the Smithsonian National Air and Space Museum in Washington, where it dwarfs the adjacent *Flyer* of the Wrights and the *Spirit of St. Louis* of Lindbergh. At the same time, the whole craft must also be light. That a plane with nearly a one-hundred-foot wingspan could weigh only sixty pounds strains credulity, but lift must equal weight (including the pilot cum engine), so designers of human-powered craft trim weight in every way they can. The extreme fragility of all such airplanes comes as no surprise. As the motto of one design team put it, "if it has never broken, it must be too heavy."

More than twenty years have elapsed since the flights of the *Gossamer*s, and the art has advanced. Another Kremer competition set a time limit that required a speed of over twenty miles per hour. A group from MIT won that prize in 1984 with a plane that had a lift-to-drag ratio of thirty-three, rather than the ten of the *Gossamer*s. In 1988 another plane built at MIT, *Daedalus*, was flown from Crete to the island of Santorini, four times the twenty-mile width of the English Channel. The trip of four hours took less than twice as long as the Channel crossing, though, because the newer plane could cruise at sixteen miles per hour.[37] The two hundred watts of power required by the craft lay below Wilkie's curve, but the rider must be both pilot and engine, and flying a slow craft in erratically moving air demands more than doing mindless battle with an ergometer.

The oddity of the human engine reflects itself in the history of human-powered aircraft. A large number of craft, dozens and dozens, managed to get off the ground. But they flew only short distances. Good designs could go farther, and improvements in any one craft resulted in ever-longer flights. That's the trouble with operating on Wilkie's curve: the shorter the flight, the more powerful the engine.[38] Flying the Kremer prize course in the *Gossamer Condor* took the best effort of the trained and gifted Bryan Allen. Shortly afterward the same plane was easily flown by Maude Oldershaw, a sixty-year-old untrained grandmother, in the first self-powered flight by a human female, but she flew for only a short time. Compare this behavior with that of machine-powered flight. To simplify only a little, if you can get aloft (and control the craft), you can fly a long way. Too much is made of the Wrights' initial flight of 120 feet, which just represented proper caution with a new machine of unknown habits. The craft made four flights that first day, each deliberately longer than the last. Only a minor mishap on the fourth—a wing tip brushed the ground—led the brothers to abandon the idea of buzzing the local coast guard station. Even now test pilots don't bother to go for distance.

Physics imposes no absolute limit on what a human-powered aircraft can do. Conversely, it provides no easy ride. Improved technology, such as materials of greater strength relative to weight, will continue to ease the burden on the human engine, but with aircraft already weighing much less than their riders, the quest for lightness yields ever-diminishing returns. Propeller efficiencies around 90 percent leave little room for improvement, and I'm not convinced that flapping wings are anything but a fantasy, something strictly for the birds. Airfoils that achieve a little more lift relative to their drag will be developed, but after a century of intense interest, airfoil design is a mature field. Of especial interest, therefore, are devices that provide auxiliary fuelless power, such as solar panels on the large wings, devices to store energy at least briefly, and arrangements to use irregularities in the way air moves as energy sources.[39] In practice, the limiting factors may be the cost of playing with a dangerous and expensive technology that lacks any great commercial or military prospect.

Interest remains, if without the media hype of the early eighties. Three new Kremer prizes challenge designers to achieve higher speed, better maneuverability, and takeoff from water. Design, construction, and testing proceed on several craft in several countries.[40]

One More Human-Powered Vehicle

Down to earth again, in several senses. Only rowed boats exceed in historical importance a certain simple and mundane human-powered vehicle, which is, not to play with suspense, the wheelbarrow. Wheelbarrows, as in Figure 8.8, are marvelously ingenious mechanical devices. They're force-multiplying levers, with a wheel in front as the pivot point, their loads borne in the middle, and their operators' effort well out between the handles. They're wheeled vehicles that move forward with low resistance and bear most of their load on something besides the

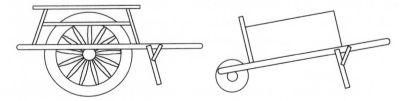

FIGURE 8.8. On the left a Chinese center wheel wheelbarrow; on the right a conventional wheelbarrow.

human torso. What load the wheelbarrow puts on the torso presses downward along the length of the spine, in as trouble-free a position as a load on the head or shoulders. Few wheeled vehicles can match its maneuverability. Can we quantify its advantages? Not with particular precision. Its effectiveness as a wheeled vehicle depends on the smoothness and hardness of the terrain to a much greater extent than does legged locomotion. Still, under favorable circumstances a worker with a wheelbarrow can move a lot more material than the carrier of a load and far more than two workers on opposite ends of a palanquin carrying passengers or a hod carrier for freight.[41]

The simplicity, versatility, and peculiarity of the wheelbarrow all need emphasis. At rest it sits on a tripod of supports, so all supporting points contact even an irregular substratum, with none of the awkward rocking of a quadrupedal table or chair the legs of which mate poorly with the floor. The rear supports lack wheels, so the wheelbarrow doesn't need brakes to keep it from wandering off on its own. Loading a single wheel stresses the ends of its axle equally, without the subtle asymmetry that requires "dishing" of carriage wheels, shifting the plane of the rim to one side of the hub. In addition, single wheels can be fixed to their axles if the builder prefers. By contrast, paired wheels fixed to the ends of an axle don't like to go around corners without some provision for different turning rates. Cars use a so-called (and complex) differential gear, and some pedaled toy cars drive just one of their rear wheels.

Let's look at the versatility of the wheelbarrow, which can go where no other wheeled vehicle dares venture. A path suffices, as does a single inclined board. No need for a road or paired tracks. Nor does a sideways tilt of the path tip the vehicle, as it would for any two-wheeled chariot, rickshaw, yard cart, or wheeled suitcase. A wheel may need a reasonably hard substratum, but one suitable for a single wheel can be improvised almost anywhere. A wheeled vehicle can climb a curb up to half a wheel's radius, but doing that takes a good jolt of power. The light wheelbarrow responds when we increase power briefly, something we do well. Just put on a little speed before the bump, and use the added momentum of the load to go up and over.

As for peculiarity, a moving wheelbarrow gets support from only two points: its wheel and whichever of the operator's legs happens to be in contact with the ground. It's therefore unstable, liable to tip over at any moment, and it demands skill and vigilance from its operator. Not just visual vigilance, the operator has to bring to bear all the proprioceptive equipment I talked about back in Chapter 5. Not for nothing do almost all wheelbarrows employ a single operator. Problems of coordination, more difficult than for multiwheeled vehicles, offset the advantage of multiple pushers for heavy loads. The Chinese use two-person wheelbarrows to some extent, but these have their wheels in the middle. So pusher follows in the footsteps of puller, and each handles both sides of the barrow in what must be the only practical solution to the problem. That need for delicate control means that few ordinary wheelbarrows can depend entirely on draft animals and none on mechanical engines.

A general point lurks here, one that applies to both living and nonliving vehicles. Stability and maneuverability turn out to be antithetical; gains in one typically entail losses in the other. The birds that fly rapidly through the pine trees of my front yard use their fine sensory equipment and fast feedback loops to offset their underlying instability. You can't sail a stuffed robin across a room as a passive glider! Commercial aircraft choose stability; small combat aircraft lean toward instability. The most extreme among the latter use computer-assisted controls to handle matters beyond the neuromuscular capabilities of any pilot. A slow bicycle is more maneuverable but less stable than a fast one. Steering a sport car takes more attention, even when you're not doing anything exciting, than steering a family sedan. Perhaps six-legged insects once walked by advancing alternate triangles of three legs and thereby avoided any tippy phase in the cycle; most abandoned that stable system far back in the past, presumably in the interest of maneuverability. Wheelbarrows, no exception, pay for their maneuverability by an especially tight coupling to their operators, one necessarily tighter than that of any other wheeled vehicles, at least before bicycles came along.

Where to put that single wheel? If the operator walks behind, it ought to be far enough forward that the person has to lift rather than force downward the handles. But the center of the wheel mustn't be too far forward, or the operator will have to lift too much weight, and weight lifted is work wasted; unfortunately, the muscular motor pays energy to produce force. That suggests a small wheel just in front of any load-carrying box. Trouble then comes from another quarter. Small wheels climb bumps poorly and thus tax one's strength and patience on rutted paths. So we compromise, putting the wheel as close as possible to the box but using a wheel still large enough to travel the intended route. The Chinese have a solution (shown in Figure 8.8) uncommon in the West: They use a huge wheel and put half the load on each side. In that way they can use wheels as much as three or four feet high, wonderfully terrain-insensitive, if a bit massive. A person can ride on one side of the wheel if counterbalanced by personal baggage on the other, or the baggage can be divided and the person perched (perhaps precariously) on a platform above the wheel. Or two people can ride, one on each side, with their bags placed (in a way familiar to all of us) on an overhead rack in the middle. A donkey hitched to the front sometimes provides assistance.[42]

About a thousand years ago wheelbarrows came to the West from China, where they had been in use for the previous thousand years.[43] Wheels themselves may have diffused east from Mesopotamia, but wheeled vehicles were over two millennia old before the appearance of this simple, serviceable, but subtle device. Students of medieval mechanics, beginning with Lynn White, agree on this Eastern origin, so we needn't just take the word of the great orientalist Joseph Needham, for whom almost everything started in China and then diffused west and east.

(Joseph Needham was as remarkable a person as you'll encounter here. Originally a chemical embryologist, he wrote a three-volume treatise on that subject when in his early thirties, gaining recognition as one of the field's greatest theoreticians. In the late thirties (his and the twentieth century's) he acquired a Chinese student who taught him classical Chinese. He spent much of World War II in China as part of a British government mission; then and there he began serious sinological work. The result has been a most amazing treatise, *Science and Civilization in China*. Seventeen volumes were published during Needham's lifetime, and several more have appeared since, the ongoing output of the Needham Research Institute at Cambridge, which he established and which I believe the Beijing government supports. The treatise itself must set some record for engaging readability, not a generic trait of treatises.)

Wheelbarrows continue to be used for short-distance transport in such places as construction sites. We in the West use them for longer-range service only under unusual circumstances; recall pictures of rural refugees carrying salvaged belongings, critical necessities, even elderly family members in wheelbarrows. The Chinese long ago used them for military transport, and the overloaded soldiery of Europe, as you'll see in the next chapter, might well have used them to advantage. Puzzlingly, their ergonomics and energetics haven't received much attention. We even make yard carts that give up two of the wheelbarrow's chief virtues. Not only do they use two wheels and thus tip over when going across a slope, but they have a crossbar that connects the handles. It forces the operator to stand behind the handles and lift a load no longer pressing directly and tolerably downward on the spine. The two wheels do compensate with larger wheels, better for bumps, and a more rearward wheel position, minimizing lifting.

BICYCLES, ROWED BOATS, pedal airplanes, and wheelbarrows test our capability for putting out power when coupled to devices unlike anything in our ancestry. In that sense, they bear testimony to the versatility of the human skeletoneuromuscular equipment, our motor machinery. Of course, as the proper iconoclast or skeptic will add, they ought to fit us well since we designed them for just that and nothing else.

CHAPTER 9

More Tough Tasks

We cut 1909, when smelt were first planted in the Great Lakes, and when a wet summer induced the Legislature to cut the forest-fire appropriations. We cut 1908, a dry year when the forests burned fiercely, and Wisconsin parted with its last cougar. We cut 1907, when a wandering lynx, looking in the wrong direction for the promised land, ended his career among the farms of Dane County. We cut 1906, when the first state forester took office, and fires burned 17,000 acres in these sand counties; we cut 1905, when a great flight of goshawks came out of the North and ate up the local grouse. We cut 1902–3, a winter of bitter cold; 1901, a centennial year of hope, of prayer, and the usual annual ring of oak.

Rest, cries the chief sawyer, and we pause for breath.[1]

ALDO LEOPOLD, ecologist

PEDALING NEITHER BICYCLES NOR AIRCRAFT HAS DRIVEN human history, and the role of endurance running has diminished since we turned to traps, weaponry, and agriculture and stopped chasing down our meals. In this day of cell phones we needn't run from Marathon to Athens to announce that the enemy has been beaten. Still, we've been our own best engines for most of our years as a species, and aerobically demanding nonrecreational activities retained importance well into the twentieth century. Running forebears equipped us to put out power; in vocational matters we've only recently turned away from that endowment.

The underlying problems of hard work remain those that human-powered flight raise in starkest form. How can tasks be designed and tools contrived to require the least total energy or the least intensive power output? How can an activity use enough muscle mass to work a human at or near maximum power for extended periods? How can the human be coupled to the task so the greatest part of that power performs useful work?

Carrying Loads

Humans kill or collect food in one place but consume it in another. Most preagricultural humans moved from place to place with their tools, material for shelter, and accumulated food. These prehistoric people did far more carrying than any other vertebrates. Only a few insects—ants, bees, cicada-killer wasps, for instance—go in for similar big-time carriage. Carrying heavy loads may be less important now, but we haven't given up the practice. Routine carrying, mostly for short distances, still occurs vocationally; standard packages of items such as fertilizer and shingles reach a hundred pounds, and small jobs don't merit pallets and fork-lift trucks. Carrying accompanies more traditional life-styles and their recreational approximations. In rough terrain, wheeled vehicles perform poorly, and humans often prefer using themselves as pack animals. Soldiers still carry; long marches with heavy packs afflict every recruit in every army. Late in the nineteenth century a British royal commission recommended a load of forty pounds for soldiers, following reports that men reached the battlefield too tired to fight.[2] But the recommendation has rarely been heeded. During the First World War the British required their infantry advancing under fire to carry as much as sixty pounds.[3]

How much can we carry? I've heard that the maximum weight a person can lift approximates the maximum that he or she can carry; if you can get it off the ground, you can lug it around. With practice, humans can transport loads exceeding their body weights. I've run into wild claims, almost always about the natives of some place off the beaten track. Porters on the south slopes of the Himalayas supposedly carry over 200 pounds for extended periods in hilly terrain and thin air.[4] Tea toters of the Sino-Tibetan border region reputedly carry individual loads up to 360 pounds.[5] Such extreme figures ask for skeptical scrutiny.

Given our splendid metabolic scope, maybe we shouldn't dismiss them as fanciful. How much power might these performances represent? We should multiply our basal rate by about 3.0 to account for walking. We include an additional factor of about 2.5 to account for the load. We put in a factor of, say, 1.5 to account for long, upward slopes. (Upward slopes that take only a few minutes can be compensated by downward slopes; recall our anaerobic reserve.) That comes to a little over 11 times basal rate, or a metabolic input of about 900 watts. That's hard to sustain for more than a few minutes. Routine vocational activity that takes a metabolic scope of 11, equivalent to 13 if we factor in altitude, demands respect. But it's not impossible. So far, so good.

To put it another way, such a performance ought to give a person a prodigious appetite since it amounts to almost eight hundred kilocalories per hour. Doing that for six hours a day (if we assume some rest and some long, downward slopes) as well as managing life during the remaining eighteen requires that a person consume over six thousand kilocalories of food each day. Lumbering, the hardest occupation for which I have a figure,[6] requires almost as much, so the estimate might just mean something. Still okay.

But to put it yet another way, things don't look so good. If the same power were invested in an energetically efficient activity (on an ergometer perhaps), it would produce an output of about 225 watts. That's about double what Desaulgiers and Smeaton estimated long ago. It's fully a third more than Douglas Wilkie's figure for sustained activity of a trained athlete at sea level, which, to allow for high altitude, perhaps ought to be decreased from 150 to 125 watts. That justifies a decent skepticism. Indeed one elaborate study of load carrying came up with a typical load for a professional Tibetan porter of 60 pounds, carried 12 to 15 miles per day.[7] The figures agree with the loads soldiers can carry during long marches, about which more below. Even so, it remains higher than what I'm told is the standard for East African porters working in hilly country, limited by contract to 35 to 45 pounds.[8]

How much should we carry? Ordinary adults increase their oxygen consumption in close parallel with any increase in load beyond body weight. That formed the basis for the factor of 2.5 just used. At a given speed, carrying your own weight will double your oxygen consumption. Thus you might think, right off, that load shouldn't matter in any energetic terms. A heavier load might mean less total time or that fewer people could complete some large task, but the cost would be the same.

The fly in that particular ointment consists of body weight. A 150-pound person with a 50-pound load carries, in all, 200 pounds. The same person with a 100-pound load carries 250 pounds. The heavier load takes 25 percent more energy (250 versus 200) but delivers 100 percent more cargo (100 versus 50). So if we consider total energy per unit distance, apportioning cargo in relatively heavier loads reduces the cost of transport. Beyond that, using heavier loads requires fewer carriers whose own food must be carried, saving both time and money. An unpleasant conclusion.

But load carrying can deviate from that linear path. Dick Taylor and his colleagues, with their motorized treadmill and oxygen-measuring equipment, didn't limit their work in Nairobi to wild animals. They persuaded humans to walk along while wearing masks to collect expired air at the energetically optimal speed for human walking, a little over two miles per hour. (Our walking has a sharper optimum than our running.) Untrained humans, male and female, gave the results they expected from data for military recruits. Any additional weight pushed up oxygen consumption and thus cost energy, and it behaved just like an increase in body weight. By contrast, tests on five African women who had been carrying heavy loads since childhood gave a different result. They carried loads of up to 20 percent of their body weights without increasing their oxygen consumption at all! With further loading they breathed harder, but they maintained their initial advantage, consuming less oxygen at any load than the other humans. Without loads, the African women did no better than anyone else, either at rest or at any speed tested.[9]

The basis of this remarkable superiority remains unknown. The women might have changed anatomically over their years of carrying loads, often up to 70 percent of body weight. Or they might subtly alter their gaits to use slower and more efficient muscle fibers. Or something else entirely. We can probably rule out an explanation based on the particular way they carry loads, shown in Figure 9.1. Several of the women were from the Luo tribe, who traditionally carry loads atop their heads. The others were of the Kikuyu tribe, who centers loads on their backs, supporting them with straps across the foreheads. Furthermore, a study done earlier by the U.S. Army found that load position made no difference. With the loads on their heads, American males took 25 percent more energy to carry 20 percent more weight, not noticeably different from how we usually carry things.[10]

FIGURE 9.1. How East African women from two different tribes carry loads.

Recently one of Taylor's former students, Rodger Kram, now at the University of Colorado, showed that the African women stood out only by the modest standards of the human species. Rhinoceros beetles do far better. He persuaded them to walk (at about a fortieth of a mile per hour) on a treadmill while carrying weights of up to thirty times their own weight. They carried the weights at only a fifth the cost of moving their own bodies. How they manage remains as mysterious as the performance of Africans whom Dick Taylor tested. Curiouser and curiouser, rhinoceros beetles don't ordinarily carry loads at all. Still, similar, if less dramatically cheap, carriage has been found in some ants, which do carry things around.[11] If we're to figure out the trick (assuming a single trick), our experimental material will more likely be insects than people.

How should we carry things? Energetics may not matter as much as other factors. I used to carry a backpack up small mountains. Only a little experience and experimentation persuaded me to carry the load as near as possible to a vertical line running upward from between my feet through and above my center of gravity. That meant, in practice, high on the back and over the shoulders, to take advantage of the way the head protrudes forward. Tipping was no problem since we have a marvelous reflex that adjusts our posture to prevent it. We mainly want to keep from stooping, the latter a recipe for backaches and other reasons for staying in bed the next day. The head strap of the Kikuyu may help, but my old backpack, World War II surplus, came with a small card that suggested using the head strap—a tumpline, the word of Algonquian origin—for only short periods lest the user get a monumental pain in the neck. How we ought to carry comes down to experience and practice as well as to biomechanical rationality. Magnificently malleable machines, we humans.

You don't want to carry excess baggage on your feet, though. That costs about five times as much as carrying a load elsewhere.[12] Taking a pound off your shoes helps as much as taking five pounds out of your backpack. The practical lesson is that lightweight shoes save disproportionately. Walking or running barefoot has at least that to recommend it.

Another possibility for economizing: We decrease the cost of walking and running by storing energy, gravitationally and elastically, for short periods. Can we use elastic energy storage to reduce the cost of carrying a load? After all, the minimum cost of moving something horizontally at a constant speed is zero, so any work done at all must represent inefficiency. Using springy poles, as in Figure 9.2, might help. Such a pole finds use in many parts of the world. It may rest on one shoulder, extending fore and aft with equal loads at opposite ends. Using them in pairs, one on each shoulder, distributes the load and allows the arms to swing naturally. Some poles run across the shoulders, with their paired loads well to either side of the body. Still others run between a person in front and another behind, with the load carried between the two, as with hand-held stretchers or shoulder poles that carry a heavy carcass home from a kill site. The poles most often used, of bamboo and other woods, don't lack for springiness.

About a dozen years ago Rodger Kram, then still in Dick Taylor's lab, looked into the possibility. He asked men to walk on a treadmill while they carried loads of about 20 percent of their body weights on individ-

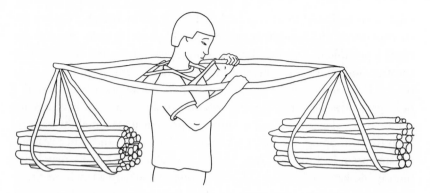

FIGURE 9.2. Carrying a load with springy poles.

ual fore-and-aft poles.[13] Even though Kram picked conditions as favorable as he could, he failed to find any saving of energy from using poles. Oxygen consumption rose with loads the same way it did for ordinary people with backpacks. So why do people use springy poles? Simple: comfort and convenience. When we walk or run, we go up and down slightly. Kram showed that with these poles, loads didn't go up and down, and peak forces on the shoulders decreased by almost half. The low position of hanging loads enables a person to shoulder a heavily loaded pole without assistance and makes balance easy on rough terrain. Also, two-person poles can bear heavy or awkward loads on terrain unsuitable for wheeled vehicles.

Slope matters a lot, as you might guess, and it works, or makes you work, just like added weight. A five-degree upward slope doubles the energy cost at a given speed, so it's equivalent to carrying a load equal to your body weight. Ten degrees triple the cost, equivalent to carrying twice your weight.[14] In practice, a person slows down when climbing. People also slow when loaded, so the suggestion made earlier that heavier is better or at least no worse oversimplifies things a little. Asked to "work hard," twenty-year old male volunteers in one study unconsciously adjusted their effort so they kept their metabolic input at about five hundred watts. That gave an optimum load of seventy to one hundred pounds—on perfectly level ground.[15] If we assume the same self-adjustment, the greater the slope, the lower the optimal load will be.

Going uphill does have one thing going for it: The negative muscular work invested in slowing your swinging appendages decreases with slope until, at about 15 percent, you do only positive work. Between that and other factors, the steeper the slope, the lower the cost of gaining altitude. At least up to a point. Above about 25 percent the cost starts to increase, and you do better if you take a longer, less steep route. Does reality match these predictions from work on treadmills in laboratories? Yes, fairly well. According to one study, mountain paths go straight uphill until they encounter slopes of about 25 percent. At least they do at altitudes below about ten thousand feet; at higher altitudes paths tend to be less steep, again what we expect from metabolic data.[16] The study limited itself to data for human footpaths, but we anticipate that larger animals pick less steep and smaller animals steeper paths—for reasons that the next chapter explains.

We still have lots of questions about load carrying. How far can we generalize from a single, if careful, set of measurements on five African women? Why were they so much better than the rest of us? Do pole carriers (not just Rodger Kram) select poles of the proper length and springiness for particular loads so the poles don't bounce up and down in resonance with walking and running movements? Finally, why do most books on human biomechanics say so little about load carrying?

Historical Interlude: Soldiering Loads

Through the First World War armies marched fully equipped for long distances. Victory could turn on those marches, as summed up in the motto of the Confederate general Nathan Forrest in the American Civil War: "Get there first with the most men." But long marches presented commanders with difficult choices, often resolved as much by custom as by rationality. A soldier needed to carry food, clothing, and bedding; in addition, he had to carry weapons and, depending on weaponry and era, ammunition and protective shielding. Pack animals and wagons presented logistical problems and, as often as not, lagged far behind. A fit human male can lift a load equal to his body weight, but he can carry it for only a short distance and at a slow speed. Any increase in one of three variables—load, distance, and speed—must be paid for by a decrease in some combination of the others.

In the early 1920s an officer in the Royal Army Medical Corps, N. V. Lothian, surveyed the history of load carrying by soldiers back to the earliest data he could reconstruct.[17] He found that commanders routinely overrated motivation and heroics and underrated physiology. Fully loaded soldiers could do only about twelve miles each day if they had to march day after day and retain their effectiveness, and the figure hadn't changed since Xenophon's ten thousand had retreated to the Black Sea from Persia. More often than not, nonfighting personnel managed most of the baggage—Caesar's *impedimentae*. The Romans divided their foot soldiers by load, with lightly loaded units used for scouting and pursuit, and they made great use of baggage trains of carts and porters; fully loaded legionnaires rarely made forced marches. The predominance of armored cavalry during the Middle Ages[18] complicated the issue, with vast numbers of noncombatants carrying the necessities for the knights. But even infantrymen had their carts and camp followers.

Beginning in about the eighteenth century, things appeared to deteriorate. Loads on combatants increased, and both dress and equipment became less practical. During the American Revolution the British not only wore bright red coats but bore especially great loads and inappropriate equipment. From Lothian's quotations: "We borrowed from Germany cross-belts which compressed the chest, but had the advantage of throwing the sword to the rear to knock against the calves and the cartridges, and to quarrel with the haversack; long gaiters which squeezed the legs and stopped the circulation in that most useful member of the foot soldier. . . ." Prussian troops fared even worse: "In one night the soldiers looked as if they had aged ten years, almost at every step lay a fainting man, and entire troops lay on the road-side. In such fashion the entire army marched in four days to Dresden, and getting there exhausted even to death found the Saxon Army fresh and lively."

What have the loads been? "Normal" has remained about 60 pounds for the past few centuries. The Prussians just mentioned were burdened with 80; occasional forced marches limited loads to about 40, with additional material brought up later. During the nineteenth century a few systematic investigations suggested that 40 to 50 might be more reasonable, on the basis of military, as opposed to humane, considerations. The numbers corresponded to a third of body weight, which Lothian recommended as a benchmark. But the old habit of overloading soldiers remained, and as noted earlier, troops during the First World War carried 60 pounds or more into combat. Perhaps few marched as far in that war, but the soldiers themselves seem to have been smaller and lighter than their predecessors, being recruits from industrial slums rather than hardier rural peasantry. Relief came only through the arrival of motor transport. A truck does far better than human or horse in almost every respect: speed; weight efficiency of fuel; capacity relative to operating personnel. Moreover, trackless wilderness has become rare. So long marches can't touch the mobility of the motorized army.

Yet we still carry loads in sacks on our backs, perhaps more often than ever. Hiking with packs retains its popularity, and the book bag has grown, repositioned itself, and afflicted ever-younger people. Hikers carrying self-imposed loads become fanatic weight economizers, down to shortening shoelaces and cutting labels out of clothing. Thirty pounds rates as a nice load, 40 something to worry about; the benchmark limit stands at 25 percent of body weight. (Of course hikers deliberately seek rough terrain, so their 40 might be as taxing as a soldier's 60.) Book bags for growing children, though, have become a bit of a problem. Too many children assume that anything that can be carried ought to be, and too many choose to bear the paraphernalia of affluence. One source recommends that school packs not exceed 15 percent of body weight, meaning that a 60-pound child should carry no more than 9 pounds, an 80-pounder no more than 12.[19]

Moving Large Objects

Ever since the Agricultural Revolution, groups of humans have felt compelled to cut, shape, and move large pieces of stone over long distances. The heroic connotation of the word "monumental" can be no recent acquisition. In every case carried stones bore symbolic significance, devoid of practical benefit as food, clothing, or shelter. At best, dealing with them might improve the social cohesion of a community. Agriculture is a seasonal business almost everywhere, and communal building projects provide organized activity during the off-season. So we shouldn't make too much of the cost of these operations in terms of lost productivity. Animal power played no big role presumably because of problems of gathering and coordinating large teams and of effective harnessing (about which more in the next chapter). How big were the stones? Monoliths in the Middle East reached at least one thousand tons; those in northern Europe and the New World, three hundred tons or more.[20]

Of the many megalithic structures of northern Europe, the most famous must be Stonehenge, near Salisbury, England. While its overall size may be remarkable, the individual stones weigh no more than about 50 tons. But they reached the site from as far away as southwestern Wales, and by one estimate, more than 30 million person-hours went into its final stage of construction.[21] More impressive is a single standing stone (called a menhir), Le Grand Menhir Brisé, erected about 1700 B.C.E. at Locmariaquer, in Brittany. It has now been toppled and broken into four pieces, but the original item, 67 feet long, weighed between 300 and 400 tons. A distance of 50 miles separates Locmariaquer from the nearest-known quarry for its particular kind of stone.[22]

How were big stones moved? The practical weight limit for carrying by an individual human is about body weight, say, 150 pounds. A group of humans, carrying an object slung by ropes from a set of shoulder poles, can manage no more than 2 tons. That takes, by one observation, about 35 men, and each contributes well over 100 pounds when the weight tax levied by the poles is factored in. Aside from any problem of coordination, that dead load tax gets worse as the weight goes up. To get an idea of this problem of diminishing returns, I tried a back-of-the-envelope calculation, scaling up 2 tons carried by 35 men to a 300-ton stone carried by a larger group of people. A straight extrapolation puts their number at 5,250, which might be arranged in a phalanx of 70 by 75. Say 75 untapered poles bear the load, and 70 people hold each pole aloft. Assuming 2 feet of pole per person and assuming reasonable values for allowable sag in the middle of the pole, the stiffness of wood, and the density of wood, I estimated that the weight of the poles would exceed that of the stone.[23] (Fewer, longer poles would exacerbate the problem.) If the poles can bear the load, the people can't bear the poles, never mind the load.

Thus the behavior of beams rules out just using more carriers. Moving loads above two tons demands more ingenuity, by both the ancestors who did it and by the contemporaries who guess how it might have been done. Some large columns went by barge. There is a description of an Egyptian column that was pulled out onto bridging over a slip. A well-ballasted barge was then slid in underneath. Removal of the ballast raised the barge up to the column, which released, pushed it down again. Logrollers can be repeatedly moved as they come free from the rear around to the front of a load. But we have little evidence of their use, which I think is surprising since I found rollers easy and effective when a colleague and I towed the trunk of a tree. One of us pulled the rope, while the other shifted the rollers; together we managed perhaps half a ton for a short distance over very irregular ground. Still, rollers are slow and give problems of alignment if used in large numbers.[24]

Persuasive evidence supports dragging, including dragging up greasy tracks on deliberately built ramps, levering columns from horizontal to vertical with ropes and upright booms, raising blocks with rocking platforms and wedges, and using dragged sledges, sometimes with lubrication. Naturally these schemes involved large numbers of workers, even discounting as obvious exaggerations post facto reports, such as that of the ancient Greek historian Herodotus (who passed through Egypt long after the pyramids had been built). One way or another, humans mainly pulled on ropes.

How do you pull on a rope? Two factors matter, one independent of muscular force and the other entirely dependent on it. If you stand exactly upright, you can't exert a steady pull of even the slightest magnitude. As you try to make the load come toward you, you tilt toward it instead. Therefore, you have to lean away from the load. If you lean only slightly, as when you open a door, the work comes not from your muscles but from your weight, as in the middle in Figure 9.3. As the load moves toward you, your center of gravity moves earthward, something

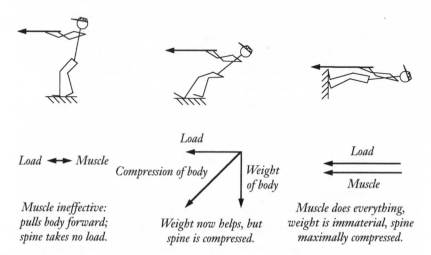

FIGURE 9.3. Pulling in various postures: when nearly upright, when leaning backward, when horizontal.

it's happy enough to do. The more you lean, the more effectively your weight pulls—if by "effectively" we mean the force it puts on a horizontal rope. The extreme, as on the right in Figure 9.3, would be to lie horizontally, with your feet braced and your body just off the ground. The formula from elementary physics implies that you then use gravity alone to pull with an infinite force.[25]

Trouble, though, comes because when you lean enough to approach the ground, the force from weight, however great, can't pull anything very far. Any minor stretchiness in the rope will eat up the entire distance change, and you'll drop the rest of the way. So your weight—that is, gravity—helps less than it might. At the same time, the farther you lean, the better your muscles can pull against the load. We're good at extending the legs and flexing the arms, especially the former; we can rise from squatting positions bearing considerably more than our own weights. A horizontal person can pull with a force exceeding body weight. In pulling postures between upright and horizontal, both weight and muscle contribute, with the best angle for leaning back determined by the relationship between weight and strength.

An interest in competitive rope pulls has generated data on how hard people can pull, at least against a minimally movable load. We do best facing the load and pulling with the rope roughly waist high. Keeping the rope low gives body weight better leverage and allows a more effective grip. While the peak force of a particularly strong and heavy man might reach 140 pounds, for sustainable force 70 pounds may be the practical maximum.[26] In the moving of heavy stone objects, even that 70 pounds must be an overestimate; the faster the load moves, the less force a person can exert to keep it moving. So we might take 50 pounds as an estimate of what one person contributes to the effort of a group pulling something along.

We'd then like to have some estimate of the resistance imposed by a load, ideally in relationship to its weight. That's no problem for hoisting something vertically. For moving anything horizontally, resistance varies all over the map, depending on the technique used for pulling. Ideally the load needn't resist at all since ideally no work need be done. Friction, the ups and downs of bumps, a bit of air resistance—all can be reduced to very low levels. The ancient Greeks boxed some things into great drums that could be rolled, and rolling resistance depends on little more than the slope, bumpiness, and hardness of the surface. That's why railroads, originally short on power, used metal wheels on metal tracks and the gentlest of grades. By contrast, dragging an unsupported and irregularly shaped mass across soft ground would give Hercules a bad time.

The pyramids used blocks weighing two and a half tons each. Although that's a little too heavy for carrying over any distance on shoulder poles, in mechanical terms it's still modest. The overall magnitude of these jobs presents the real problem. For the task of quarrying and transporting over two million blocks, each weighing five thousand pounds, one can be sure the designers of the great pyramid of Cheops (Khufu) gave some thought to laborsaving technology. After all, even slaves or off-season farmers must still be fed and sheltered. Furthermore, speed mattered. Doing the job in twenty years meant moving a hundred thousand stones each year or nearly three hundred per day, perhaps as many as five hundred per day if laborers had to be released for planting and harvest. The stones of the chambers within the pyramids, some as much as sixty tons, presented the more severe mechanical challenge. But even these pale before blocks of a thousand tons and obelisks of five hundred tons that have been moved with little more than levers, wedges, inclined planes, sledges—and human muscle.[27]

Still, why use blocks of even a couple of tons? Among premodern structures only the Great Wall of China exceeds the bulk of the pyramids of Egypt (and, incidentally, those of the Maya in the New World), and it is built of sixty-pound bricks, either dried or fired.[28] I think both Egyptians and Chinese designed rationally, but the cost accountants of the pharaohs faced a different scaling rule from those of the emperors. Where stone gets cut into blocks, fabrication cost should follow surface area. The larger block has less surface area in relation to its volume, less than a quarter as much if we compare five-thousand-pound cubes with sixty-pound cubes. If cutting is the rate-limiting step, because of either constrained access space in quarries or slow cutting, then bigger is better. After quarrying, the Egyptians carefully trimmed rather than crudely cracked the blocks of the pyramids, and they probably did that by steadily feeding sand into the saw blades' grooves. It could not have been fast, and the lifetime of the pharaoh limited the duration of the project. Maybe we ought to focus more on the labor involved in cutting and shaping the stones of the pyramids than on the labor needed for their transport. For bricks, on the other hand, surface imposes no cost beyond that of any mortar used to join them. So the overall cost ought to scale closely with volume, and large size buys no advantage. Under these latter circumstances, sixty-pound bricks make better ergonomic sense; even they might be a little larger than optimal, given the deliberately hilly course of the Great Wall.

When human power alone does such great deeds, we too easily overestimate the strain on the individual laborers. Not that hauling big stones around comes easy. Rather, the task may be so demanding of force that it can't work a person at anything close to the limit of aerobic power. For moving large statues, the Egyptians seem to have used a drummer to synchronize the tugs of the army of workers, implying that while they may have pulled often, they weren't pulling steadily. Better by far to be employed moving megaliths toward Stonehenge than, as in the last chapter, to be a convict rowing a sixteenth-century French galley.

Clearing Land and Dismembering Trees

Agriculture feeds more people on a given amount of land than does hunting or grazing, but starting a farm isn't for the fainthearted. Unless you're into irrigation, any land suitable for planting will already be dense with plants. The basic game thus consists of replacing what nature plants with what you plant. Grassland sounds ideal, but plows equal to the task of breaking prairie sod have been around for less than two centuries. Before that, pastoral, nomadic societies rather than farmers and the city dwellers they supported occupied the great grasslands of North America and Eurasia. Through most of our agrarian history, we've done serious farming (again, excepting irrigated land) mainly on ex-forest. How, then, to make forest into ex-forest?

First we have to get rid of the trees. Both trees and crops live on sunlight, and shade-grown crops yield too little relative to the labor we invest. So we start by chopping down the forest. In North America our mythological giant Paul Bunyan created the prairies with the aid of his ax, ox, and crew of loggers. Legend aside, we looked at axes two chapters back, where we noted the clear effectiveness and possible energetic superiority of short-stroke stone axes for felling trees.

But I wonder about the full relevance to Stone Age life of the test of old against new axes. Felling is only the first operation in clearing, and removing the felled trees makes a lot more trouble (and, believe me, gives much less gratification) than felling. Gravity now helps little, and branches have to be removed before a trunk can even be rolled. Even for walking across the land, much less for planting crops, felled trees must be removed. Better to kill the trees first, and to kill them when they're leafless so their corpses will intercept little sunlight. (A newly killed tree often retains dry leaves for a long time; autumn leaf fall depends on much more than leaf death.) An agricultural system lacking horse- or tractor-drawn plows ought to find dead trees no great obstacle. For a few years limbs and sometimes trunks will fall, but that's a minor nuisance. Standing trunks might even provide support for crops, such as many legumes, that cling and climb. Alternatively, once dry, standing trunks can be burned in place, with their ash fertilizing the soil.

So how to kill trees? Nothing easier, just girdle them, meaning you cut a circumferential ring down to the cambial layer just beneath the bark. A girdled tree promptly dies, and short-handled stone chipping axes should shine as girdling tools. The Indians of eastern North America routinely girdled trees and planted their crops amid standing trunks.[29] Indeed we have worldwide evidence of ancient girdling, and it still finds use in many third world countries.

Girdling, though, produces no construction material and little fuel wood. That's where axes come in, and that must have been the impetus for the invention and spread of the efficient North American polled ax, that full-body aerobic tool. Eastern North America was heavily forested, winters were cold, and fireplaces were inefficient. Sparse populations and poor transport conspired to make the area wood-fueled and wood-housed as well as encouraged wooden (corduroy) roads and wooden bridges. Benjamin Franklin, describing his new "Pennsylvania" stove in the mid-eighteenth century, complained about the distances from which fuel wood had to be hauled to Philadelphia a century after its founding.[30] Years later we built gigantic railroad trestles using wood that was cut and shaped locally. In its various versions, Paul Bunyan's great ax remained well into the nineteenth century the first-stage tool for harvesting wood.

At every step we invest energy when we convert standing trees into usable lumber. Not only must a tree be cut down, but it must be cut up, transported, and cut up further. Axes did well for felling and for a lot of ancillary trimming and clearing, and they still play a role in these activities. Modern axes may do better than ancient ones, but the ax's place has been largely usurped by the saw. Just as the role of axes has diminished, so have those of other percussive tools, such as adzes, once used to convert the round surfaces of logs to the flat surfaces needed for building, whence the expression "hew close to the line." Saws have swept the field. One might guess that the great advantage of saws over axes lies in their superior suitability for motorization. That may be the case, at least in part. Data for nineteenth-century forestry distinguish between exports of hewn and sawn wood,[31] meaning ax-felled tree trunks and sawed-up lumber. At least in that century, axes felled and trimmed, and the trunks then went on to nonportable motorized sawmills.

Power saws may have a venerable history, but human-powered rather than motorized saws deserve much of the credit for the change from ax to saw. By the start of the twentieth century the terminological distinction between "hewn" and "sawn" wood had become anachronistic. We're all familiar with the ordinary carpenter's saw and the hacksaw. A few of us know compass saws, coping saws, and miter box saws—just to touch the diversity of handsaws. The heroes of the present story are none of these delicate items, but rather huge bucking and felling crosscut saws suitable for use with entire trees, such as the ones shown in Figure 9.4. These consume any amount of power a human can generate, with a motion much like rowing, and they consume that power for long

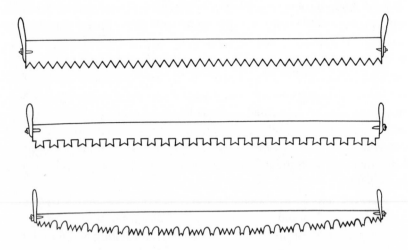

FIGURE 9.4. The crosscut saw (bottom) and its antecedents, a conventionally toothed one (top) and one with M-shaped blades (middle). The latter comes from an illustration in a medieval manuscript.

enough to put the sawyer well out along Wilkie's curve. Too little attention has been paid to the historical development of these big saws; somehow Paul Bunyan's ax carries superior symbolic value.

The crosscut timber saw underwent as revolutionary an improvement in nineteenth-century North America as the ax had a century earlier. First used for bucking—that is, for cutting up a felled tree preparatory to moving it—the saw then displaced the ax for felling itself. For bucking, axes waste more of the useful length of a laboriously felled trunk than do saws. In addition, in bucking, the sawyers cut downward, relatively easy, while in felling, they have the harder task of forcing the saw sideways. However, by the 1880s big crosscut saws had become so good that they could fell trees five times as fast as axes.[32]

Understanding why these modern crosscuts worked so well requires a look at how saws in general work. The basic idea of drawing an edge back and forth goes back to the Stone Age. The edge could be serrated, as in most wood saws, or it could be straight, with some abrasive grit fed into the groove, or kerf. As mentioned earlier, the Egyptians and others cut stone with bronze saws using this last trick. Serrated tools do better for wood, but not just any sharp and jagged edge will be effective. The teeth must make a kerf a little wider than the thickness of the back of the blade, or else the friction of wood against blade (binding) will tax the process. For this reason, teeth splay slightly outward, one to the left and the next to the right. Then the right number of teeth must contact the wood. If too few, they dig in and catch; if too many, they don't press into the wood far enough to cut much. The thicker the board or log, the more coarsely toothed must be the saw. Having too many little teeth creates another problem. Sawdust gets trapped between the teeth and then carried out of the kerf in them; if the volume of sawdust from a stroke exceeds the volume of the gaps between the teeth, it will work its way upward and make the saw bind. Big teeth need more space between them so that they can take the debris from a longer cut. Saws for logs require very coarse teeth, as was known (by the irrefutable evidence of illustrations) by the Middle Ages and probably far earlier.

What was not appreciated until the nineteenth century was the special problem imposed by a wide kerf. The thicker the log, the longer must be the saw; to get the sawdust out effectively, it has to be twice as long as the cut. To be stiff enough, a longer saw needs a thicker blade (although external bracing, as in a bow saw, helps somewhat); a flexible blade won't press evenly against the bottom of a long kerf, so the kerf won't be planar, and so the sawyer will curse the binding blade. (Cut a thick trunk with a bow saw, and you'll see what I mean.) A tree-cutting saw needs a thick blade and makes a wider kerf. In this wider kerf, dragging slightly splayed teeth doesn't work so well. If the teeth are as wide as the kerf, they make too much contact with the wood to penetrate decently. If they're narrow, they cut grooves running along either side of the kerf and leave wood standing in the middle. A wide kerf requires some kind of chiseling action. This much has long been known. Log saws with (inverted) M-shaped teeth appear in drawings from the Middle Ages, one of which is reproduced in Figure 9.4.[33]

The breakthrough consisted of separating cutting and chiseling functions by alternating splayed cutting teeth with unsplayed teeth that chisel. Figure 9.5 shows the basic action. The splayed cutters, a trifle longer (in practice about one sixty-fourth of an inch, or 0.4 mm), make two parallel grooves, and the chisels (called rakers) scoop out the crosswise fibers in between. Between the cutters and rakers, large cutouts in the saw (gullets) provide space for the sawdust, which consists of fibers as long as the kerf is wide. A properly sharpened and well-adjusted crosscut saw advances at a remarkable rate. Each stroke may deepen the kerf by five or ten millimeters, and the gullets emerge from the log completely filled.

I haven't learned how, where, and when this separation of rakers and cutters began. It seems to be North American and mid-nineteenth cen-

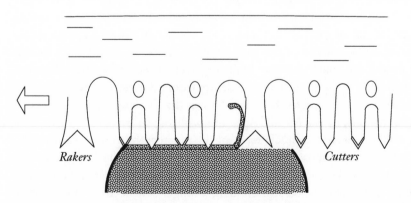

FIGURE 9.5. The action of the cutter and raker teeth of a crosscut saw.

tury. Some hint comes from the etymology of the word "raker," the (more or less) novel tooth. An old tool called a raker was used to remove loose mortar when bricks and stones were being repointed; like a pair of raker teeth on a crosscut saw, it chiseled in both directions. That meaning makes the bigger dictionaries, disappearing only in the most recent. "Raker" as a short, chiseling saw tooth appears in unabridged American dictionaries back into the nineteenth century but not in those (even, shamefully, the giant *Oxford English Dictionary*) of British origin.

Modern crosscuts persisted in American commercial logging operations until at least the late 1940s. Two-person saws in the East ranged between five and eight feet in length, and in the Northwest up to ten feet. Special saws of up to sixteen feet cut giant redwoods in California. They were pulled by sawyers who had to stand on springboards mounted on the trunks to get above the flared bases of the trees.[34] Using these latter must have been heroic. A crosscut saw must be moved by at least the radius of the trunk with each stroke in order to empty the gullets in the middle. Crosscutting by hand doesn't just use hands and arms; you have to brace the feet and contract every large muscle above them. Think of making a four-foot-long stroke while standing on a short and (to judge from photographs) none too secure platform.

However done, cutting across tree trunks by hand demands power. By one set of measurements, a group of loggers in India worked at nearly 500 watts.[35] Recall that this is the metabolic rate that twenty-year-old American males spontaneously picked when they were asked to do "hard work" on a treadmill with different loads and grades.[36] What did the Indian loggers accomplish with their 500-watt expenditure? They could cut 200 square centimeters per minute, which translates into cutting across a 20-inch (or 50 cm) trunk in ten minutes. I find that speed impressive but credible. My wife and I, when we were weekend sawyers in our forties, could at best do only about half that rate cutting though white oak, a fairly dense wood, and we worked under near-ideal conditions: freshly tuned saw; vertical bucking cut; a single cut rather than repeated cuts. Aldo Leopold gets it right when he ends each paragraph of his account of the bucking of an old oak with the refrain "Rest, cries the chief sawyer, and we pause for breath." Aerobic work is crosscutting, probably more so than any other activity on a traditional farm. As the old saying (old saw?) goes, the firewood warms you twice.

In one of my less successful forays into woodcutting, I combined the basic action of rowing a racing shell with the operation of a two-person crosscut saw. As we've known at least since the work of Henderson and Haggard, rowing with immobilized feet and a fore-and-aft sliding seat permits maximum aerobic output and a good distribution of effort among the big muscles. Besides improving muscle action, I planned to circumvent one of the less handy aspects of the crosscut. While it works well for cutting up a big, well-braced trunk, it has an awkward way of rolling instead of cutting anything small and movable. A conventional sawhorse just tips over, and with sufficient bracing it's no longer portable. Building the device shown in Figure 9.6 gave no problem, although the thing had to be almost twelve feet long to accommodate my six-foot saw. (The Mark II version would have folded about hinges in the middle.) Skateboard wheels made fine rollers for simple wooden seats. A V crotch with beheaded nails kept the log in place and prevented it from being rolled by the passing sawteeth. The device worked wonderfully, slicing a five- or six-inch trunk in less than a minute, but I had given insufficient thought to a most mundane factor. Any trunk thin enough to be heaved onto the rig took no time to cut through. Then the sawyers had to detach their feet from stirrups, stand up, move the trunk, sit down, and reattach their feet. Reloading took as long as cutting and destroyed the pleasure of the experience. Self-loading? There would go any pretense of simplicity and portability.

Powered saws appeared early in the twentieth century, but only the advent of lightweight gasoline engines and the invention of the modern saw chain (in the 1940s) gave them sufficient advantage to displace crosscuts. Crosscuts still predominated for felling in the late 1940s in the American South,[37] but the region had especially low labor costs, and

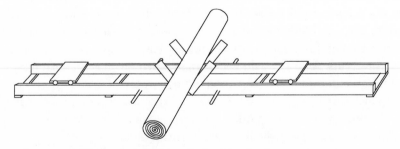

FIGURE 9.6. My rowing machine for bucking logs.

southern yellow pine probably taxed these saws less than would hard-woods or the bigger western softwoods. The demise of crosscut saws, at least as amateur tools, should occasion a little sadness. They're safe, quiet, odorless, and, when well adjusted and sharpened (itself a satisfying activity), only about five times slower than a chain saw. My wife and I made cuts of up to thirty inches across. That exceeds the capacity of the household fourteen-inch chain saw, which can't handle more than twice its length and becomes ever more cumbersome as it approaches that limit. Energetic cost? I think we should shift the entry to the asset side of the ledger. What productive house-and-garden activity can provide such good aerobic exercise? A fine thing for a sunny fall or winter day!

Making Lumber

These days most felled trees end up at lumber mills or pulp mills. Pulping wood, to make paper, has been done for a mere two centuries and has never depended on human muscle. Milling lumber, though, has a longer and more diverse history. Portable lumber mills find occasional use even now, but we usually carry (or drag) wood to a sawmill some-where beyond the forest. Cutting long pieces of wood lengthwise takes a lot of energy, and sawmills have been mechanically powered for fewer than a thousand years anywhere. How can (or, perhaps, could) muscle power make logs into lumber?

The traditional tool was the adz, now something even rarer than the crosscut saw. An adz takes the basic ax blade and rotates it about its long axis so it cuts crosswise to the arms of the person wielding it—like a pick, hoe, or mattock. An ax blade itself may serve as a crude adz. Stone Age axes did so, as did the axes of some American Indians who could acquire metal blades only through occasional trade on unfavorable terms and needed adzes to hollow out wooden boats and make totem poles. Adzes, either short- or long-handled, chip away at the surface of a log, slowly converting it from a round to a rectangular cross section. Adz work takes as much skill as strength and more of either than it does power. I still see occasional adzed beams, the treasured and exposed "rough-hewn" tim-bers of a few old buildings and more new ones willing to pay for the "ye olde" ambiance. Adzing makes only one beam out of each section of trunk, so adzed supportive beams and columns have often been com-bined with sawn lumber for floors and walls. Adzes may be cheap and portable, but adzing takes time. That's why log cabins have half-round logs; why go to the labor of flattening sides that needn't be fitted together?[38]

Saws hold even greater advantage over adzes for lengthwise cutting than they do over axes for felling and bucking. Even so, one is awed by the very idea of making boards by hand-sawing big logs lengthwise. But we've been doing just that since the ancient Egyptians. Wherever metal saws have been available, some version of the same human-powered sawmill has been used. This repeatedly invented contraption is called a pit saw, and Figure 9.7 shows one version. Typically, the log extended over an actual pit, and the saw moved up and down, driven by one sawyer in the pit and another standing atop the log or on a platform above it. Few tasks sound less appealing than operation of a pit saw; each sawyer had to perform an awkward motion at a high power level again and again and again. The sawyers pushed and pulled directly on the saw; at least I've not run across a picture or description of a pit saw that used any linkage to permit more convenient movement.[39]

Human-powered pit saws still find use in third world countries. But pit saws were among the first things motorized, with water-powered sawmills in Europe going back about a thousand years. The way a mill

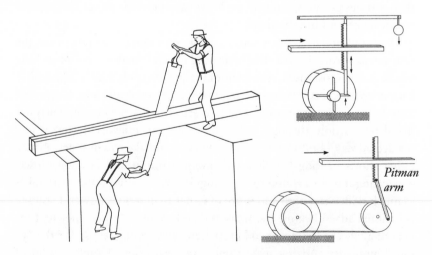

FIGURE 9.7. Pit saws: On the left, one saw hand- (or, better, body-) operated; on the right, two run by waterwheels. Of the latter, the top one is medieval, the bottom a more modern version with a pitman arm.

wheel drove a pit saw deserves a few words. The topside sawyer gave way to a springy pole, a weight and a pulley, or a saw-stretching frame, a device that kept the saw blade straight and tensed. In place of the bottom sawyer, at first a set of cams on the waterwheel pushed the saw up. The saw made as many strokes for each turn of the mill wheel as the wheel had cams, as in Figure 9.7. The arrangement had the useful effect of giving stroke rates higher than the slow turning rate of the mill wheel. But it must have given the saw a jerky motion as well as forced it in only one direction. Later a strut was extended either from the periphery of the mill wheel or from a faster wheel that was pulley-driven by the mill wheel. That strut drove a saw in a framework both up and down. What we're looking at is a device that converts rotational into reciprocating motion, the reverse of what a crank does. Cranks came late, and these anticranks came even later. Their use in sawmills dates from the fourteenth or fifteenth century, well after water-powered mills had become common.[40] More powerful sawmills used several saws in the framework, so a group of boards could be cut with each pass of the log; gang saws they were called. Fancy versions automatically moved the log forward by a predetermined distance between successive strokes.

While circular saws and band saws long ago replaced pit saws, we retain a curious connection with these fine old machines. We use the term "pitman arm" for a rod attached near the edge of a wheel, the anticrank that converts rotary motion into push-pull motion. For years I assumed that the term could be traced to someone named Pitman (and I would have capitalized it as I would a Pratt truss or a Prony brake). But no, the name memorializes the countless men who stood in the pits and pushed up and pulled down on lumber-milling saws. Pitman arms, incidentally, can be handy things. I once used one to make a small electric motor (well geared down) rock a cradle. It worked much better than the saw I described earlier.

WITH THE WISDOM of hindsight, we can recognize two concurrent lines of historical change. The stories of axes and crosscut saws show how humans have increased the efficacy with which we use our muscles. Better tools, tools that take less fastidiousness and more muscle, let us labor harder and work more efficiently (again, not the same thing.) As sawmills and sailing ships show, we've at the same time relieved ourselves of muscle-powered work with all manner of laborsaving devices. At first glance the two lines of change contain a contradiction, something for historians to explain. In a larger sense, though, there's no contradiction. Both represent ways to increase the productivity of human labor, which translates into wealth, however distributed or reapplied. Increased yield relative to effort motivates mightily, and in the end the efforts of the Luddites to keep labor from being displaced have always come to naught.

CHAPTER 10

Bringing Animals to Bear . . .

The great machine stood between the high, hive-shaped stacks. The ten horses were standing hitched to the five long wooden sweeps of the horse-power. The driver stood on the board platform in the center with his long whip in his hand. The pitchers had climbed the stacks with their forks, the handles polished by long contact with hard hands. . . . A deep growl, like that of a bulldog magnified fifty diameters, filled the air, and as the cylinder gathered speed, it rose from a bass to a baritone, and then to a tenor of a volume which sang over four square miles of haze-obscured prairie. The feeder looked up at the pitchers, saw the man who pitched to the machine with his next bundle ready to fall on the table, saw Frank with his bandcutter's knife ready to slice softly through the band of it, and then he moved the first two sheaves gently over between the open lips, deftly twitched their butts upward, and the great operation was on.[1]

HERBERT QUICK, *The Hawkeye*

O NLY THE OLDEST OF US AFFLUENT AND MOTORIZED FIRST world people recall a time when animals other than humans worked for a living. "Forty acres and a mule" went the unfulfilled promise after the American Civil War, with the latter as important as the former. Paul Bunyan had his blue ox, Babe, and Alexander sat upon his horse, Bucephalus. Bedouin came with iconic camel, Saami (Lapp) with reindeer, Inuit (Eskimo) with husky dog. Llama, yak, elephant, and water buffalo defined yet other cultures. After an absence of several chapters, nonhuman animals return to this account, here to trade a share of our work for food, shelter, care, and (at our pleasure, not theirs) access to mates.

In relation to the number of species around us, we've domesticated precious few animals, even if we take advantage of lots of others, and only several of those domesticated contribute mechanical work. Consider our birds—mainly chickens, ducks, pigeons, turkeys, geese, ostriches, and peacocks. Most get eaten. Although pigeon racing persists as a hobby, only a few pigeons have worked their muscles on our behalf as message carriers. The same goes for the small mammals: Cats and ferrets serve as vermin exterminators; rabbits and guinea pigs serve only at the table. Beyond these, we raise two kinds of insect, bees for honey and beeswax and silk moths for silk fiber. All that I've mentioned joined our economy ages ago. We've begun again to expand the range of creatures that we breed and raise in captivity—we can now buy farmed bison and ostrich in local markets—but most of our recent activity has been aquacultural. If we're talking about making animals work, only bigger than average mammals merit attention.

The Players

Our working animals thus make up a small minority of a small minority. While both their numbers and their diversity have declined over the past century, they were once central to our agriculture, industry, transportation, and traditions—religious and secular. Who are these domesticated coworkers? These days we need to start with a list, here arranged (as is the wont of the biologist) by lineage.[2]

BOVIDS
Domestic goat: *Capra hircus*   55–210 lb (25–95 kg)
Together with sheep, our first domesticated herbivore, perhaps by 7000 B.C.E., from *Capra aegagrus*, the bezoar goat, in southwestern Asia. Goats are especially tolerant of variations in food source and climate. While their small size works against their use for power, goat carts find occasional use, and goats sometimes work as pack animals for hikers in mountainous country.

Domestic cow and ox: *Bos taurus*   1,000–2,000 lb (450–1,000 kg)
Separately, in central Eurasia and southern Asia, as early as 6200

B.C.E., from the now-extinct aurochs (*Bos primigenius*). Zebu cattle in Africa probably represent yet another domestication of a different sub-species of aurochs.

Water buffalo: *Bubalus bubalis*   550–1,200 lb (250–550 kg)

From southern Asia or China, about 4000 B.C.E., from the wild *Bubalus arnee*, now rare. Water buffalo still find wide use in southern Asia as both draft animals and milk suppliers.

Yak: *Bos gruniens*   700–2,200 lb (330–1,000 kg)

In the general area of Tibet and the Himalayas, from *Bos mutus*, the wild yak, now very rare. The time of domestication is unknown. Yaks work well as pack animals, even at altitudes above eighteen thousand feet.

EQUIDS
Horse: *Equus caballus*   400–2,000 lb (175–930 kg)

In eastern Europe, probably Ukraine, about 4000 B.C.E., from the wild horse (*Equus ferus*). Horses have been less changed by human manipulation than most other domesticates and easily readapt to the wild, as did the horses lost by the Spanish that then formed the basis of the horse-based buffalo-hunting cultures of the Indians of the North American plains.

Donkey: *Equus asinus*   about 550 lb (250 kg)

In the Middle East, perhaps Egypt, around 3000 B.C.E., from the wild ass (*Equus africanus*). Donkeys can be ridden or used to pull loads, but their small size is often a drawback. Their main role has long been the fathering of mules. Today's donkeys are even smaller than their forebears or wild asses[3]; I presume they're just large enough to mount mares.

Mule: *Equus caballus* x *Equus asinus*   about 1,100 lb (500 kg)

Sterile hybrids of maternal horses and paternal donkeys, probably first developed in the Middle East before 1000 B.C.E. Mules have notable stamina, carry or pull heavy loads well, and work well in warm climates. (The opposite hybrids, hinnies, turn out to be less useful.)

CANIDS
Domestic dog: *Canis familiaris*   3–200 lb (1–80 kg)

Our first domesticates, going back at least fifteen thousand years, but from wolves (*Canis lupus*) that had separated from the modern wolf lineage about a hundred thousand years ago. Perhaps no domesticates have been more modified by deliberate selection. Humans and dogs probably teamed up as hunters in several places in the Northern Hemisphere.

PROBOSCIDEANS

Asian elephant: *Elephas maximus*   5,500–12,000 lb (2,500–5,500 kg)

Probably first tamed by about 2000 B.C.E., perhaps in the Indus Valley of India. Breeding hasn't changed Asian elephants much; both their intelligence and tractability are native traits. Their large size makes them attractive for royal mounts, military use, and ceremonial events. While inefficient for most of our tasks, they perform well for high-strength pulling tasks, as in logging (Figure 10.1), for which teams of equids would be awkward.

African elephant: *Loxodonta africana* 6,000–15,000 lb (2,700–7,000 kg)

Larger than the Asian genus. African elephants have not been widely used. They're reputed to be harder to tame, but the Romans (and, most famously, Hannibal) used them in warfare. However, a good authority, Juliet Clutton-Brock, notes that when these sensible animals encounter arrows, they turn around and often damage their own army.

FIGURE 10.1. Elephant and rider pulling a log, to get some sense of relative sizes.

CAMELIDS

Llama: *Lama glama*   300–350 lb (135–155 kg)

Probably the first camelids domesticated, from guanacos (*Lama guanicoe*), in South America by about 4000 B.C.E.. Llamas are too small to ride, but they make excellent pack animals: surefooted and tolerant of high altitude, limited water, and extremes of temperature.

Bactrian camel: *Camelus bactrianus*   650–1,500 lb (300–690 kg)

The two-humped camel, which originated in Central Asia, about 2500 B.C.E., probably from the wild camel (*Camelus ferus*). Both domestic camels are marvels of endurance, carrying substantial packs with minimal food and water in hot deserts. They can be used for plowing or pulling carts, but their long legs make such activities awkward.

Dromedary: *Camelus dromedarius*   650–1,500 lb (300–690 kg)

The one-humped camel, a more graceful beast, which appeared in Arabia about the same time as the Bactrian from an unknown ancestor. It shares the special utility of the Bactrian camel but can run faster.

CERVIDS

Caribou (North America),
reindeer (Europe): *Rangifer tarandus*   130–700 lb (60–318 kg)

So similar that we can't tell when the wild reindeer were first bred in captivity. They've clearly been milked, eaten, ridden, and used as pack animals for millennia in Scandinavia and adjacent areas of Russia by the Saami (Lapps).

Moose (North America),
elk (Europe): *Alces alces*   450–1,800 lb (200–825 kg)

Tamed, milked, and ridden in Scandinavia. Whether elk were truly domesticated (including breeding, for instance) in antiquity remains uncertain. They are currently being successfully domesticated as draft animals in Russia.

WHAT CAN WE make of this short set of species?

- First, even this list exaggerates the diversity. African elephants and elk have never been important domesticates, and many of the others have had only local significance.
- They're all quadrupeds, leaving little uncertainty about where to place a load—at least the purely gravitational load of the pack animal. All must be loaded above the quadrilateral formed by a line connecting the feet. Otherwise the animal might tip at rest, or the load would be poorly divided among the legs. I wonder if a quadrupedal walker (perhaps a centaur) could learn to load such bipeds as ostriches or ourselves.
- All are relatively large, aerobic mammals, comparable with or larger in size than us. Birds (except perhaps ostriches) don't get big enough to do much of our pulling and carrying, while present-day reptiles lack adequate aerobic capacity, being sprinters at best.
- Except for dogs, they're all herbivores. For basic ecological reasons, food for carnivores costs more than food for herbivores. While some North American Indians used them to carry or pull light loads, dogs have been important only in the far north, where the Inuit, in particular, lack good access to vegetation.
- All these herbivores can digest cellulose, or they at least can persuade some personal protozoa to do the digestion for them. We can't do the trick; humans and most other mammals and birds (except during our bouts of voluntary starvation) focus on the seeds, fruits, and storage organs of plants. So our beasts of burden not only eat foliage and other cheap vegetation but compete with us for fodder just to the extent that a field might produce wheat instead of hay.
- As far as we know, not one of these animals was first domesticated as a working creature. Except (again!) for dogs, all began by supplying, in various combinations, meat, milk, fur, leather, fiber, manure, and some lesser products. Most, for instance, have at one time or another been milked. Domestication may not be trivial, but it's a lot easier than coaxing animals to contribute the power of their muscles to tasks of our choosing. We didn't put the cart before the horse.
- We can discern a similar sequence in the acquisition of mechanical tasks by these domesticates. All began by carrying loads, and all remain usable as pack animals. The larger ones were then ridden, which is to say that one or more of us served as a load. Finally, most were persuaded to pull loads. People carrying and pulling demand

finer control by us or more cooperation from the animal, but perhaps more important, they require fancier ancillary technology. You need a bit and bridle and perhaps a saddle or their equivalents for a rider alone to control a mount, and you need a harness if the animal is to pull. More about these critical devices shortly.

The Cost of Size

A working dog weighs at least a hundred pounds, an elephant perhaps ten thousand, so our animals span a hundredfold range, from a size comparable to us to that of the largest terrestrial animal. Not that we couldn't use still-smaller mammals. Tiny electrical motors run our lives; the computer and its partners on my desk use at least half a dozen. Nothing prevents me from connecting a generator to the activity wheel of the cage of a pet rodent and using the output to keep the battery of my laptop computer or electric razor charged up. Nothing, it seems, except the bother. In fact, some basic physiology weighs in against the utility of small birds or mammals for replacing the local utility company. This issue of size is worth a little attention, although it involves a thicket of relevant factors.

One might guess that small creatures have an intrinsic advantage. An ant can lift many times its own weight, suggesting that it has some basic muscular advantage. But that advantage proves illusory, one of the tricks played on our intuition by the way the rules for scaling work. Indeed, that last word, "work," holds the key. A muscle develops force in proportion to its cross-sectional area. The smaller animal simply has more muscle cross section relative to its volume or weight than does the large one just as a matter of geometric scaling. If we assume no great change of shape, a smaller object has more area in relation to its volume than a large one—whether outer surface area or muscle cross-sectional area. So the ant develops lots of force for its size. But the work a muscle can do (as opposed to the force it can exert) depends on its volume—cross section times length—and volume remains volume, so the larger animal can do as much work for its size as the smaller. The ant does lift a relatively heavy weight, but it doesn't lift it as far. Work (or energy) is force times the distance the force moves something, and in the more relevant terms of work, the ant enjoys no advantage. Nor would a mouse.

You might then guess that size plays no role, with the smaller creature no better off but no worse off either. After all, the muscles of a mammal of any size make up about 40 percent of its overall mass, and at least roughly, muscle is muscle, whoever grows it and in whatever size it comes. That's true enough, but bear in mind that both the power and efficiency of a muscle depend on the speed at which it shortens. Recall also that we're unable to develop anything close to our maximum muscle power for any significant length of time. Our cardiovascular capacity hits its limit when we use but a fraction of our musculature. So we have bases yet to touch.

Perhaps we can sharpen the question by asking how, given the resources of a large zoo, we might carry a given load with the least investment in fodder. Should we use one elephant or should we divide it among a hundred dogs or thirty-five llamas or twenty-five donkeys or seven horses? All of these can carry loads of (at least roughly) the same fraction of their body weights. Furthermore, all use energy at about the same rate for carrying a load as for getting their own masses around. So we need no distinction between loading and self-loading.

The bottom line: Bigger is better. Cardiovascular factors, basal metabolic rate and metabolic scope in particular, tell the tale. Basal metabolic rate doesn't follow body mass but instead increases more slowly when mass goes up. Consider the dog and elephant again, the latter a hundred times more massive. The basal metabolic rate of the elephant will be not one hundred but only about thirty-two times greater than that of the dog.[4] In other words, relative to its mass, the elephant manages its personal activities on about a third of the energy needed by the dog. If (and we'll get back to the point) each can carry the same fraction of its weight or can pull a plow or cart with a force that's the same fraction of its weight, the elephant wins by a factor of three.

Several other factors increase the bigger animal's advantage, although none bears much on the specific comparison of dog and elephant. Metabolic scope, the factor by which basal energy consumption can be raised, increases with body size. Dogs have an exceptionally high scope, and we don't have good data for elephants, so we have to revert to idealized animals of one hundred and ten thousand pounds. On average, the larger can expend energy in vigorous activity at a rate a third higher than the smaller, compared to its basal rate.[5] So the bigger animal not only spends less to keep its engine idling and in repair but can work harder than the smaller relative to that base line cost.

Furthermore, longer legs swing back and forth more cheaply than shorter ones. If we assume that swinging legs operate as pendulums (and make a few other assumptions), we conclude that the bigger animal doesn't have to swing its legs through as wide an arc to walk at the same speed. A shorter arc and lower frequency of swinging mean that muscles can contract at lower intrinsic speeds (again, shortening speed relative to muscle length). As we noted earlier, lower intrinsic speeds mean cheaper and more efficient operation.

In practice, some potential complications don't make much difference. For one thing, large animals go only a little faster than small ones. Long ago A. V. Hill predicted that provided they are of similar shape, all animals should run at the same speed; his reasoning was based on the idea that most of the work an animal does goes into moving its legs back and forth.[6] Secondary factors, such as differences in shape between small and large animals and in stride-to-stride energy-conserving mechanisms, give the larger animal that slight advantage. But in the size range of interest here, speed just doesn't vary with size in any systematic way. The biggest mammals clearly don't outdistance ones of somewhat lesser size; a horse with a lightweight jockey goes a little faster than a greyhound but more slowly than a sprinting cheetah, long-legged camels and giraffes are not famously speedy, and elephants charge at no remarkable rate.[7]

Besides, top speed has little bearing on what we get from a beast of burden. What matters more is the maximum walking speed of an animal, the speed at which it switches from a walk to a trot. Thanks to the efforts of R. McNeill Alexander and his coworkers, we have a good formula for that transition point.[8] They've shown that the switch occurs at a speed that's about 2.2 times the square root of the distance between hip and ground—with speed in meters per second and distance in meters. For a typical human, that's equivalent to going a mile in about twelve minutes. If our paradigmatic one-hundred- and ten-thousand-pound animals have the same shape, then the larger will have legs 4.6 times longer than the smaller (the cube root of their hundredfold mass difference). It can walk faster by the square root of that 4.6, or about twice as fast. No big deal, considering their vast difference in size—and considering something else.

To walk or run faster takes more power; experience tells us that we expend more energy per unit time when we go faster. Less self-evidently, speed doesn't much alter the cost in energy of going a given distance. Going twice as fast costs about twice as much per unit time, meaning that it costs just the same per unit distance. When they pick speeds, animals care enough about cost to pick the ones that are most economical.

At these best speeds that rule about constant cost per distance holds. Dan Hoyt, with Dick Taylor (whom we met in connection with metabolic scope two chapters ago), measured the energy cost per unit distance for ponies trained to run at different gaits and speeds on a treadmill. They then watched what the ponies did in their paddock and found that the animals picked speeds for each gait—walk, trot, and gallop—that gave the minimum cost for that gait. Of equal significance, those three minima turned out to be the same. The upshot: For a given animal, the cost of self-transport doesn't depend on how fast it goes.

One-half of the bottom line: Cheaper maintenance metabolism and more efficient use of muscle each give large size an advantage; cardiovascular and muscular factors coconspire. Consider, to put some numbers on the argument, the price animals pay for hauling themselves around, their minimum cost of self-transport. (Remember that the cost of load carrying tracks the cost of self-transport.) From the work of a number of clever and patient investigators, we have data for animals ranging from ants to horses; Figure 10.2 summarizes an enormous effort. The ant may do the heavy lifting, but the horse wins hands down in the long haul. It

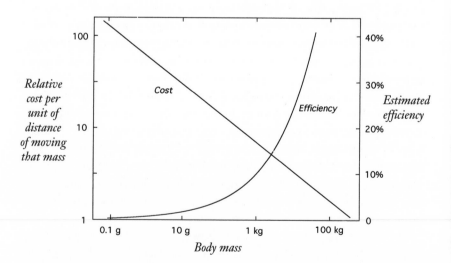

FIGURE 10.2. How the minimum cost of self-transport varies with body weight. Note the geometric scales on left and bottom (for which metric units work best). The data come from R. McNeill Alexander (1999).

does better by a factor of almost a thousand in energy expended to carry a given mass a given distance.

The other half of the bottom line: We mustn't forget that the bigger animal walks faster, and as the adage goes, time is money. The trucking company needs half as many trucks if each truck can make its trips twice as fast. So the horse not only gets there with greater physiological efficiency but saves by getting there faster as well.

Finally, a few data from creatures closer to home and hearth. In one investigation, some strong dogs pulled a sledge while the force and speed with which they pulled were recorded. The dogs could sustain a power output for ten minutes of about 4.5 watts per kilogram of body weight. That's similar to the rate of 3.7 watts per kilogram found in measurements on pulling horses or on rowing Yale crewmen.[9] To put it another way, dogs (and horses) don't get far from Douglas Wilkie's curve. Yes, size matters, but within the narrow range of us and our working animals other factors blur the force of any general rule. Forget gerbil- and hamster-powered technology, though.

Size, Slope, and City

So far we've been on the level—no walking or running uphill or down. For plenty of places, level is laughable. I grew up in a hilly town along the Hudson River, and I still recall being startled at an age of about seven by the flatland of New Jersey. I now live in Durham, a city in the gentle hills of the Carolina piedmont. Durham came about as a post–Civil War artifact, a place where two railroads crossed. It lacks any salient geographical feature, such as a mountain, river, or harbor, and you can head away from it in any direction. Its layout strikes the newcomer as peculiar and confusing. The conventional rectilinear gridwork of streets occurs in small patches in between old radial arteries that meander as if intent on disorienting the stranger. These peregrinating pathways can be distinguished on a map by name as well as by appearance: As Chapel Hill Road, Hillsborough Road, Roxboro Road, Old Raleigh Road, Old Oxford Road, they tell where they go. As you can see from Figure 10.3, the layout reverses what you find in more recently evolved cities, where the curved streets of residential areas form patches among broad, straight avenues intersecting at right angles.

Ground-level experience reveals that our radial roads don't go up and

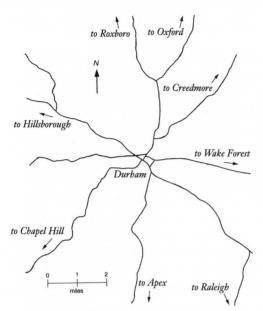

FIGURE 10.3. The radial roads leading to and from the center of Durham, North Carolina, from the 1920s. Many of these have since been straightened or rerouted.

down the way the secondary streets do. Their routing attempted to avoid the ubiquitous slopes of the piedmont without, as is now the custom, digging and filling. Why this antipathy toward slopes? Dismissing it as pure cultural artifact would be a mistake. Muscle mechanics, not misanthropy, made the mess.

Consider what happens when an animal goes uphill. To the ordinary costs of locomotion and body maintenance you now add the work of lifting body mass, of work against gravity. That work varies with the height to which the mass is lifted; as important here, it varies with the mass itself. And that work incurs the expected metabolic cost; it's proportional to body mass. But as we've just seen, the cost of getting from one place to another, the cost of self-transport, doesn't increase in proportion to body mass but does so more slowly. The cost of body maintenance, basal metabolism, also increases more slowly than body mass. So the larger animal faces a greater cost of going uphill relative to both self-transport and body maintenance. Dick Taylor and his collaborators made proper measurements, so we're not just arguing hypothetically. A chimpanzee running up a fifteen-degree slope consumes about 50 percent more oxygen than when it runs on the level. A mouse incurs much less difference—it cares less about the grade of its path—but it uses oxygen (in relation to its mass) many times faster than the chimpanzee.[10]

A given slope slows a horse and wagon more than it slows a horse alone, it slows a horse more than it slows a person, and it slows a person more than it slows a dog, relative to the speed at which each travels on the level. Durham was a market town, and farmers coming into town had an easier journey if their horses, mules, or oxen didn't have to pull loads uphill. Even slight slopes slow or stop a horse that's pulling a loaded wagon. By contrast, a squirrel climbs a tree nearly as fast as it runs on the ground. You notice something analogous when running or riding a bicycle: you become aware of slopes that pass without notice when you walk. Here the main culprit, though, is speed. You ascend faster, so the work of lifting happens over a shorter time and thus demands a higher level of power input and output.

You can see the antithesis of our road-building tactics by watching, as I did once in Panama, a line of leaf-cutting ants, each carrying a piece of leaf homeward along a prepared avenue. The ants route their roads so they take the shortest distance; they often reject going around in favor of up and over. To my eye they don't just follow a compass direction; they circumvent irregular obstacles when crossing over them would lengthen their path. But, I should add, even small creatures—maybe not as small as ants—do expend extra energy when slopes get extreme, and the origin of that cost remains obscure. Cockroaches can be persuaded to run on treadmills (Bob Full, who runs the polypedal lab at Berkeley, proves particularly persuasive). They turn out to consume twice (but only twice) as much energy climbing a forty-five-degree slope as they do walking on the level and three times (but only three times) the energy with the treadmill set at the full vertical ninety degrees.[11]

Thus the big animal enjoys a significant advantage in overall cost of transport, a slight advantage in speed, and a disadvantage when sloping paths must be followed. We should be impressed that Hannibal brought the Carthaginian elephants over the Alps. We also have a better sense of why the great leader anticipated by the prophet Isaiah was expected to "make the rough places plane."

Carrying Loads

We saw in the last chapter that a human, admittedly one not expected to enjoy the experience, could carry loads of up to about half of body weight. How much load can we put on our various beasts of burden?

Maybe they can carry more, relative to their weights, since they divide a load among four legs and have an obvious place to put it: on top or on the sides of their backs. Or maybe they must carry less since, except for horse and dog, we beat them in metabolic scope. Figures for safe (or humane) loads differ widely even for specific kinds of animals, nor should we expect precise numbers since many of the animals vary a lot in lean weight. In general, a reasonable load appears, on average, to be no more than about a third of the animal's weight. Donkeys can do a bit better, carrying perhaps 50 percent.[12] I calculate a figure of over 60 percent for a llama from published numbers for maximum loads and maximum body weights, but I suspect that either one of the figures is incorrect or the two are in some way incompatible.[13]

In this sense, we're better pack animals than our pack animals. Still, the calculus contains complications that have been chewed on by military planners through the ages. A downside to using pack animals comes from their inferiority on grades, beyond considerations of surefootedness. In one study, two donkeys (called in the study burros, in the Spanish-derived vernacular of the American West) and two humans walked with loads of a third of body weight. The donkeys carried added weight somewhat more cheaply relative to their body weights than the humans, although these particular humans did less well than others for whom we have data. On gentle (2 percent) grades, though, the humans did better than the donkeys. The humans, of 80 kg average weight, had to put out 50 percent more power to ascend than to descend such a grade, but the donkeys, of 174 kg, needed to double their output to do the same. Slopes, once again, are worse if you're big.

Thus, on reasonably level terrain that affords good footing, large pack animals do better than do we—if the measure is weight carried in relation to food consumed. That's the consequence of their lower (relative) basal metabolic rates. By the same measure, pack animals of our size or a little larger don't do as well as we do. That's the consequence of our ability to carry a larger load. Yet again other matters complicate the calculus. Consider what changes if the load carrier must carry its food. The ability to get energy from cellulose enables an animal to eat food that others reject, but it doesn't translate into lightweight food. For that you need a high-fat diet, something more suitable for the palates of people and dogs. Dogs of course are too small to live cheaply, and it takes many of them to move a substantial load. Both factors offset their impressive metabolic scope and power output when pulling. Humans do better. According to one comparison, a horse eats as much as six men, but six men can carry far more than one horse can carry and a little more than a horse can haul.[14]

That brings up the other way we use animals to move things. We add wagon or carriage and ask the animal to pull rather than to carry; the horse of the previous paragraph hauled much more than it could carry. So we have to worry about wagons, and wagons care more than animals about what's beneath them. A horse, shod or unshod, may be less versatile than a goat or a dog, but it's far less fastidious about footing than even the best of wagons. Moving a wheeled vehicle along a level and unyielding surface costs little; that's why railroads preceded motorcars by almost a century. Their operating cost, though, rises rapidly with increasing bumpiness and softness. So we can't fall back on a few globally applicable figures to compare the cost of transport by a draft animal with that of a pack animal. In the American West, nineteenth-century settlers traveled long distances in Conestoga wagons pulled by horses or oxen, crossing the Rocky Mountains on trails whose only maintenance was continuous usage. But that country consisted of dry plains and sparsely forested hills. In the heavily forested Appalachians of the East, a century earlier, the scenario had differed. A recent biography of George Washington describes how General Braddock's army inched its way for a hundred or so miles from the edge of settlement to its defeat at what's now Pittsburgh, Pennsylvania. Washington, still too young and provincial to be heeded, "urged Braddock to think about employing pack trains of horses instead of the slow, heavy Conestogas that required building a wide, graded roadway. Horses would be more suitable for narrow trails through the wilderness. But Braddock was addicted to moving vast quantities of artillery equipment and ammunition by wagon train, even if it meant laboriously building roads ahead of him."[15]

The persistence of pack animals reflects the absence or poor quality of roads rather than tradition or a lower level of technology. Even in supposedly advanced northern Europe, so poor were the roads that pack animals remained important until just a few centuries ago. In medieval England, transportation of the lord's crop to market was a requirement of an estate's peasants, and records indicate that long trips to carry wheat, rye, oats, and other crops were routine. The powers that were regarded transport by pack animal as about half as effective as cart hauling, and they paid accordingly.[16] A fifteenth-century English woolen merchant still could not rely on wheeled transport, as we find in one account: "Then the great bales were carried on the backs of pack-horses by the ancient trackways over the Wiltshire and Hampshire Downs, which had been used before the Roman conquest, and thence through Surrey and Kent to the Medway ports by the Pilgrims' Way."[17]

So the picture is mixed. Armies have preferred human transport to the extent possible, but they've rarely relied on it entirely. Having carriers, soldiers or animals, that can then serve as fighters sounds better in theory than it usually works in practice. But soldiers wear out—the agony of "de feet" rather than of defeat—and neither pack-carrying nor cart-hauling horses make good mounts for cavalry. The classic peddler carried his pack, but the miner in the American Southwest led his burro; climbers reach summits unassisted, but we guess that trekkers adopt llamas for something other than their attractive eyelashes.

Working in Place

The Old World ancients discovered that wheels could be used for something other than turning out pottery or easing the movement of a sledge. They also discovered that a domesticated beast of burden could be used for something other than carrying a load or pulling a wagon. They figured out that animals could turn wheels that were fixed in place, wheels that could do a lot of useful tasks. That's the basis of a technology that lasted for millennia but that has receded into insignificance during the past century. I'll use the term "treadmill" for these revolving devices, but I mean to include a wider variety of things than the term usually encompasses. They have in common their fixed position, rotational motion, and whole-animal muscle power; most look nothing like the fancy exercise machines now popular. Humans as well as domestic animals turned treadmills, showing, I suppose, how little distinguishes a slave or convict from any other source of involuntary living labor.

We can distinguish three basic treadmill designs; Figure 10.4 gives them in diagrammatic form. The apparent oldest uses a vertical shaft with a horizontal extension. An animal walks in a monotonous circle and thereby cranks the shaft. Since the animal itself rotates, the device needs no true (and relatively modern) crank, such as a bicycle pedal. The vertical shaft may be connected directly to the business end of the machine—

FIGURE 10.4. Three kinds of treadmill: An animal walks in a circle; a person climbs a rotating, inclined disk; several people climb within a cage that operates a crane.

for, say, grinding grain between stones—or that shaft may drive a horizontal shaft using a pair of what we now call bevel gears, with the horizontal shaft raising and lowering a chain of buckets that descend into a well.

A second treadmill takes the form of a huge, hollow wheel within which domestic animals or humans walk and climb, much like the exercise wheels in rodent cages. By contrast with the first design, it asks its motor to climb a slope rather than to pull a load. The greater the load on the treadmill, the greater its resistance to turning. The animal will then have to climb a steeper part of the wheel to keep it going. The increasing slope of the inside of the wheel ensures that the animal's output can match itself to the load. Rodent cages notwithstanding, the wheel works best with bipeds, so humans have usually served as motors. As a cage (it's sometimes been called a cage wheel) it makes no demand that the driving creature be domesticated or even harnessed. One account speaks of bears as power sources as well as of the excellent service given by goats.[18]

An alternative wheel, once of penal popularity, asks its power sources to walk up the outside. That sacrifices the nicely self-adjusting feature of the cage and can be worked only by humans, but it allows use of a smaller wheel.

The third design also makes an animal climb against gravity but asks it to walk up a moving platform of constant slope. This of course is the familiar treadmill of fitness centers. It needs a platform that can support an animal but that can still be flexible enough to go around two revolving drums, so it's a more complex machine. However, a constant slope makes the device less finicky about who drives it. Load matching can be done by changing the slope, but it can't be done by the motor creature itself. This third design came into use more recently; nineteenth-century American farms made great use of it. Like the cage wheel, it demanded next to nothing in the way of harnessing. Ironically, harnessing, on which the earliest design depended, presented a particular problem for the ancients.

Many other designs—or at least variants on these basic ones—found use. A person could sit on a plank and push out with the feet near the periphery of an inclined disk. Or an animal could walk up along that same periphery. Or a person could pull downward on crossbars on the outside of a revolving vertical wheel. Or he or she could pull on a slack chain hung over such a wheel. The makers of modern exercise equipment might do well to peruse the compendiums of great Renaissance engineers such as Georgius Agricola or Georg Bauer (1494–1555), Agostino Ramelli (1531–1608?), and Fausto Veranzio (1551–1617), all three, by the way, available in modern editions.[19]

We know only a little about the effectiveness and efficiency of old treadmills, either as a group or in comparison with one another. I have found no book on their fascinating history and diversity, and they deserve their Boswell, or at least their Dava Sobel or Stephen Ambrose. One guesses that few would win any prize for efficiency, yet they must have been effective enough to be worth the effort, often considerable, that went into their construction. While ancient, they came well after wheeled vehicles and potters' wheels. They're probably of Middle Eastern or Indian origin; even Joseph Needham doesn't argue for Chinese use before about the sixth century c.e.[20] Since their main use in that part of the world was and remains lifting water (as the sakieh), their need for a bevel gear to drive a bucket chain may have delayed their invention.

The Greeks may have used an analogous treadmill to turn the upper stone of a rotary grain mill by about 400 B.C.E. By the time Pompeii was buried, in the first century C.E., the Romans were using slaves and donkeys to power highly effective grain mills, such as the *mola asinariae*, shown in Figure 10.5.[21]

The revolving cage wheel may be a Roman invention. The best surviving account of Roman engineering, that of Vitruvius, talks about it, and there are allusions to it in less specialized sources, such as Suetonius. Unambiguous images occur on bas-relief sculptures, and remnants have been found at Pompeii; a version of one of the latter has recently been reconstructed.[22] As with grain mills, humans and donkeys provided the power. Such wheels provided the motors for the cranes that lifted blocks into place when tall structures were constructed. Centuries later the same treadmill cranes helped a more technologically savvy culture build the great medieval cathedrals.[23] According to one source, a cage wheel 16 feet in diameter and 8 feet across accommodated 6 to 8 men, who could lift 1 ton a distance of 27 feet 40 times per hour—600 foot-pounds per second of power.[24] Dividing that output among 8 workers, we can calculate a power output per person of just over 100 watts. That attests to both decent efficiency for the machine and considerable effort for the workers (who may not have worked steadily). Of course, if the power requirement were less, the operators could have used fewer workers; we come back to John Smeaton's eighteenth-century datum that a laborer could work at an output of around 90 watts.

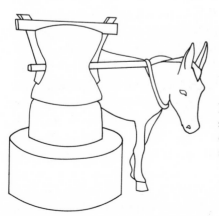

FIGURE 10.5. The grain-grinding *mola asinariae* of the Romans. This one is being turned by a donkey in ancient harness.

Raising water has always been the main task of stationary, muscle-driven engines, and we still use hand-operated pumps. But that often involved more than the short distances needed for wells, rivers, and the leaky hulls of wooden ships. In one Roman mine a cascade of eight pairs of treadmill-powered scoop wheels raised water almost a hundred feet. Nor were grinding grain and lifting stonework the only other tasks asked of these engines. Medieval Europe powered a wide variety of machinery with treadmills, including sawmills, pile drivers, dough-kneading machines, dockside cranes, and bellows. From a bit later we have reports of such odd things as a dog-powered wheel that turned a roasting spit.[25]

As far back as the eighth century C.E. the Chinese used human-powered treadmills to power paddle-wheel boats, an element of Asian technology to which Joseph Needham gives his usual attention.[26] The arrangement mates two effective devices. As we've seen, treadmills can achieve respectable efficiencies. At the same time, drag-based paddle wheels, if not as efficient as lift-based propellers, do much better than paddles or oars for all but small boats. Oar and oarsmen don't scale up well; hull length that might accommodate oars increases by a lower factor than hull surface that incurs drag, and oars get heavy and clumsy with multiple oarsmen. Making wide, high paddle wheels raises only minor mechanical problems, and they should couple easily to treadmills low and amidships.[27] They're as good as oars and better than screws for shallow-draft vessels navigating in confined places; with a pair of side wheels turning in opposite directions, a ship can turn around in its own length. The earliest ocean-crossing steamboats used paddle wheels; with the inefficient, coal-burning steam engines of the time, that says something about their propulsive efficiency. Lovely, large paddle-wheel ships steamed up and down the Hudson River into the 1950s; they provided the most pleasant form of transportation I've ever experienced.

The nineteenth century marked the heyday of treadmill technology, but in two contrasting contexts: alliteratively, prairie and prison. Beginning in the 1880s, combines, pulled by up to forty horses, reaped and threshed wheat as they moved through the fields. But in the preceding decades threshing had depended on large, stationary threshing machines, typically horse-powered, as in Figure 10.6. These were designed to be broken down and moved from farm to farm by their crews. The quotation with which this chapter begins describes threshing day on an Iowa farm. The great machine arrived late the day before and

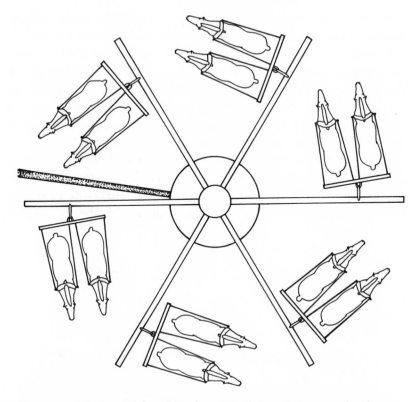

FIGURE 10.6. A nineteenth-century large North American sweep threshing treadmill, from above. In this particular design, each animal must step over a radial drive shaft once for every turn of the mill. In others, the drive shaft extended outward above the animals.

was assembled before dawn; after working all day, it was dismantled and moved (perhaps by its own horses) to the next farm. Interestingly, by using circling animals, these big, sophisticated machines reverted the oldest of treadmill arrangements. A popular magazine from the post–Civil War period, the *Prairie Farmer*, was filled with references to treadmills and advertisements for their sale. Some, even larger than the one in *The Hawkeye*, employed up to twenty horses to drive their rotating sweeps.

But most nineteenth-century American devices were smaller, general-purpose machines powered by one to four horses. These often used the third of the treadmill designs, two rollers and an inclined belt. "Belt," while descriptive, may mislead the modern reader: crosswise wooden boards, hinged side to side, formed a moving platform that slid on greased ways between the rollers. A one- or two-horse model looked like an open horse trailer, a horse-drawn one, of course, as in Figure 10.7. It worked much like a tractor with a power takeoff, pulled out of the barn and positioned wherever needed for power-demanding chores, such as sawing logs. These multipurpose units could be purchased as late as 1890, and their use persisted into the twentieth century.

So did the use of the horizontal mill with radial sweeps. For some tasks the two competed. Sweeps were simpler, bigger, and lighter relative to their size or potential power, but they required more goading of the driving animals, and many designs required that the animals step

Figure 10.7. A nineteenth-century small North American portable treader for general use on a farm.

over a horizontal drive shaft once in every revolution. Treads were more compact, as well as easier and quicker to put into action, but they were harder on the animals' hooves. They were advertised as being twice as effective, not an unreasonable claim.[28] On an inclined belt or vertical wheel, an animal works by lifting itself; attached to a sweep, an animal pulls. Yes, carts achieve carriage more cheaply than do packs, but that's just the advantage of wheels. The bigger the animal, the better lifting itself becomes as a way to extract power; pulling remains almost indifferent to size. That advantage of large size comes down to what we earlier viewed as a disadvantage: the large animal's greater cost of going up an incline compared with merely going forward.

Not that the treadmilling devices came only in large, horse-powered versions. Household (or dooryard) versions for sale included a butter churn the manufacturer of which asserted that dogs, goats, sheep, and even children provided suitable power sources. Treadmill technology reminds us that mechanization needn't imply muscle-displacing motorization. It may also save labor by improving the coupling between muscle and task and by facilitating the use of animal instead of human muscle.

But the technology began to decline before the end of the nineteenth century, mainly because steam power displaced horses for stationary engines. We don't give this early steam-powered technology enough attention. The engines may have been less fuel- and weight-efficient than internal-combustion engines, but they proved practical for use far from centers of population. A steam-driven external-combustion engine operates at much lower temperatures than any internal-combustion engine, so its construction takes neither fancy metal nor precision machining, and its repair is similarly forgiving. Furthermore, it can be adapted to run on anything combustible. I recall seeing a steam-driven snowplow, pressed into emergency service by our county on one occasion in the mid-forties, making its way on firewood. The wood came from the people whose streets were belatedly being cleared; presumably wood had been stacked along roads when steam plows were the norm. Farm museums in North America still display huge steam engines, self-propelled tractors that could get from farm to farm but were intended for use on arrival as stationary power sources, just like the sweeps and treads that preceded them.

The other big use of treadmills was penal. For that matter, the great

Oxford English Dictionary, whose *T* section was compiled at the turn of the twentieth century, recognizes only a penal application for treadmills. Punishment and profit: what a nice combination. Prisoners were put to work, sometimes exceedingly hard work. Treadmill punishment could not have been uncommon. Thomas Henry Huxley, famous defender of Darwin in the latter part of the nineteenth century, assumed he'd be understood when complaining about an adversary, "I would willingly agree to any law which would send him to the treadmill."[29]

Brian Cotterell and Johan Kamminga, in their book on preindustrial technology, note that early-nineteenth-century Australian convicts worked up to twelve hours per day at a power output of 70 watts and that some claimed to prefer hanging to working on the treadmill. That's no small output, comparable (given the errors in such numbers) to Smeaton's benchmark 90. If we assume a 25 percent efficiency, it corresponds to a metabolic rate (above basal) of 280 watts; sustaining it for twelve hours would require almost three thousand kilocalories of extra nutrition.

Nor was the treadmill just a British "correction" device. A recent history of New York City[30] mentions its use in the early nineteenth century, referring to an American treatise of 1824 with the informative title of "The History of the Tread-mill, Containing Its Origin, Construction, Operation, Effects as It Respects the Health and Morals of the Convicts, with Their Treatment and Diet, also a General View of the Penitentiary System."[31] The piece provides yet another source of power data. On this New York treadmill, shown in Figure 10.8, prisoners climbed, eight at a time, on treads protruding from the wheel, the latter a little over 5 feet (1.57 meters) in diameter and turning three times each minute. If we assume that a typical prisoner weighed 132 pounds (60 kilograms), the prisoner must have worked at a power of almost 140 watts. Since the normal duty cycle allowed each prisoner to rest a third of the time, the sustained output would have been a little over 90 watts, sustained, according to the report, for up to ten hours a day. Again Smeaton's datum receives confirmation, as does the unpleasantness of the regimen.

Of course, used judiciously (I use this last word in a nonlegal sense), a treadmill could be a good thing for people who, being incarcerated, lacked aerobic exercise. It combined exercise and punishment for inmates with useful tasks, such as grain grinding for prison bakeries. The report on the New York treadmill makes much of its noninjuriousness,

Figure 10.8. The treadmill of the New York City prison at Bellevue, as it looked in 1824.

and the assertions concerning the health of the inmates forced to use the treadmill don't seem unreasonable. Additional food did get provided for them at a quantity that appears sufficient. But the punitive character of the treadmill remains evident, or so one can guess from its claim of rapid attitude adjustment of obstreperous or recalcitrant inmates. Remember, most of those inmates were mere vagrants, not Lizzie Bordens.[32]

Even if rare, treadmills exist yet. A colleague saw one in operation in North Africa during the 1950s, and another was mentioned at about that time in use on Ibiza, one of the Balearic Islands in the Mediterranean.[33] Still more recent is a report of a treadle pump in common use for irrigation in Bangladesh, a pump run by a person who continuously climbs what looks like a Stairmaster exercise machine.[34]

CHAPTER 11

Bos versus *Equus*

With oxen I have succeeded everywhere; but with horses I have not been so successful, particularly in marshy places, where the fillies being full of gross humours were much more troubled with the strangles than those bred in high situations. Other disorders fell upon their eyes and limbs . . . .

. . . if a colt has a defect, you sell him for a trifle; and should any accident happen, such as breaking of limbs or the like, you must kill them . . . yet there is some resource with oxen; for if it so happen that you cannot sell them to the butcher, their meat will serve for the use of the farm.[1]

PROU DE MONROY, 1796

TWO GROUPS OF ANIMALS DOMINATE THE PULLING GAME: Among the equids the horse-donkey-mule team; among the bovids the ox–yak–water buffalo team. Familiarity and fashion favor the equids. They retained a role in urban transport in the West after oxen—castrated bulls, sometimes called bullocks—disappeared into rural obscurity. Among people unimpressed by fast automobiles, equids retain to this day a peculiar cache as racers; they merit their own museum, the International Museum of the Horse.[2] They still grace our literature and humor; we have equestrian statues and even horse grave-yards. Historical primacy, though, goes to the bovids. Balaam may have ridden his ass, but it neither plowed his field nor pulled his wagon. These tasks demanded a harness, an easier fit for bovids than it proved to be for equids.

Neither of course pulls anything in nature. Carrying a pack or rider differs only in quantity from hauling along the animal's own viscera or perhaps a fetus. By contrast, pulling uses an animal's locomotory equipment in a way far from any possible basis for its evolutionary success, as far from biological normality as is bicycle riding for us. We shouldn't be surprised that persuading an animal to pull might be tricky. What's surprising is that it became such a routine, widespread, and persistent practice.

Harnesses

To this day great numbers of bullocks pull carts in Asia. One estimate has seventy million of these bovids working in India alone, with an aggregate potential power almost equal to the country's electrical capacity.[3] The connection between cart and bullock couldn't be simpler. For a one-animal cart, a cylindrical crossbar, little more than a pole, serves as the yoke and connects the front ends of the fork, as in Figure 11.1. The fork and yoke can be dropped over the animal, which then pulls with the yoke bearing on the front of the fleshy, fatty hump between its shoulders, a hump particularly well developed in these derivatives of zebu cattle. Although a rope usually encircles the animal's neck, it's little more than a necklace, insurance against the yoke's flying up and off if the cart hits a bump or the load shifts rearward. To use animals in pairs, the yoke extends to either side of a single pole that runs forward from the cart. The arrangement has performed well for millennia; its main drawback is ulceration of the bullock's hump from the pull of the yoke.

FIGURE 11.1. A bullock cart as used in southern India.

Bullocks or oxen give good service as draft animals. But they plod along, short-legged creatures that move slowly even when lightly loaded. A horse can pull harder and move faster, and with a light load it can move much faster for long periods. People attend bullfights but not ox races, and no form of cow cavalry has gained military sanction. Nonetheless, as draft animals bovids came first. Whereas equids have been part of our domestic entourage for perhaps six thousand years, only for the past thousand or so have they contributed much as draft, as opposed to pack, animals. On only one occasion, so it seems, did we bring together the pieces necessary to make them work as proper draft animals.

That history implies some difficulty in making equids pull. Horses evolved in the New World, and the natives had at least two opportunities to capitalize on their ability to pull loads. In the first instance, the Paleolithic hunters that arrived after the last Ice Age found horses on the plains of North America; the horses became extinct shortly afterward,[4] as likely as not hunted down by the humans. They might have been domesticated, as happened a little later on the Eurasian steppes, but they weren't. In the sixteenth century the Spanish expanded north from Mexico, and enough of their horses went wild to reestablish the species as part of the North American fauna—after an absence of about ten millennia. This time the natives did domesticate them, but only as mounts and pack animals. In the Old World, horses, donkeys, and later mules served as mounts and pack animals for several millennia before they were harnessed. For several thousand years thereafter they pulled only fast but light military and racing chariots.

That the ancients cared so much about horses while using bovids to pull their heavy loads reemphasizes the difficulty. Cavalry based on lightweight, horse-drawn chariots were first developed in Sumeria and Babylonia (and, at about the same time, in China). Horses ridden for warfare came later for both Persians and Greeks, at least in part because the horses of Eurasia were too small. Larger animals from North Africa provided the basis for the mobile, mounted Alexandrian army along with lessons in horsemanship from the Persians.[5]

War chariots abound as motifs on Greek vases, so we know how horses drew them. A one-horse chariot had a pole that extended upward from the front of its platform and across the back of the horse. But two- and four-horse versions appear more often. In a four-horse chariot, shown in Figure 11.2,[6] the pole went forward between the pairs of horses, so it looks less ungainly. But it shared the same peculiarity, a high front end connected to yoke and throat straps. There's no lack of evidence, historical and experimental, that a horse cannot pull effectively when hitched this way. The classic evidence comes from measurements by a retired French cavalry officer, Richard Lefebvre des Noëttes, in 1931. He found that hitched with such a yoke and strap, a horse could pull only about a quarter to a third as hard as a horse with a modern harness.[7] Keeping the chariot light provided a partial solution, even if to our eyes the use of four horses for such minimal vehicles looks silly—like using a huge engine in a car and then omitting all but the low gears. Traversing big bumps took large wheels, which the chariots had, but their wheels were so lightly constructed that flat spots developed if the wheels were left mounted.[8] Athena needed a special subgoddess to remove and hang up the wheels from her chariot at night. According to late Roman law, a two-horse or two-mule vehicle wasn't supposed to convey more than two people, each with about the carry-on baggage we now take on airplanes—about 400 pounds or 180 kilograms all together.[9]

FIGURE 11.2. The hitching arrangement of a four-horse Greek chariot, as pictured (and simplified here) on a vase at the North Carolina Museum of Art in Raleigh.

Following Lefebvre des Noëttes, we blame the inadequacy of the classical hitching system on the absence in equids of the high shoulders of bovids. Thus the yoke had little to bear on, and the throat strap took most of the load. Yet the attachment of the throat strap to the yoke meant that the strap pulled upward as well as backward. It pulled at a right angle to the neck, about the worst possible orientation for an animal that breathes through a pipe running down the lower side of its neck. So the horse choked if the load increased beyond a relatively low value; with a light load it could race forward, while more heavily loaded, it became immobilized. But that was only part of the problem.

Two other factors must have exacted penalties. For one thing, the yoke and throat strap attached the load too high, as seen diagrammatically in Figure 11.3. For their weights, horses have longer legs than do oxen. Consider how you pull on a rope. You do best pulling with it low—waist or hip high. Yes, the reliefs we have of great teams of ancient Egyptians pulling sledges show ropes over their shoulders. But their upright, forward-facing posture as well as the position of the ropes suggests ceremony rather than routine labor. If you must pull on a rope over your shoulder, you lean far forward in order to push backward on the ground with a combination of weight and muscle, something mentioned a few chapters back. Your forward force balances the rearward force of the load. Quadrupedal horses don't lean as readily as bipedal people, reducing the utility of that option. Horses behave more like bodies whose ori-

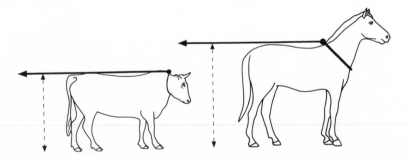

FIGURE 11.3. The lines of action of the forces on ox and a horse with high yokes. Note the greater lever arm tending to lift the front of the horse. The animals have been drawn to have the same torso lengths and the same foot placements.

entation is fixed. The higher the location of the rearward force on such a body, the more the body tends to tip over backward. That means pulling the horse's forelegs off the ground, something muscles can counteract only as part of the horse's weight. Try it yourself: pull back on a string attached at different heights to a toy horse (or another quadrupedal structure, such as a chair) that stands on a carpet.

The other penalty comes from force concentration. A strap loads a line around the front of the horse. So a given load exerts a high stress: force divided by the area over which it's applied. Sitting on a board feels better than sitting on a narrow rail. Your weight doesn't change, but the board spreads it over a wider area and thus lowers the stress. The yoke of an ox presents the same problem, but the ox's fortuitous anatomical peculiarity, its padded hump, reduces the force concentration. We don't often harness ourselves to haul loads, but when we do, as did the ill-fated Scott expedition to the South Pole, we use some gridwork of straps across our chests.

A chest strap instead of a yoke and throat strap worked better, and it saw use in both China and the West. It got the load lower and may have reduced tracheal constriction. But it still concentrated the load. The solution to the problem came to Europe, most likely from Asia, in the early Middle Ages, in the latter part of the ninth or the early tenth century. It was the solid but padded modern horse collar, depicted in Figure 11.4. The collar alleviated all three problems: It minimized the force on the horse's trachea, it lowered the height of the force, and it distributed

FIGURE 11.4. A horse with a modern collar; the load is better distributed and borne lower.

the force over the various bones of the horse's pectoral girdle. It did something else of great importance. It could be attached to the load by lengthwise traces as well as by poles, so that one horse could be harnessed in front of another.[10] Road width might limit parallel hitching to two horses; no such limitation faced a serial arrangement. Serial hitching mattered in rough country, and the street designs of some cities still reflect its use. Salt Lake City, in the American West, and Ballarat and Bendigo, in inland Australia, got their broad avenues so long teams could make U-turns. Lengthwise traces do fine for pulling, but since one can't push on a rope, wagon trains can't easily back up!

Our language preserves the distinction in hitching technique. Picking verbs, we "yoke" oxen but "harness" horses; choosing collective nouns, we speak of a "yoke of oxen" but of a "team of horses."

More Horsy, and Other, Technology

Effective hitching depended on another piece of technology, something even less familiar to modern, urbanized (or at least petroleum-powered) people. Traces rather than poles allowed serial hitching, but they're not the whole answer for parallel hitching. Which pulls if two horses or oxen are hitched side by side, each separately attached to a cart or plow? The laggard need bear no load at all! Also, how does one turn without asking the outer animal to take the bulk of the load? The solution comes in the form of a whippletree (or whiffletree), something Lynn White called "a grubby but important element in the history of land transport."[11] A whippletree (see Figure 11.5) looks like a dichotomously branching genealogical or phylogenetic diagram. Crucially, each animal always pulls on the end of a crosswise bar. With it animals can pull with comparable effectiveness even if they are not exactly abreast. The relative lengths of the bars can be chosen for teams of odd numbers of animals or animals of varying capabilities. In addition, the whippletree acts as a shock absorber, decoupling the cart from sudden jerks by the animals—more important with horses than oxen. It also equalizes the stresses on the traces when the cart turns, playing a role now assigned to the differential gear of an automobile.

Whippletrees came into use in the eleventh century, a little later than the horse collar. As more modest appliances, they're harder to trace, but one appears as a detail in the border of the Bayeux Tapestry, a linen

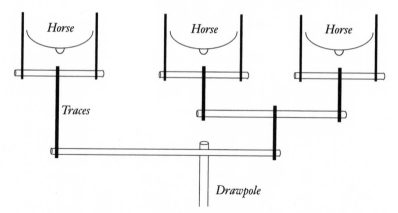

FIGURE 11.5. A whippletree, from above, for three horses of equal ability.

embroidery 231 feet long and just short of 20 inches wide. The tapestry, now displayed in the former Bishop's Palace in Bayeux, France, depicts scenes from the Battle of Hastings, the key to the Norman conquest of England, and was made between 1066 and 1077.

A third momentous device, the horseshoe, came into use during this same period as well. A consistent difference between the feet of bovids and equids forms the basis of our present taxonomy. We put bovids in the order Artiodactyla, all of the members of which have even numbers of toes, two or four, on their feet. The bovids, specifically, have two toes, bearing the split hooves upon which their biblical edibility depends. We put equids in the Perissodactyla, the members of which have a variable but usually odd number of digits and bear the largest part of their weight on their middle ones. All equids of the present and recent past ("Recent" in paleojargon) have just that single functional digit, corresponding to our third finger, and all walk on hooves that correspond to our fingernails and toenails.

The foot of a cow, as in Figure 11.6, represents a more versatile, less specialized design than that of a horse. For that matter, the whole locomotory setup of a cow is less fancy than that of a horse—in the basic features, such as shorter legs and lower metabolic scope, that come from their ancestors, not their breeders. An ox can walk pretty much wherever it or we prefer. But horses ran across the dry steppes of central Eurasia and North America, where they evolved their strange, specialized toenails. Unshod horses went wild in the American West, and even now

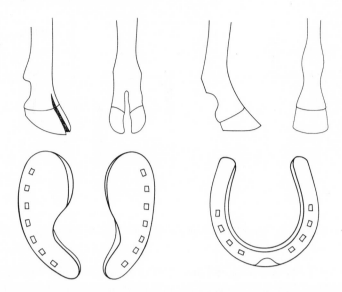

FIGURE 11.6. The feet of ox and horse, with the shoes for the toenails of each. An ox wears eight shoes, one for each functional toe.

feral horses live happily on the sandy barrier islands that form the coast-line here in North Carolina. If what's underfoot isn't too wet or rocky, they do fine. But stone-paved roads or nice damp soil tend to crack or rot hooves, and iron footgear for horses goes back to the Romans and the pre-Roman Celts.[12] Allusions to nailed, as opposed to laced or otherwise attached, horseshoes, though, go back only to about 900 C.E. They tell all too little about the origin of our odd but effective way of nailing bits of iron to a horse's toenails. The utility of horseshoes has never been disputed, especially for the heavier animals bred during the Middle Ages. Indeed the sizes of antique horseshoes record that increase between the thirteenth and fifteenth centuries.

Yes, oxen do get shod, and the figure shows the pair of shoes for a single ox foot. Ox toes splay out a little as they touch the ground, something a single shoe on a foot wouldn't allow. An ox therefore needs, in all, eight shoes for its load-bearing toes. But the practical performance of oxen depends much less on whether they're shod than does that of horses.

We'll leave two further developments, stirrups and saddles, for a subsequent chapter since their use is as much military as civilian and because they're less important to a story that's mainly about pulling. Of more relevance here is the cultural transformation that took place in the early Middle Ages. The widespread adoption of horse collars, horseshoes, whippletrees, stirrups, and saddles formed a critical part. Indeed the economic efficiencies resulting from introduction of these devices may underlie the other aspects of that transformation. The horse of the Middle Ages could pull with four or five times the force of its Roman predecessor, and it could keep pulling longer.[13] Lynn White makes power the centerpiece of a revolution in which cultural change permitted technological change, which in turn permitted further cultural change. Power in this context means power in the physical sense in which we've used it here and mainly muscle power—never mind kings, armies, and shifting political lines.

The Romans cared little about nonhuman sources of power. Of course they used oxen for plowing and mules and donkeys for other chores, they put sails (if inefficient ones) on their ships, they ran simple waterwheels, and some of their treadmills relied on nonhuman animals. But these contributed only a tiny fraction of Roman power. A slaveholding society, some have guessed, stifles technical innovation in general and laborsaving innovation in particular. What's the incentive, with such cheap labor? I (not alone) wonder about this facile explanation since even slaves must be fed, clothed, and housed. Furthermore, slaves represent a capital investment, something that labor for hire does not.[14]

Perhaps Roman uninterest in laborsaving technology reflected not the cheapness of labor but the lack of incentive to innovate among the laborers themselves and among others, such as overseers, who were close to the tasks. In other words, maybe the population that could profit from economies of labor didn't overlap with the population that might have created laborsaving devices. Whether evolutionary or technological, most innovation comes in small increments, and potential increments may be mainly evident to those hewing the wood and drawing the water. Granted that only a tiny fraction of laborers and small-time craftspeople have a creative turn of mind, what must matter is if that tiny number have opportunities for making and then profiting from an improvement.

However we explain it, the Middle Ages saw greater elaboration and proliferation of laborsaving technology: of oarless and thus more labor-efficient sailing ships, of waterwheels, and of windmills. The Romans knew how to make water-powered grain mills, but they remained uncommon. By contrast, the 1086 Domesday Book of England, a kind of national inventory, lists 5,624 such mills. Windmills may have originated in Persia or Afghanistan, but the modern form, with a horizontal axle well off the ground, was invented in northern Europe during the Middle Ages. It served the same purpose as the waterwheel in places, such as the Low Countries, where the streams didn't drop enough for waterwheels to work. Even the design of carts and carriages improved, so not only were horses more effective at pulling, but they needed to develop less force relative to the payloads they moved. Thus advances in employing horses as draft animals came in the context of a technologically diverse and widely disseminated shift away from reliance on the power of human muscle.

Many (but not all) subsequent historians have endorsed Lynn White's point that the Middle Ages was a period of major technological advance, both indigenous and imported, rather than the dark period of pre-Renaissance stagnation presumed by people looking at artistic and literary output. Pipe organs, mechanical clocks, trousers, felt making, skis, soap, wooden barrels, alcohol distillation, the magnetic compass, gunpowder, printing, wheelbarrows, improved mining and metalworking methods—the list gets lengthy! In the West, at least, technological change in the five hundred years after 800 C.E. dwarfed change in the thousand years preceding it. Never mind political fragmentation, artistic rustication, or religious homogenization.[15] Or the schoolbook "Dark Ages," as nothing but knight time.

About Plowing in Particular

Almost all hominids eat plants. About ten thousand years ago humans began to improve their access to plants with agriculture, initiating the most multifaceted social revolution in the history of the species. Early in the game we must have realized that we could improve our harvests if we didn't just drop seeds on unprepared ground. Nature may do so, but nature gets horrible yields compared to her investment in seeds. Seeds get eaten by birds and rodents, they blow or wash away, they fail to find rooting access, and so forth. Beyond yield maximization, agriculture boils down to an attempt to divert to your own metabolism and biomass the greatest fraction of what your seeds produce. Better by far to prepare the soil, loosening it so the seeds can fall beneath the surface and then get covered, so rain will be better absorbed, so preexisting weeds don't preempt soil and sunlight, so granivorous animals aren't encouraged.

How to prepare the soil? Wooden digging sticks have been widely used, and they're effective. They do best where seeds are planted in little hills, where a field must only be cleared and not tilled as well. The fragility of even fire-hardened wood had to limit the metabolic demands of the task, however unpleasant a long period of uninterrupted soil breaking might have been, much as we surmise for felling with Stone Age axes. I find doing the equivalent task with a four-tined pitchfork strenuous, but my steel tool is tough and forgiving, so I can work as hard as I'm able. Agriculture in the New World remained based on digging sticks of one form or another, and lots of primitive agriculture still relies on hoelike derivatives of digging sticks.

In most situations, plowing does a much better job at controlling weeds, adjusting the texture of the soil, and getting rid of the stubble of the previous crop. How to plow? Human power rarely works well. Joseph Needham mentions a Chinese plow that's pulled by a person in the front and guided and pushed by another behind, but it's suitable for only the lightest and shallowest plowing. Humans as a rule can't power a plow that goes deep enough to do much good unless they're combined into unwieldy teams. Or unless they employ force-advantaging accessories. Both Chinese and Western illustrations show human-powered plows that run back and forth across fields on ropes. Workers on either side take turns turning the radial spikes of drums onto which the rope winds.[16] We have no evidence, though, that the scheme loomed large in either culture.

Here's where a large draft animal makes a job not just easier but possible at all. Big draft animals do the job well; modern tractors just go faster, pull several plows simultaneously, and take less upkeep. Wherever such animals have been available, meaning almost everywhere in the Old World, they've pulled plows. The failure to domesticate an animal large enough for plowing may have been the prime impediment to the spread of complex societies in the New World.

Animals still pull plows, even if they're no longer economically significant in first world countries. I recall seeing horses plow small plots in the 1940s in the Hudson Valley. When I first arrived in North Carolina in the mid-sixties, I rented an apartment from an elderly couple who maintained a vegetable patch behind them. Each spring a man arrived with mule and plow and did the patch, which then produced a profuse crop of wonderful tomatoes. Not that we pilfered; since the crop always exceeded their ability to consume it, our landlady made the rounds, urging them on us. Since then powered garden tillers, rented for a day or purchased, have displaced mule and plow.

Much human history turns on plowing. It in turn depends on three interrelated factors: how we designed plows and what we built them of, the types of soils we could till, and how we've used animals. Up through Roman times plows were light affairs, adequate for the friable soils of the Mediterranean region and the Middle East. But the Romans had to plow their fields twice, the second pass at a right angle to the first. They then harrowed them to break clumps further and to work in the seed. The Romans put iron cutting tips on their plows, which must have permitted a sharper edge and thus easier pulling than a completely wooden plow, and they used iron tines on their harrows. Oxen of course did most of the pulling.

The Middle Ages saw major changes in the technology of plowing, as in so many other things. As the great French historian Marc Bloch drew to our attention, iron plows became stronger, heavier, and effective enough so that a single plowing sufficed. Of greater importance, the plows could then manage the heavier, moister soils of northern Europe and thus bring previously marginal land into cultivation. Not only did this land need more force in order to be penetrated, but the wetter climate meant that it had to be plowed more deeply for sufficient drainage.[17] While drought makes more news than drowning, roots breathe oxygen just as we do, and the water in saturated fields can become suffocatingly anoxic. Unlike the appurtenances that put horses to work, the use of heavy plows spread (or, one might say, penetrated) slowly, beginning in pre-Roman northern France and Britain.

These heavier plows acquired some clever accessories. Consider what happens if you force a stick through relatively loose soil: You get a groove flanked by two ridges. But if you do this in dense, moist soil, you may, for all your effort, get neither; the groove simply closes up behind you. Using an asymmetrical plow or holding the plow at an angle helps by throwing to one side most of the soil that's raised. By the thirteenth century a typical plow had a so-called moldboard behind and to one side of the plowshare itself.[18] A moldboard, shown in Figure 11.7, pushed the soil away from the cut. As always, forces must balance, so a moldboard on one side demanded a bearing surface, called a landside, on the other. How to maintain the proper depth with these heavier and deeper-

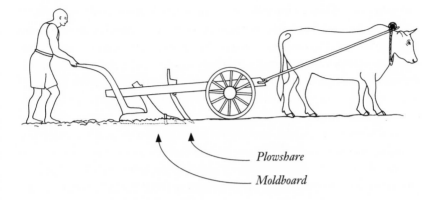

FIGURE 11.7. Ox, plow, and plowman from medieval northern Europe.

cutting plows? The medieval plowman used either of two devices. A plow might have a wheel in front—that is, running ahead of the plow-share. By rolling along the unbroken surface, the wheel ensured that the plow would penetrate only to the depth set by the relative heights of wheel and share. Alternatively, a sliding board (footboard) in front could fix the depth. Finally, horses came into use for pulling plows although, as we'll see shortly, that didn't displace oxen.

Implementation of these new tools had profound effects. More food could be produced as more land was brought into cultivation, and that food could be produced with less labor. More effective crop rotation may have increased productivity, although historians differ on the extent of any increase.[19] In all, the changes allowed a lot of human labor to do something other than eke out a living directly from the soil. They may also have permitted greater consumption of animal muscle: fat and meat.

One measure of the wealth of a self-sufficient society might be the fraction of the population that does something other than produce calorie-providing crops for its members. In contemporary terms, we're talking about wheat, rice, potatoes, and a few other carbohydrate-rich seeds and storage organs of plants. By such a measure (and including oats, millet, and barley), medieval Europe outscores the Roman Empire. Northern Europe, once its heavier soils could be handled with efficiency, may have led the world, and leadership by this measure might, as the biologist speculates, have fueled both its complex feudal society and its technological creativity. A larger fraction of the population could build cathedrals, operate mines, make consumer goods, serve in armies, write and copy manuscripts, and so on. At the start of the nineteenth century, before the next revolution in agriculture, a farmer could feed three or four nonfarmers. That figure may have been stable over the preceding half millennium or so, but it must have far exceeded what could be reached by classical civilizations.[20]

In first world countries we now do ten times better, and we enjoy the resulting material and cultural wealth. At that low labor level, though, so few people work the land that even major changes in their numbers no longer have much impact on the nonfarming labor force. Increase in per capita wealth must now hinge on other changes in what things we do and how we do them.

The results of that medieval agricultural revolution may persist in other ways: in the northward shift of the cultural and political center of gravity of Europe and in the way that European civilization outdistanced that of India or China. As Lynn White put it,[21] the question of why the East stagnated is moot. India and China progressed at a normal rate. What happened was that the velocity of Western technological change increased; to put it in the analogous physical terms, medieval acceleration produced a higher velocity. Most likely the great leap in productivity in northern Europe after about the year 1000 reduced the normal constraints on population growth and increased urbanization, and both forces promoted that acceleration. Even the Black Death of the fourteenth century may have helped by initiating a period of relative labor shortage. For a critical period—since population growth eventually eats up most gains—Europe found itself atypically resource rich and with a special incentive for disseminating laborsaving technology.

The North American adds a grace note. Even after centuries of incremental improvement, the heavy plows that brought new land into productive use in northern Europe couldn't cut it in the tallgrass prairies of the American Midwest. Bringing farming to the prairies in the mid-nineteenth century took another technological jump: John Deere's plow, made entirely of steel, right back to the moldboard. At least this change wasn't left unsung. The American composer Virgil Thomson wrote music, still played, for a documentary film of 1937, *The Plow that Broke the Plains.*

Horse versus Ox Economics

The great latifundia—agricultural estates—of imperial Rome ran on oxen and slaves. Mules and donkeys helped, but just a little; in aggregate power only slaves approached oxen. One Roman landowner, who died in 8 B.C.E., left seventy-two hundred oxen and four thousand slaves.[22] Economics gave little incentive to use horses; in an ox yoke, a horse could pull only four times what a human could, and the horse cost that much more to feed. By contrast, the medieval demesnes and peasantry could make good use of horses; thanks to horse collar, horseshoes, and such, the animal could pull fully fifteen times as hard as a man.[23] Still, the acre got its definition in the Middle Ages as the area that could be plowed in a day by a competent plow-man with his team of oxen. In 1620 the Pilgrims took with themselves to Massachusetts on the *Mayflower* cows, oxen, and pigs, but no horses.[24] Question: why did oxen persist as draft animals? For persist they certainly did, and to the extent that draft animals remain useful, they remain in use.

Nor are oxen the sole persistent bovids. Data from the government of India, conveniently for us, separate working cattle and buffalo from animals raised for milk and meat. Its data allow us to compare the muscle power of bovids and equids in a country still largely dependent on such muscle, a country, moreover, of great climatic and agricultural diversity. In 1966 working bovids numbered about 75 million, or roughly 1 per 8 humans. Of these, buffalo made up about 10 percent. Equids amounted to only 2 million, divided almost equally between horses and donkeys. (Additional work came from 1 million camels and 30,000 yaks, each important in specific regions, and 60,000 mules.)[25] That 75 to 2 bovid-to equid ratio should correspond to the ratio of available muscle power, since horses are larger than oxen, but donkeys are smaller.

Horses and oxen don't form elements of incompatible technological complexes like automobiles and airplanes or E-mail and snail mail. Instead their use resembles that of sailing ships and oar-driven ships—down to reports of the two in common harness. But for all but the smallest vessels, sail decisively displaced oars and paddles, while we've noted that horses and oxen have been coworkers for a millennium. The practical choice involves a complex of factors, some biological and basic, others geographic and socioeconomic and thus context contingent. In recent years that choice has come in for attention in studies of medieval life in northern Europe, where both played significant roles.

First, there's food. An ox eats about the same amount each day as a horse, but it's less fastidious about its fodder, being content with hay and general pasturage. Horses need some grain as a fuel additive, traditionally oats but now more often corn, a New World contribution. Oats cost more, in part because raising a given quantity takes more acreage. Depending on time and place, pasturage may or may not represent a real cost; with ample meadows, oxen can just be turned loose. Similarly, the cost of oats, although always higher than hay, can be hard to account. In the typical three-field rotation, oats, barley, and legumes are the spring crop in one field; wheat is the main autumn crop in the second; the third remains fallow. Oats, then, don't compete for space with wheat, the main cereal crop. Nor in most places do humans eat oats in quantity. Samuel Johnson may have described oats as "a grain, which in England is generally given to horses, but in Scotland supports the people." But Scotland had a much lower population than its oat-eschewing neighbors.

Second comes plowing force and power. The horse pulls about as hard as the ox, but it can move 50 percent faster while pulling, and it can reportedly work several hours longer each day. Since power is force times speed, 50 percent faster means 50 percent greater power output. That 1:1.5 ratio has been a remarkably robust datum, unchanged for almost a thousand years despite both biological (breeding) and technological (harnessing) advances. Speed makes a real difference to a farmer since with faster plowing it takes fewer teams to complete the task within the proper season. For harrowing, speed may matter even more. Faster harrowing may be better harrowing, since with greater momentum the projections of the harrow strike the clods with greater effect. Harrowing may be lighter work, but that's less relevant; it just means harnessing fewer animals to a harrow than to a plow. Where a demesne maintained both oxen and horses, the horses did the harrowing.[26]

At the same time, the ox has an edge for zero or very low-speed pulling, whence the expression "strong as an ox." At zero speed, power output is zero, and force, the ox's forte, determines the choice between engines. The lower speeds demanded by rougher ground favor oxen, an important factor when new land was being brought into cultivation (assarting). In addition, wetter ground favors oxen, and during the Middle Ages the animals saw greater use in the wetter parts of England. Because wheeled plows worked better on drier ground—in wet soil a wheel sinks and clogs—horses more often pulled wheeled than footplated plows.

Third comes hauling, the other use of draft animals. Neither demesne nor peasants could let their animals off the hook when plowing and harrowing were done. Oxen drew heavy *plaustra*, while horses drew lighter *carecta*, both easily maneuvered two-wheeled carts. The oxcart carried more but bogged down more easily, even if oxen excelled at unbogging. The horse cart went much faster, giving it a big advantage. Getting to market faster permits one to get to a market that's farther, and with horse-drawn transport markets could be larger and fewer. That then would tend to lock in a shift to horses, which (but possibly for some other reason) happened in many places. The speed difference when hauling exceeded the speed difference when plowing; a fourteenth-century account has horses traveling more than twice as far in a day.[27] On the Silk Road to China, a distance an ox-drawn wagon could do in twenty-five days, a horse-drawn one could do in ten to twelve.[28] But again multiple factors come into play. In an emergency the horse-drawn vehicle could speed up far more; as we've seen, few mammals of any size or lineage can match the metabolic scope of a horse. The ox, though, could pull a wagon that weighed 50 percent more, offsetting some of the horse's advantage, and its adventitious feeding offset still more.

The advantages of horses in harrowing and hauling—and, incidentally, as pack animals—gave them greater versatility. That gave them greater appeal for peasants, who owned only a few animals, than for the demesnes, which maintained major menageries. So a consistent difference between peasant and demesne superimposed itself on a gradual rise in the penetration of workhorses. At the Norman conquest, horses made up over 30 percent of the work animals held by peasants but only 5 percent of those on demesnes. By 1600 the numbers were 70 and 50 percent, and English agriculture had become predominantly horse-powered.[29]

Despite the results of this marketplace test, oxen retained advantages. They were less high-strung and easier to train, and they were less susceptible to disease than horses. They represented a lower capital investment since they were cheaper to buy and their harnessing equipment cost less. At least in England, a final factor operated. Only the English, it seems, followed Pope Gregory III's injunction against eating horse-meat.[30] The old ox could be recycled with greater profit than the worn-out horse. Even in Brittany, though, echoes of culinary choice can be found; this chapter began with the comments of Prou de Monroy, estate owner from Brittany, as he admonished the ever hippophilic but never hippophagic English.

What of more recent history? Northern Europe may not be India, but remember that oxen accompanied the early emigrants it sent to North America. A continuing literature comparing oxen and horses points up the lack of closure for the issue. Horses have retained their superior place as cultural symbols, for which their speed and military utility must take much of the credit. In the English language oxen trudge alongside only two iconic figures, the fourteenth-century English Piers Plowman and the tallest of the figures of nineteenth-century American tall tales, Paul Bunyan. Oxen have their defenders, but as I read them, they sound just that: defensive. Thus Prou de Monroy, in 1796, felt compelled to make the case for oxen, which were anything but an innovation in agriculture. Still, he did provide data. He found the horse's advantage for plowing more than offset by its additional costs; his bottom line gave a 9:5 expense ratio against a 3:2 labor ratio— 80 percent greater cost for the same 50 percent greater labor that's cited by modern medievalists. Moreover, in the end, he noted the superior value of ox dung as manure.

Half a century later R. L. Allen's book on American domestic animals engaged in no overall advocacy, but it did point out (defensiveness, again) how easily the ox can be undervalued: "The horse and mule are fed with their daily rations of grain when at hard service, and if the spirit of the ox is to be maintained, he should be equally well fed, when as fully employed. Great and permanent injury is the result of [poor] feeding and severe toil, exacted from the uncomplaining animal. His strength declines, his spirit flags, and if this treatment be continued, he rapidly becomes the stupid, moping brute, which is shown off in degrading con-trast with the more spirited horse, that performs, it may be, one half the labor, on twice his rations."[31] Allen continues with comments that echo Prou de Monroy on the superiority of horses for harrowing and drawing wagons on account of their speed, their offsetting greater initial cost, the greater cost of their food and harness, and their lower salvage value. He also cites contemporary English data showing that whereas twelve horses can do the work of twenty oxen, oxen work out to be cheaper in the long run.

Halfway between Allen's day and ours, in 1923, oxen still played a role in the United States. A general book on logging notes the greater use of horses and mules, but it once again takes note of the ox's advantage for strong, low-speed pulling, here for skidding along newly cut logs: "Oxen stand heavy pulling day after day better than other draft animals and also require a minimum of attention because only one feed per day is necessary if the animals are turned out to graze at night. They are slow on short hauls but they can be loaded more heavily."[32] Skidding freshly cut logs the short distance to the nearest road and truck demands an unusual combination of force and maneuverability on irregular ground. In southern Asia it's the only activity in which draft elephants play a major role.[33] In developed countries it may have been the last stand of draft animals of any kind. Despite the case just made for oxen, horses were the last to go. In a 1953 book on techniques for forestry we find the following (noting, of course, the final pun): "For [skidding] there is still nothing to beat the horse. It has quite enough power, can maneuver better than a tractor, and also possesses a useful amount of 'horse sense.' Light tractors can be used in place of horses, but do not get in and out of trees so well, and are baulked by rough, rocky slopes that the horse takes in stride."[34] By the end of that decade the standard works mentioned the horse only as an anachronism, rapidly giving way to tractors (not farm tractors but tank-tread vehicles). A decade later no mention of horses remained; the plug that pulled had itself been pulled.

This fluctuating economic advantage seems historically curious. Not that the case is unique, merely unusual. I can point to the coexistence of diesel and gasoline versions of internal-combustion engines, but I should then note the displacement of external by internal combustion. More interesting are the secondary consequences of relying on one or the other engine. The slower ox means that the plowman should live closer to the field, while the faster horse, by allowing labor to live farther away, generates a settlement pattern with fewer hamlets and more villages. The spacing of centers of population in the American Midwest contrasts sharply with that of the Old World, from which the immigrants came. But the Midwest was overwhelmingly horse-powered, with the same general-purpose engines used for road vehicles as for farm work. My wife's grandmother, in high school in rural Illinois in

1895, could drive to school when a horse wasn't needed for work on the family farm.

Human power may have done the tasks of antiquity, but by the Middle Ages we had persuaded animals to take over most of the burden; by time of the Norman conquest, northern Europe had got 70 percent of its power from animals, with humans, watermills, and windmills sharing the residual 30 percent. Only recently have we lost contact with this animal-powered world, which retained its dominance for a thousand years. Even if attenuated, it can be seen still in third world countries. Perhaps over-population, pressing land to produce plants for human consumption, threatens its continued existence more than does technological advance. Feeding engines on fossil plants doesn't threaten our food supply as directly as does feeding animals on freshly grown plants. I use the word "threaten" not out of nostalgia, first world hubris, or personal preference for the smell of manure over that of petroleum residue. In poking around for material for this chapter, I ran across a study published in Sri Lanka in 1981 as a joint effort of the Agrarian Research and Training Institute, of Colombo, and the Reading University Farm Power Study Team, of Britain. Here's the first paragraph of their conclusions: "The observed costs of tillage and threshing by animal and tractor power to the national economy . . . present an overwhelming case for expanding the proportion of paddy land tilled by buffalo and for reducing current levels of investment in tractors. To plough an acre of land by draught animals is far less expensive to the national economy than tractorized plowing."[35]

Competing Variables

A general issue pervades this comparison of oxen and horses. It comes from the physics that underlies the biology that underlies the economics. Often the bottom line depends on two variables that get multiplied together. The electrical power consumed by an appliance equals the voltage difference between the paired wires coming in times the current that flows through it. The energy we can get from an impounded body of water equals the volume of the water times the height of the dam. The energy you get from your food equals the quantity you eat times the calorific content of each portion. The amount of gasoline you consume on a trip equals your rate of consumption times the length of the trip—either gallons per mile times miles or else gallons per hour times hours. Sometimes we find it helpful to call one variable the extensive variable (current, volume, distance, time) and the other the intensive variable (voltage, height, energy per volume, consumption rate).

In all such cases, the same final factor can come from any combination of the two contributing variables. A light bulb in an automobile operates at 12 volts rather than at the (North American) household's 120. The bulb in the car therefore needs ten times as much current to produce the same light (with the same power) as one in the house, and a 3-volt flashlight needs another fourfold increase in current. Higher dams give more power for the same amount of water coming into their lakes. And so forth.

We've seen many devices that can adjust the mix of the two variables. As they fix the mix, levers, gearboxes, and pennate muscle hookups get around the limited conditions under which our engines—electrical, internal combustion, muscular—give their best power outputs and efficiencies. For domestic animals serving as power sources, the relevant variables are force and speed. Relative to its own weight or power input (food) the ox does better at force but worse at speed than the horse. Borrowing a term from the electrical engineers, I might say that the ox is a lower-impedance machine and the horse a higher-impedance machine. The distinction—or, really, the continuum—will help us see a forest through a lot of trees in the next chapters.

CHAPTER 12

Killing Tools: The Big Picture

I tell you that in the arts of life man invents nothing; but in the arts of death he outdoes Nature herself, and produces by chemistry and machinery all the slaughter of plague, pestilence, and famine. . . . When he goes out to slay, he carries a marvel of mechanism that lets loose at the touch of his finger all the hidden molecular energies, and leaves the javelin, the arrow, the blowpipe of his fathers far behind. In the arts of peace Man is a bungler. There is nothing in Man's industrial machinery but his greed and sloth; his heart is in his weapons.[1]

GEORGE BERNARD SHAW, *Man and Superman*

𝔍𝔣 𝔞 𝔪𝔞𝔫 𝔬𝔣 𝔤𝔬𝔬𝔡-𝔴𝔦𝔩𝔩 𝔩𝔦𝔨𝔢 𝔪𝔢 𝔠𝔬𝔲𝔩𝔡 𝔤𝔢𝔱 𝔥𝔦𝔰 𝔥𝔞𝔫𝔡𝔰 𝔬𝔫 𝔱𝔥𝔦𝔰 𝔫𝔢𝔴 𝔟𝔬𝔪𝔟 . . . 𝔥𝔞! 𝔍'𝔡 𝔰𝔥𝔬𝔴 𝔱𝔥𝔬𝔰𝔢 𝔴𝔥𝔬'𝔡 𝔩𝔦𝔳𝔢 𝔟𝔶 𝔱𝔥𝔢 𝔰𝔴𝔬𝔯𝔡. . . . 𝔍'𝔡 𝔡𝔯𝔬𝔭 𝔱𝔥𝔞𝔱 𝔟𝔬𝔪𝔟 𝔞𝔫𝔡 𝔍'𝔡 𝔣𝔬𝔯𝔠𝔢 𝔭𝔢𝔞𝔠𝔢 𝔯𝔦𝔤𝔥𝔱 𝔡𝔬𝔴𝔫 𝔱𝔥𝔢𝔦𝔯 𝔟𝔩𝔬𝔬𝔡𝔱𝔥𝔦𝔯𝔰𝔱𝔶 𝔱𝔥𝔯𝔬𝔞𝔱𝔰![2]

WALT KELLY, *Pogo*

HUMANS—MALES IN PARTICULAR—CAN PRESENT HAZARDS TO other animals. We represent a more immediate peril than any competitor for food or space. While our nutrition doesn't demand that we eat animals, most human cultures prefer to do so when given the opportunity. Even where farming and commercial fishing supply ample meat, hunting and sport fishing persist as recreational activities. Nor do we represent only interspecific danger. We humans lack a strong taboo against killing other humans to whom we're not related or socially connected. That behavioral quirk (shared with some chimpanzees), in combination with our toolmaking ability and our extensive social organization, puts us far ahead of any other animal in potential for intraspecific iniquity.

Let me hasten to add that I'm neither advocate nor apologist. Biology may saddle us with unfortunate abilities and propensities, but we have sufficient cultural and behavioral wiggle space to offset any atavistic predilections—if we take deliberate steps to do so. The majority of humans now eat little or no meat; indeed they cannot, since our numbers have risen to a level that precludes substantial carnivory. Most developed countries—in sharp contrast with my own—have given up capital punishment. But murder, lethal terrorism, and warfare persist. While Bernard Shaw may have been misanthropic, his view of human creativity can't be dismissed as misplaced misanthropy.

This is the first of two chapters about tools and rules for muscle-powered killing. That's how we most often did it until the advent of chemical explosives six or seven hundred years ago. The devices and techniques that revolution unleashed may economize on power, but first and foremost, they facilitate subduing or killing prey or victim with less direct encounter by hunter or aggressor. Not that weapons need have started as distancers. Some paleontologists suspect that early humans scavenged as much as they hunted, competing with jackals and vultures rather than with lions and wolves. Perhaps hand axes for smashing bones and scrapers for coaxing off loose flesh first aided that competition rather than served as immediate, aggressive weapons.[3] I like the idea—along with anything else that might diminish our human hubris.

However the practice began, muscle-powered weapons have long been critical for humans bent on killing animals. Bare hands can do the job, but ours seem poorly designed for it. Dexterous hands that can be accelerated to high speeds have to be light in weight, with their driving muscle well uparm of their own positions. Such hands alone thus pack little punch in bareknuckle battle with large prey. Our versatile feet do well for walking, running, and climbing, and they can even do some manipulating and swimming, but they can't kick like hooves. Our upright postures may give us good cross-country vision, but in raising our centers of gravity, they deprive us of enough maneuverability to count on catching small animals. However good at running, we're poorly equipped for the kill that follows the chase.

The Methods of Myomaniacal Mayhem

The choices are wide. Bash with blunt objects; poke with sharp ones; throw or shoot off projectiles. Do so as a standing individual; do so while moving forward on a horse; do so as a coordinated team of artillerymen. Throw stones of sufficient size and density to transfer a destructive level of momentum. Or supplement them with material of biological origin, such as wood, sinew, hair, and even bone; the "jawbone of an ass" only exaggerates. Scrape together sufficient metal to capitalize on its still-greater toughness and density. At the outset we need a scheme to impose order on this diversity, doing what the biologist's taxonomic system did for domesticated animals, two chapters back.

So we begin with a taxonomy of muscle-powered weaponry. We first divide weapons by whether or not their use involves storing energy—other than the kinetic energy of weapon and weapon wielder just before launch or strike. Thus we thrust with a sword, adding no more than the kinetic energy of some moving body mass to that of the sword. But we store potential energy in the elastic deformation of a bow, energy that we put in slowly but that comes out in a great rush as the arrow accelerates forward. Then we further divide weapons by whether they're portable, small enough to be carried by an individual on foot or mounted, or nonportable, ones that must be dragged or wheeled into position or else assembled on the spot. Here, then, together with a few comments, is that hierarchical scheme:

I. Weapons that don't accumulate potential energy

A. Portable weapons

Stones, whether wielded directly or thrown. A hand holding a stone can hit with greater effectiveness than a hand alone. For one thing, the increase in mass provided by the stone may more than offset the decrease in velocity it costs the appendage. For another, the more rigid surface of the stone hits with greater force than that of a fist, quite aside from reducing skin abrasion. A thrown stone can do as well or better, even if it has to share momentum at launch with the still-moving arm. Using stones for pounding, hitting, and throwing must antedate the emergence of our species since our immediate nonhuman forebears had forelimbs similar to our own and since chimpanzees do these things.

Spears, including lances, knives, and swords, again whether wielded directly or thrown. For a given speed at impact, the weapon with the more acutely angled front will penetrate further. Only when we try to damage something harder and more brittle than any animal—a stone fortification, for instance—does pointiness lack value. Few naturally-occurring objects have proper points; fewer still combine points with the right mass to maximize the momentum an arm can impart. So humans long ago moved beyond careful selection and began reshaping and de novo fabrication. Old modified stones vastly outnumber old human bones. Even if that reflects in part their greater chance of preservation, it still attests to the antiquity and ubiquity of the practice.

Thrown sticks, such as boomerangs. Wood, especially if dry, is a material of no great density—typically about six times less than stone. A wooden projectile of a given mass will therefore have about six times the volume of a stone one and, if the same shape, 3.3 times the surface area and 1.8 times the diameter. That makes two kinds of trouble. Surface area translates into drag, a tax on its potential range, and greater diameter imposes greater difficulty for handling, at least in objects of sufficient mass for worthwhile throwing. No culture hunts or makes war with hand-heaved croquet balls.

But wood, even if dry, exceeds in toughness almost every mineral; it resists cracking whether in native or fabricated shape. Cracking a stone produces a fairly smooth new surface; cracking a piece of wood gives jagged ends and lots more surface. Making that additional surface demands energy, so cracking wood takes more effort. Stone is harder and denser; wood is tougher and more easily reshaped. That facilitates the use of oddly shaped pieces of wood, especially elongate pieces. Throwing an elongate piece, a stick, has at least two points in its favor. You grasp a stick near one end, not in the middle. So for practical purposes it extends your arm, with its center of gravity farther from your torso than even the longest of your fingers. When swung, as in Figure 12.1, a stick (unless very heavy) can move faster than your hand, and when released, it can therefore go farther than any hand-held ball. The other advantage comes from its size and from any rotational motion; a stick cuts a wider swath. Aimed at a group of birds, a thrown stick has a better chance of hitting one than a thrown stone of the same mass or volume. That rota-

FIGURE 12.1. Throwing a throwing stick. Throwing sticks for hunting didn't return like boomerangs.

tion can also pack a wallop of momentum well beyond that of the stick's forward motion. I once hit myself with a returning boomerang, one poorly thrown and with its momentum almost spent. It still left an impressive bruise.

Launched sticks and spears, including atlatls. You can take this arm-extending trick a step farther, grasping a second stick that has no function other than extending your arm. Spear-throwers—we most often use the native North American word "atlatl" or "atatl," less often the Australian "woomera"—have been around for around twenty thousand years. Without question, these devices, one of which is shown in Figure 12.2, greatly increase the distance a lightweight spear can travel. They enjoy strong advocacy from a group of ardent amateur atlatlists, who back their claims of power and accuracy with tests and competitive matches.[4] I think, though, that we ought not forget that bows and arrows have replaced atlatls almost everywhere that both were used. These spear-throwers persisted mainly among native Australians, who seem not to have known much about archery. Still, the practiced atlatlist can sometimes achieve sufficient distance and accuracy to disarm (at least) skeptics. The main disadvantage of a spear-thrower comes from the energy that's wasted imparting momentum to a part of the system that's not launched preyward. But the wastage needn't be severe since

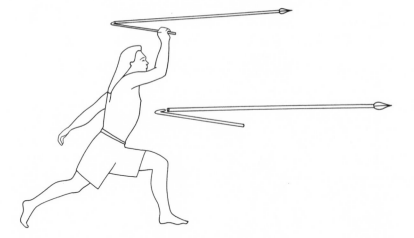

FIGURE 12.2. Throwing a spear with an atlatl. The spear has a concavity in its rear end against which the spear-thrower pushes.

the spear-thrower weighs less than the spear and since its center of gravity lies closer to the hunter than does the center of gravity of the spear.

One omission strikes me as odd. We've thrown spheres and we've launched spears with spear-throwers. Where are the arm extenders that would give spheres, most likely stones, some extra velocity? What I mean are devices analogous to lacrosse sticks. We learned the sport of lacrosse from native North Americans, but I find no reports that they used an equivalent item of weaponry.

Slings. The agency of Goliath's demise hasn't played a large role as a muscle-powered weapon. Slings do work, and like atlatls, they have their advocates and aficionados.[5] Again like atlatls, they work as arm extenders, but slings (Figure 12.3) take only tension, so they have to be whirled, and they stay extended only by the centrifugal effect. On the other hand, an adequately strong tensile element can be lighter than one that must withstand compression, twisting, or bending, so a sling imposes merely a slight penalty for its own weight. Trebuchets, about which more shortly, used very large slings but took a bad weight penalty elsewhere.

Blowguns. No other weapon makes analogous use of the muscles of the torso. The trick with a blowgun (Figure 12.4) consists of expelling a large amount of air from the lungs as forcefully as possible. You use the

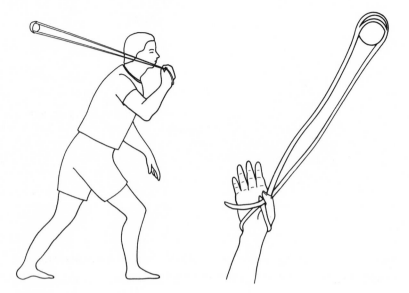

FIGURE 12.3. A sling in operation. Releasing one string sends the missile on its way.

muscles between the ribs (the intercostals), the muscular diaphragm that separates the lung cavity from the abdominal cavity, and the various muscles of the belly's wall, such as the rectus abdominus. An effective blowgun must be long because the projectile picks up speed only while it moves within the tube. What limits length is the volume of exhaled air, in practice a little less than maximum exhaled volume, what physiologists call the vital capacity of the lungs. Thus the fatter gun must be shorter, with an inverse relationship between a blowgun's cross-sectional area and its length. That limits the size of the projectile. The blowgun's capacity suffers a further limitation because maximum expiratory pressure drops as the speed of exhalation increases. Accounts of blowguns often refer to them as local weapons, then note that "local" encompasses the whole Amazon Basin and a large number of other cultures elsewhere.

Battleaxes. For some reason, these play only a small role in our myths and images, but Beowulf and his kinfolk swung them with memorable effect. By and large, they're heavy, cumbersome hand weapons that gain their advantage from having their centers of gravity well outboard of the

FIGURE 12.4. A blowgun in operation. A dart is shown in inset, with the fluff of plant material on the rear that provides a seal in the blowgun and stabilizes the dart in flight.

ax wielders' arms. A well-struck battleaxe must have been potent, but it couldn't have matched a sword for agility. Perhaps a battleaxe does best when aided by gravity, as when a mounted hunter or a cavalryman does battle with a prey or foe beneath his level—like a polo mallet or even a hockey stick. In such use it blurs our distinction based on energy storage, but the warrior in battle wasn't worrying about that.

B. Nonportable weapons

Battering rams. These too blur that distinction just a bit. If it hits with even modest velocity, a large mass can damage the brittle masonry of fortifications. It might be moved forward by a group of running humans, or it might be swung forward as a pendulum where the additional machinery can be brought up to the fortification. So we're aggregating devices from the simplest pole with reinforced end to elaborate siege engines. Besides combining high mass and low speed, all depend on a coordinated group effort under difficult circumstances, defenders dropping varieties of unpleasantness from above. The obvious countermeasure consists of reducing the brittleness of walls, using either an outer layer of thick wood or some berm of earth. But flammability limits the first, and the difficulty of making an unclimbable vertical wall does the same for the second.

II. Weapons that store up energy before firing

A. Portable weapons

Bow and arrow. The origins of archery lie, as one says, shrouded in the mists of antiquity. Bows and arrows found use almost everywhere that humans penetrated; Australia and a few places where no appropriate materials were available provide rare exceptions. So simple is the basic device that differences are as likely to reflect local tradition as independent origin. Thus we can't be sure where, when, or how many times archery was invented. The weapons themselves may not survive, but skeletons of large animals with embedded arrowheads and deposits of arrowheads alone attest to their wide use. Arrows and arrowheads weigh so much less than spears and stone spearpoints that telling the difference involves little uncertainty.

What makes archery so good is the bow's ability first to store the energy put in as the archer slowly bends it and then to release the energy rapidly when the archer lets go. The bow needs to be made of a material (or a combination of materials) that can first achieve a high level of elastic energy storage relative to its weight and then shift a large part of that potential energy to the kinetic energy of the arrow. In the normal jargon, it needs high values of both strain energy storage and resilience. Woods vary in these properties: In northern Europe, yew does especially well; in North America, ash and osage orange make the best bows. Bending a bow stretches the side facing the target and compresses the side facing the bowstring. Various ancients recognized that better bows could be built if several materials were combined. That meant putting a material, such as sinew and tendon, that does especially well in tension on the target side and a material, such as horn, that performs best in compression on the bowstring side. Of course that gets the craftsman into the sticky business of making dissimilar materials adhere even when severely stressed.

Crossbow and gastraphetes. In conventional archery, the bow is drawn while held in a position close to that of its final aiming. Simultaneous drawing and aiming may improve the rate at which it can be fired, at least a little, but it doesn't maximize the working mass of the archer's muscle. Nor does it allow much adjustment of the rate at which energy gets put in; you can't draw a much stiffer bow by drawing much more slowly. Two different muscle-powered individual weapons, weapons of deceptively similar appearance, addressed this limitation.

Of the two, the crossbow diverges less from conventional bows, although it came into widespread use later. It consists of a small bow attached to a crosspiece, with a slider that runs along the crosspiece parallel to an arrow's position, as on the left in Figure 12.5. What's special is that the archer pulls the slider back with some winching, cranking, or levering arrangement until the slider engages a catch. The archer next inserts the projectile (the bolt) in front of the slider's catch, takes aim, and then springs the catch with a trigger to send the projectile on its way. For their size, crossbows develop great power. In addition, they can be kept ready to fire without continuous muscular input, and they can be used, if a little awkwardly, on horseback. Their low rate of firing suggests greater utility for hunting than warfare; Coleridge's ancient mariner dispatched the fateful albatross with one. Nonetheless, they enjoyed a long period of military favor because their bolts could penetrate shields and body armor.

The gastraphetes of Figure 12.5—the name means "belly bow" and

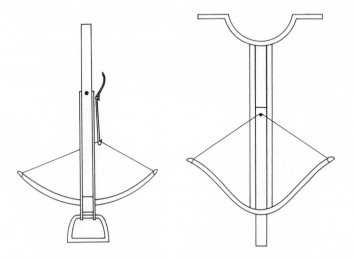

FIGURE 12.5. Crossbow (left) and gastraphetes (right). The crossbow's operator put a foot in the loop to brace the bottom of the crossbow while pulling up on the drawstring. The gastrophetes' operator put the upper end of the weapon against the belly and pushed downward to bend the bow.

refers to the way it was braced against the operator's abdomen as the bow was bent—shares the ability to be drawn, cocked, and then held in readiness. Where it differs is in the way it bends its bow. Instead of drawing back the string and the ends of the bow, the operator pushes the center of the bow forward. That's done by bracing a curved crosspiece on the rear end of the stock against the belly and pushing forward with both hands on the center of the bow. The name of the weapon reflects its Greek origin, probably at Syracuse (but not by Archimedes), on Sicily.

B. Nonportable weapons

(The term "catapult" has been applied to each of the following weapons, often to distinguish it from the other. So I think it best to use the name as a generic for both—that is, for any large, muscle-powered engine that throws heavy projectiles.)

Ballistae. These must constitute the height of technical sophistication among the weapons of the ancient Greeks, Romans, and other disputatious Mediterranean civilizations; Figure 12.6 shows a reconstruction. What we know of ballistae comes mainly from a construction and operation manual by a Roman engineer, Vitruvius. He assumes, though, various bits of commonly understood detail, and modern analysts have had to guess (and argue about) the details.

Ballistae store energy in torsion, the way a model airplane with a rub-

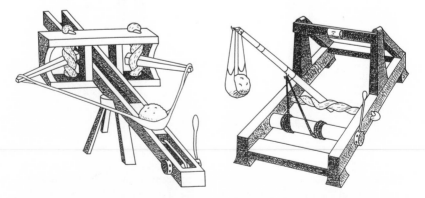

FIGURE 12.6. Ballista (left) and onager (right), as they're usually shown. I think each should have handspikes on both ends of its capstan (winder), and the handspikes should be longer.

ber band–powered propeller stores up the input for its flight. But it does so with a much less stretchy material. Rotating rearward (like the ends of a bow), each arm twists a bundle of tendon. Upon release, the bundle untwists forcefully, swings the arms forward, and sends a projectile on its way. Tendon may sound like a peculiar choice for an energy-storing material—unless we recall some odd details from previous chapters. We use our tendons to store energy from stride to stride when we run, as do kangaroos when they bound. So nature has worried about the matter, and tendon can store a remarkable amount of potential energy, an amount greater (relative to mass) than spring steel. At the same time, its resilience exceeds 90 percent, meaning that it returns in mechanical form nine-tenths of the energy that stretched it. By contrast, wood comes with no such biological imperative. Trees of excessive resilience might sway dangerously in winds. It's probably just fortuitous that some dry woods (different stuff from living wood anyway) have adequate resilience to serve as bows. But tendon is the Achilles' heel as well as the key element of the ballista; it's all too perishable, and replacing its bundle takes time.

Two artillerymen probably winched back the receptacle of the projectile, putting in the energy that triggering would subsequently release. It's hard to see how more than two men, one on each side, could have pulled on the radial spikes of the winch, although groups might have taken turns and thereby shared the work of turning. A predecessor (also in Figure 12.6) used a single swinging arm that twisted a single bundle of tendons. Its name, onager, alludes to an equid, now rare, that was infamous for kicking.[6] Replacing its massive arm with the relatively light pair of arms of the ballista got rid of most of that postshot kick, both a hazard to the machine and a dissipation of energy that might have been put to better use.

The largest of the ballistae could, we think, send a ninety-pound (forty-kilogram) stone about 400 yards.[7] Stones of that size have been recovered in great numbers from sites of ancient battles, such as at Carthage. Since archers of the time could shoot 350 yards, the longer range kept the artillerymen safely distant. Still longer range, with necessarily lighter projectiles, would not have held much advantage. As we'll see when we look into the rules for shooting projectiles, going for greater distance would have degraded performance in other ways as well.

A little more about that bundle of tendons. Tendon may be high in strain energy storage and in resilience, but it doesn't stretch by much. If it did, the contraction of a muscle might stretch it more and move bones less. You can't stretch tendon the way you stretch the rubber band that powers a toy airplane. The long swing of the pair of arms can't be asked to stretch tendon by more than, at most, 10 percent, or the tendon would break. Twisting a bundle does the trick, applying lots of force to each bit of tendon, but not extending it by much. Twisting a loop of almost inextensible cord shouldn't be undervalued. You might at some time want to reglue the crosspieces beneath the seat of a wooden chair. Just apply the glue, fit the pieces, loop a cord between the chair's legs, stick a crosspiece through the loop, and twist it. You can apply great pressure without marring the legs, and you don't have to buy an expensive furniture clamp. You can even pull your SUV out of a ditch using the same device if you happen to have a strong rope and if the world cooperates by providing a pole or tree in the right place. Or if you've a streak of fanaticism, you can build a ballista; nylon rope makes an adequate substitute for tendon.

Trebuchets. The classic version of these great medieval siege engines, one of which is shown in Figure 12.7, stored energy gravitationally rather than elastically; a trebuchet worked like a walker rather than like a runner. Pulling the projectile's long arm downward raised a weight. Triggering let the weight descend, which swung the arm upward and forward. At the end of the arm, a long sling slung the load farther upward and around and released it (ideally at least) at the right angle. People as respected as Lynn White[8] have claimed that the performance of trebuchets surpassed that of ballistae, but professional chauvinism afflicts even medievalists. Too much differs between the two for some single judgment to mean much. In particular, trebuchets were much larger. It took many men to pull their arms down, and the weapons could throw projectiles at least as big as a dead horse. A trebuchet could be used, as a ballista could not, to toss a corpse that carried some infectious disease over the walls and into a besieged fortification. No allowance needed to be made for a resilience of less than 100 percent; with gravity you get as much potential energy out as you put in. No longer was operation dependent on a scarce and perishable component.

But severe disabilities counterbalanced the trebuchet's advantages.

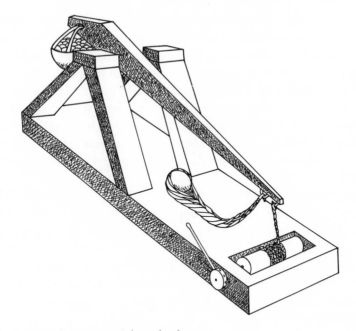

FIGURE 12.7. A counterweight trebuchet.

Most of the energy of the descending weight went into accelerating not the projectile but the weight itself and the throwing arm, which had to be a massive thing if it was to last for more than a single shot. Not only did it have to withstand the load imposed by the projectile, but after discharge it had to be stopped without destroying itself or any other part of the machine. The slider that held the projectile on a ballista put only tensile loading on the rest of the weapon (except for the two swinging arms), and tensile loads can be taken by light cord or cable. By contrast, both the projectile and the counterweight bend the arms of a trebuchet, and bending imposes a much more demanding kind of load; for the same load, a protruding beam has to be much thicker than a cable. This limitation may have set the maximum range of trebuchets, apparently a lot shorter than that of ballistae. Maximum range depends on the speed of launch, and that can be no greater than the top speed of the sling on the end of the throwing arm. Moreover, the faster the throwing arm, the harder it resists stopping. Even with its lower launch speed, the kick of a trebuchet must have been tremendous.

In this world of high-tech microelectronics, a huge but simple machine exerts a certain thumb-your-nose-at-modernity fascination—not to mention the peculiar pleasure of throwing large items long distances. So people build and even compete with trebuchets.[9] According to several reports, one Hew Kennedy, in England, has built a trebuchet that gives great satisfaction as a piano pitcher. Maybe, though, one should start on a smaller scale, throwing typewriters and mechanical calculators, both of nice mechanical complexity and low market value. Or old computers, restoring the original meaning to the neologistic "crash."

An earlier version of the trebuchet stored no energy but instead depended on a large group of people who gave a synchronized pull on ropes hanging from the end of the beam opposite the sling and load. Pulling on the ropes sent the missile on its way. This traction trebuchet originated in China and saw its first Western use by the Arabs at the end of the seventh century. The classic counterweight trebuchet, the one that gets all the attention, was Mediterranean in origin and dates from the twelfth century.[10]

The Heavy Hand of Physics

We've met the players; what are the aims of the game?

- You want to maximize the chance of hitting the prey, which means being able to aim effectively; alternatively you want a rate of hitting or firing that's sufficient to offset poor aim.
- You want your weapon or projectile to make enough of an impact when it hits to kill or incapacitate the intended prey.
- You want to minimize risk to yourself by smiting the prey from the greatest possible distance consistent with a decent chance of hitting and with adequate impact.

These interacting objectives demand compromises, the nature of which depends on the interplay of the underlying physical variables, on the rules of the game.

SPEED, SIZE, AND RANGE

As we're advised by state highway patrols, speed kills. As we're further informed, stopping distance increases disproportionately with speed. Nothing matters more for muscle-powered weaponry than these rules—and not just as analogies. Let's begin with a look at how that second warning applies to projectiles.

To simplify things, imagine throwing spherical projectiles at the upward angle that enables them to reach the greatest possible distance. To simplify further, imagine throwing them in a world in which their drag can be ignored, one in which gravity imposes the only external force that matters. In such a world, the upward angle that gives the greatest range remains at forty-five degrees for any projectile size or launch speed. Neither the shape nor the density of the projectile matters at all. We throw these projectiles at different speeds (in practice on a calculator or computer), and we watch how far horizontally each goes (on the calculator or computer). Here, in tabular form, are those ranges:

Launch Speed		Maximum Range	
20 mph	8.9 m/s	27 ft	8.1 m
40 mph	17.9 m/s	107 ft	32.7 m
60 mph	26.8 m/s	240 ft	73.3 m
80 mph	35.8 m/s	429 ft	130.8 m
100 mph	44.7 m/s	669 ft	203.9 m

Notice that doubling the speed, from twenty to forty or from forty to eighty miles per hour, quadruples your range. Faster means much farther, just as when stopping a car. Only the source of force differs: the gravitational pull of the earth instead of the friction of the brakes. An animal might throw projectiles at such speeds; the fastest corresponds to that of a baseball thrown by a fastball pitcher and remains below the initial speed of a well-driven golf ball, both muscle-powered activities.

Launch speed (often called muzzle speed, the term redolent of the smell of chemical explosives) needs a word of explanation. A projectile never exceeds the speed at which it leaves arm, bow, or gun's muzzle.[11] So all the acceleration that gets it up to speed has to happen before release. As Newton established with his second law of motion, force equals mass times acceleration. If you have a fixed amount of force to work with, you can't give a heavier projectile as great an acceleration as you can a light one. And vice versa. But acceleration is not speed, however we confuse "fast" cars with those of notable acceleration. Speed equals acceleration times time. If you can prolong the period of acceleration, you can achieve a higher launch speed and a greater range. The gun with the longer barrel, other things being equal, shoots its projectile farther, as noted earlier for blowguns. Handguns use large bullets to compensate for their short barrels. The subject may sound familiar; it came up earlier when I talked about how small and therefore short-legged animals had to achieve high accelerations if they were to be good jumpers. With equal forces at full draw, the bow whose arrow gets drawn back farther will shoot it the greater distance.

When we put drag into the mix, nothing remains so neat and tidy. Drag (whether in air or water) brings in fluid mechanics, a subject of such indecent messiness that physicists mostly leave it for engineers, practical people who can't just sweep complications under the rug. In a draggy world, shape matters. Airplanes but not spacecraft should be streamlined, and the launch angle that gives the longest trajectory drops below the textbook forty-five degrees. Size matters as well. The small projectile has more drag-experiencing surface relative to its volume or mass than does the large one. For many years I've taught biological fluid mechanics, so I've had to contend with the behavior of projectiles as small as jumping fleas and explosively ejected fungal spore clusters. For class use, I set up a computer program that makes reasonable approximations of drag and then computes trajectories. Its numbers put the issue in quantitative terms.[12]

Consider a set of spherical projectiles differing in size. We'll assume that these are three times as dense as water—a reasonable value for rocks that might be thrown or shot off. We'll assume, further, that all are launched at 56 miles per hour, or 90 kilometers per hour, or 25 meters per second. For each projectile, we ask the computer to figure the maximum range and the launch angle (above horizontal) that produces that maximum range. How can we view the effect of drag? We might look at drag as a tax on maximum range (here 209 feet, or 63.6 meters), a "drag tax" expressed as the percentage of its dragless range that a projectile can reach in a sea-level atmosphere. For such projectiles, again in tabular form:

Diameter	Mass	Launch Angle	Range	Drag Tax
200.00 mm	12.6 kg	42°	62.8 m	1.3 %
63.00 mm	0.393 kg	41°	57.4 m	9.8 %
20.00 mm	0.0126 kg	40°	46.9 m	26.3 %
6.30 mm	0.393 g	38°	32.8 m	48.4 %
2.00 mm	0.0126 g	35°	17.1 m	73.1 %
0.63 mm	0.393 mg	30°	5.9 m	90.8 %
0.20 mm	0.0126 mg	17°	1.5 m	97.6 %

(The metric system works best for this geometric series, in which each line considers a stone a little less than a third of the diameter of the one above. To put the masses into pounds, multiply each by 2.2; to put the ranges into feet, multiply each by 3.3.)

A rock 8 inches across (200 mm) that weighs 28 pounds suffers a negligible 1.3 percent loss of range to drag, and even a small rock, one of 2.5 inches (63 mm) that weighs a bit less than a pound, still loses only around 10 percent of its range. Also, drag bothers such spherical projectiles many times more than it would if they were streamlined; for drag, one can't do much worse than a sphere. But as we go smaller yet, things deteriorate. A quarter-inch (6.3 mm) pebble goes only half as far in a world with drag as in one without; David must have stood close to Goliath. A barely visible sand grain of a fifth of a millimeter drops after launch, I might say, like a stone. Drag matters more than does gravity, so speed after launch diminishes rapidly. The projectile has to get its distance while it still has some speed, so low launch angles work best. Moreover, for these small spheres (for reasons we won't get into), shape makes little difference, so streamlining can't help much.

Thus size buys range and (as we'll come to) impact. We're big enough creatures to heave rocks without paying too much for their drag. But we're just big enough. Quite a few animals, creatures as diverse as ant lions (insects that dig ant-trapping pits) and rabbits, throw projectiles in connection with rapid digging. One caterpillar throws its lumps of excrement (frass) to avoid alerting predators to its presence.[13] Besides us, though, few use projectiles as weapons; the great apes are about the size of it. Not, I'd say, a credit to primate intelligence; it's more a combination of size and requisite dexterity. Raccoons certainly have the dexterity, but they just as certainly lack the size. Ants carry sand grains, and I can imagine an antlike insect with some elastic storage system analogous to that of a jumping flea. But such an insect's projectiles would be in an 80 or 90 percent tax bracket, not in our 1 to 10 percent range.

You may catch the odor of a problem within this problem. If an animal with a fixed amount of energy wants to get some distance, it might trade size of projectile against speed of launch. Smaller size, higher launch speed. But for spherical rocks going around fifty miles per hour and below an inch or so in diameter, smaller size gives shorter range even at the same launch speed. So some of the increase in range from an increased launch speed can't be realized. Further down in size, any gain in launch speed from smaller size gets more than dragged down again by the atmospheric tax collector.

We noted another advantage of speed, killing power, but this one takes a bit more of an explanation. We'll first have to back off the immediate issue and develop some additional points of mechanics.

MASSY LAUNCHERS

When you send a projectile on its way, how much effort do you waste setting the rest of your weapon in motion? Throw a spear, and the whole thing goes into action. The only wastage comes from the moving mass of your arm, but that moving mass does still matter. Use a spear-thrower, and the mass and speed of the moving spear-thrower constitute a tax on the work you've done and the momentum you've generated, but the spear now moves much faster than your arm. A good battering ram moves little wasted mass, only that of the mass of the ropes that hold it suspended as a pendulum. A blowgun wastes only the trivial mass of some moving air. A bow weighs more than the arrow, suggesting serious inefficiency, but the bow doesn't move as fast as the arrow, so its momentum stays modest. The ballista shares this advantage, but the otherwise similar onager, with its solid throwing arm, doesn't; the kick for which it's named reflects that waste. The trebuchet looks terrible, with those great swinging arms of mass and speed comparable to that of its projectile, not to mention the momentum of the descending weight. The direction of motion of the various parts of the weapon doesn't make any difference: One arm of the trebuchet goes upward and forward; the other goes downward and backward. Bow staves move forward; guns recoil backward. Whatever the direction, energy goes to waste putting mass into unnecessary motion. You do best if your muscles impart motion to the business end of the weapon, the projectile, and to nothing else.

We need some measure of this second tax. The ordinary ratio of projectile mass to total moving mass won't quite do the job because it makes no allowance for the speed of the rest of the weapon; it would make no allowance for leverage, the speed and distance advantage that the bow gives to the arrow. A kind of fake mass simplifies the job; it's the mass that would impose the same tax if it traveled at the same speed as the projectile. Since in none of these weapons does any part go faster than the projectile, this "virtual mass" never exceeds the actual moving mass of the rest of the weapon. Equipped with this new variable, we can use as a measure of the weapon's mass efficiency the ratio of projectile mass to total effective mass, projectile mass divided by the sum of projectile and virtual masses. A perfectly efficient weapon would be one in which no mass except that of the projectile was put into motion. Or we can put the problem in terms of a "mass tax," the percentage loss relative to the same weapon with zero virtual mass. (This talk of masses has an odor of Marxmanship.)

How do the various weapons compare in terms of this mass tax? A little data can be gleaned from published reports; the table below combines them with the guesswork necessary to put all the data into comparable form.[14]

	Mass Taxes
Throwing a 0.6 kg spear from the hand	40%
Throwing the same with a spear-thrower	17%
Roman ballista	30%
Trebuchet	90%
Bow and arrow	50%
Blowgun	less than 14%

One might have expected better of the bow and arrow since the bow moves very little when launching its arrow. The virtual mass of a long bow runs around 5 percent of its actual mass—impressively low. That's offset by the fact that the bow, at perhaps a pound and a half, far exceeds the weight of the arrow, running around an ounce.[15] Values for bows and arrows range from about 30 to 60 percent, with impractically massive arrows suffering the lowest tax rates. For hand-thrown projectiles, the lighter the projectile, the worse the tax, since the throwing arm represents an unchanging mass.

This business of mass efficiency rationalizes some otherwise peculiar phenomena. An efficient weapon transfers to the projectile almost all the kinetic energy put in by its operator. What if no projectile gets launched? That can spell trouble. Beginning archers are urged not to release the string of a bow without shooting an arrow; unless the string is slowly moved forward and the energy absorbed by the operator's muscle, the bow may break. Not that a slow forward movement by the archer should present any problem; recall that we exert more force and achieve greater efficiency when we ask our muscles to produce tension while lengthening. I suspect that unloaded ballistae imposed the same risk, but with their greater mass efficiencies, trebuchets should have faced less of a problem. Still, whatever stopped the movement of a trebuchet's arm would have felt some additional load if no projectile had been launched. Closer to the here and now, you can strain your decelerative muscles when going through the motions of throwing but not launching any projectile. Beyond any sense of accomplishment, you feel better when you hit a baseball than when you swing and miss.

But you can swing a baseball bat or a golf club with impunity, so greatly do their masses tax us. For that matter, the rules of baseball (or cricket) assume that more times than not the batter will fail to hit the ball. Unlike, say, lacrosse sticks, bats must be massive since they persuade their projectiles to move by hitting them; you need a good supply of momentum to transfer in the collision between bat and ball. A baseball bat outweighs a baseball sixfold, and most of the bat goes about as fast as the ball; that translates into a mass tax of around 85 percent.[16] The driver weighs seven or eight times as much as a golf ball, regulated at 1.62 ounces; allowing for some weight in the handle rather than the head puts the tax at about 70 percent. To feel the importance of keeping the moving mass of weapon far above the projectile's mass, try hitting a tennis ball with a badminton racket. These high mass taxes explain why no impact device analogous to baseball bat, golf club, or tennis racket makes our list of historically significant weapons.

ENERGY AND MOMENTUM

We begin with a paradox. Among the chief variables up to this point have been force, work, energy, and power. Muscle uses energy to produce force and do work, and its work can be viewed as its energy output. Power expresses the rate at which it works: energy divided by time, as either power input or output. To these we now add a most important variable for mechanical problems, something called momentum that we've used earlier only in a loose, metaphorical sense. Momentum and energy, specifically kinetic energy, combine the same two variables, mass and velocity, but they do so in different ways.[17] On that paradoxical difference hangs a great weight of significance for weaponry.

Kinetic energy, energy by virtue of motion, combines mass and the square of speed: one-half times the moving mass times its speed squared. Double the speed, and you increase the kinetic energy fourfold. To state it another way, if you can put no more than a fixed amount of energy into something, doubling its speed requires that you drop its mass fully fourfold. On the other hand, if that something is a projectile, doubling its speed increases its range fourfold; as we've seen, the range of an unpowered missile, if we assume insignificant drag and a shot at the best upward angle, varies with the square of launch speed. So for a fixed input of energy, the range of an unpowered missile increases by the same factor as its mass decreases.

Momentum, though, combines mass and velocity as simple multiples with no square of anything; it's what Newton called the quantity of motion. Consider, again, what happens when for a given amount of energy, speed is doubled. Twice the speed and a fourfold reduction in mass mean a twofold reduction in momentum. At the same kinetic energy, going twice as fast means going with half the momentum. Why should that matter? Simply because the effectiveness of a missile in its final collision with prey or target depends on its momentum rather than on its kinetic energy. In more intuitive terms, hitting with half the momentum means hitting half as hard. Decreasing mass buys greater range, but at a price in the momentum of impact.

Why should momentum, not energy, be so momentous? In any

collision, overall momentum does not change, so the combined momentum of projectile plus target after impact equals the sum of their momenta before impact. The greater the momentum of the projectile, the greater is the change in momentum it imposes on the target. The greater that change of momentum, the more damage inflicted. The big, slow projectile makes better use of the same energy input than the small, fast one because it hits with greater momentum. The lower-impedance trebuchet (to use a term raised in the last chapter) looks better than the higher-impedance ballista. In a sense, you can't win. While slower movement does more damage, it achieves a shorter range. For efficient conversion of kinetic energy to momentum, nothing looks better than battleaxes and battering rams; with zero range, you can raise mass until you reach some other limit. In other words, the momentum you achieve relative to the energy you put in, yet another kind of efficiency, goes up just as speed goes down. (From this viewpoint the ideal speed is zero. That, though, puts both momentum and kinetic energy at zero and achieves total impracticality. We shall ignore this last paradox since the weapon won't ever reach the target anyway.)

Encapsulation: If a fixed amount of energy gets invested, then range trades off against momentum with a vengeance. Increasing speed by a factor of two so as to go four times as far requires that projectile mass be dropped by a factor of four. As a result, momentum drops by half. A table will help clarify all this jockeying of interrelated variables; for simplicity, let's ignore the units and put things in relative terms.

Kinetic Energy	Speed	Max. Range	Mass	Momentum
1	1	1	4	2
1	2	4	1	1

A look at momentum and its conservation in any collision makes sense of something that will reappear in the next chapter. Imagine charging forward on your horse, long lance pointed ahead, and making contact with another rider. The collision transfers much of the momentum of your lance and you to the opponent; since your mass is unchanged, your velocity must decrease. Unless your horse slows abruptly at just that instant, you are ignominiously pushed off its wrong end. Unless, that is, your feet fit into stirrups and you're leaning forward. In that case you can transfer the force of impact to the horse, which can pass it along to the earth. To put it in terms of momentum, to your mass in the collision the horse's can now be added, so you have far more momentum to contribute to the impact. At the same time, you have a massive horse with which to share the change in speed on impact, so it changes less; more mass means less speed change. With saddle and stirrups, rider and horse become a unit, and being horsed does more than provide a high and fast, if precarious, perch. Saddle and stirrups put a rider a leg up on the opposition.

IMPACT

Enough about a projectile's launch and trajectory. Neither matters as much as the impact at the end of its flight. Here again some curious but ultimately enlightening bits of physics come into play. We speak of the force of impact, but how does momentum in flight translate into force at impact? Momentum turns out to correspond not to force but to something closely related, a variable that hasn't made much impression outside books on physics: impulse. Impulse, to simplify just a bit, is force times time, the time over which the impact occurs. Nothing happens instantaneously, but a rock contacts a masonry fortification for less time than does an arrow penetrating an animal. For a given momentum, less time in contact means a more forceful contact. In warfare, peak force against walls may count for a lot, so round rocks and cannonballs have an advantage. In disabling prey, the higher force from a shorter contact time matters less than penetration and internal injury, so pointy projectiles have long been dominant. No absolute distinction can be drawn, but the general difference between weapons for warfare and weapons for hunting must trace to the contrast between the large and brittle artifacts of human culture and the smaller and softer creations of nature.

Brittleness? A well-struck wall crumbles. Only a few craniums, turtle shells, and the like are at all prone to cracking, and these turn out to be less brittle than they look. Size? The smaller the creature, the less it cares about acceleration. Force, again, is mass times acceleration. If your mass is low, suffering an injurious force takes more acceleration. A rock that hits a squirrel (unless it squashes it against a tree or another rock) just sweeps it along with little injury. It transfers little of its momentum to the squirrel, so the impulse (force times time, again) isn't great. We hunt small prey best with small, fast, sharp-pointed projectiles, which will penetrate rather than move the prey—arrows, bullets, and so forth—counting on fatal internal injury rather than on simple smashing.

We might look with the eyes of prey rather than predator, target rather than aggressor, and ask how to minimize a blow. A hard shell or cranium provides good resistance to penetration, but it leaves one vulnerable to the blunt object, mainstay of murder mysteries. Hard armor affords a sense of security, even if often illusory, so we've dressed in it, carried it as shields, and draped our horses with it. Every so often, though, a better idea surfaces. The boxer rolls with the punch, minimizing the force of its impulse by maximizing contact time. The catcher holds out a soft glove and lets it come backward as the ball enters, repeatedly stopping balls that exceed ninety miles per hour, balls that would cause severe injury if they hit the head or torso. Effective armor, even if it doesn't sound like armor, has been made from basketry and from layered feathers. The Spaniards in Florida found padded cloth more effective than chain mail in stopping Indian arrows.[18] Earthen battlements often afford better protection than stone walls, as found out by both sides in the American Civil War, Stonewall Jackson's image notwithstanding. The notion tells us something about armored organisms such as armadillos, hard beetles, turtles, and extinct glyptodonts: They must be more vulnerable to sharp-toothed or sharp-clawed predators, if less so to kickers or punchers.

CHAPTER 13

Wielding the Weapons

From 1918 to 1924 I was badly schooled in a California military academy that operated at the technological level of the Spanish-American War. I learned to ride bareback and have detested horses ever since. My enthusiasm for the stirrup was confirmed by the more advanced stages of cavalry training. Since the spear was never widely used in North American armies, I am no lancer. I am, however, probably the only living American medievalist who has ever taken part in a charge at full gallop by a line of cavalry with sabers bared. We yelled like Comanches less to terrify the hypothetical foe than to encourage ourselves in the face of the possibility that a horse might stumble. Our stirrups were a notable consolation. Those who doubt that the coming of the stirrup opened new possibilities in mounted warfare are invited to ride stirrupless in strenuous cavalry maneuvers.

LYNN WHITE, medievalist

Powering the Weapons

Bare hands may empower us as manipulators, but for physical power they need help. So we pound with stones, chop with axes, and stab with big knives. Effective weaponry more often than not depends on energy and power and only indirectly on force. But we can't forget that we're dealing with interdependent variables.

Force times speed equals power. If our power output were fixed, then achieving enough speed for a projectile to go the distance would set the force we'd have to exert. But muscular systems don't work that way; the best output power comes at a particular combination of contractile force and shortening speed. These in turn depend on a host of other factors, mostly connected with the nature of the load. In almost every case, the unaugmented arm and torso produce too much force and too little speed to kill large prey with projectiles while you keep your distance from teeth, claws, horns, and hooves.

How can we get more speed at less force while maintaining power output? For simple weapons that store no energy, the easiest solution consists of extending our arms, what slings and spear-throwers accomplish. Onto the physiological complexities we now stack layers of external complications: How much does reaching sufficient speed compromise output? How much does greater speed compromise accuracy? How can a projectile of minimal mass distribute that mass to give it enough strength and stiffness? How can the mass and structure of the overall device be matched to the mass and shape of the projectile it throws? No mere matter, the design of a near-optimal sling or spear-thrower. The difficulty of quantitative analysis should impress us with the empirical achievements of our ancestors.

Still, those ancestors managed slings and spear-throwers as much by skill in operation as by either sophisticated design or clever use of physics. Moreover, in almost every instance weapons that could store up energy ultimately displaced nonstorage types. Our power output, if not quite fixed, still remains severely constrained, and energy storage amounts to power amplification. Power, once again, is energy per unit of time. Even a low level of power going in accumulates lots of energy over time; that energy can burst out at a high power level during launch. That's a fine thing when a small group of hunters shoot arrows at boar or bison—as long as enough arrows can be loosed in the brief encounter so the prey falls rather than hastens out of harm's way before a second volley.

By contrast, warfare most often depends on the repeated operation of a weapon, with its maximum repetition rate a key criterion of its martial quality. Where muscle provides every watt of power, the limitations of muscle power constrain the whole system. Looking at what muscle can do gives us an important, and neglected, perspective on old weapons of war.

Consider a large ballista, a device of magnificent range and efficiency, that shoots 90-pound (40-kilogram) stones 400 meters. As noted in the last chapter, that maximum range depends only on launch speed, which must have been about 140 miles per hour (63 meters per second). We get the kinetic energy of the projectile by multiplying its launch speed (squared) by half its mass—almost 80,000 joules per shot. If we assume a mass tax of 30 percent, a loss caused by the imperfect resilience of cow tendon of 10 percent, and miscellaneous other losses (friction, wasted motion) of perhaps 10 percent, the artillerymen will have to put in about twice as much energy as comes out as the projectile's kinetic energy. That's around 160,000 joules per shot. We should also assume splendidly fit artillerymen since we're told that they were given special training and food. Maybe they could have sustained outputs of 200 watts each. How often could two artillerymen manage to crank back the projectile in its holder? Watts are just joules per second, so we divide 160,000 joules by 400 joules per second for the two artillerymen; as a result, we get 400 seconds. That's 6⅔ minutes.

A pair of athletic artillerymen could thus get off a shot about every seven minutes. From the descriptions of ballistae, simultaneous use of more than two people seems unlikely. Pairs may have alternately loaded projectiles in front and cranked at the rear, but that would have sped things up only a little. Power—maximum human power output as given by the Wilkie curve—may have limited the effectiveness of ballistae as severely as did the supply of cow tendon. While ninety-pound projectiles could do only a little damage to fortifications, we can see why neither Greeks nor Romans built the larger ballistae described (hypothetically) by Vitruvius.

Trebuchets, for all their grotesque size and inefficiency, circumvented the ballista's power limitation by permitting an almost unlimited number of artillerymen. Without knowing (even roughly) how many persons operated a given model, we can't do an analogous calculation. But we can still recognize a problem of power. As mentioned in the last chapter, the original traction trebuchets depended not on a counterweight but on groups of men pulling down at a signal on ropes from the beam. These could throw 130 pounds (60 kilograms) a vulnerably short ninety meters.

How does that stack up as a human-powered task? If we assume the proper 45-degree pitch and no drag, it required a launch speed of 68 miles per hour (30 meters per second). For a 130-pound load, that's a kinetic energy of 27,000 joules ($\frac{1}{2}$ x 60 x 30^2). If each person pulled downward for 3 feet (almost 1 meter) with a force of 50 pounds (223 newtons)—problems of coordination and location probably precluded greater force—each supplied 220 joules. Provided there was perfect efficiency, the weapon still required more than 120 artillerymen; in practice perhaps 200 were needed. After all, few of the artillerymen would have pulled without impediment and in the optimal direction. On the one hand, a traction trebuchet could accommodate big groups of people; on the other, it had to use large numbers to achieve its combination of load and distance. The Muslims at Sind (in India) in 708 C.E. reportedly used 500 people to operate such a weapon. The bottom line should come as no surprise: Humans can't heave big rocks decent distances without energy storage, but at least a traction trebuchet could fire often. At a modest 100-watt power level, a person can supply 220 joules every two seconds. Since launch and reloading must have taken longer than that, power imposed no practical limit for traction trebuchets.

A modern calculation suggests that a counterweight trebuchet could shoot a 500-pound (225-kilogram) projectile 260 meters, four times the weight and three times the distance of a traction trebuchet. That would have required an initial speed of 110 miles per hour (50 meters per second) and a kinetic energy of nearly 300,000 joules. To do the same, a traction trebuchet, even one working at perfect efficiency, would have needed almost 1,400 people. So the counterweight trebuchet economized on labor, and it reinvested that saving in a combination of missile weight and distance. Still, pulling down a 10-ton counterweight again and again could not have been easy. In reducing the number of operators while maintaining a reasonable repetition rate, the counterweight trebuchet must have reintroduced power as a practical limitation. Of course it became monumentally heavy. Storing energy gravitationally takes far more mass than storing it, as did a ballista, elastically—unless one hoists a weight to an impractically great height.

What about the longbow, allegedly the crucial weapon in the English victory over the French at Crécy in 1346? We know a lot more about longbows than about catapults of any kind. In particular, we know how far they were drawn and with what force, making calculations simpler and less uncertain. For a simple bow, which these were, the energy put in is half the draw distance, 14 inches or 0.35 meters, times the maximum draw force, 70 pounds or 300 newtons—a little over 50 joules.[3] If we take the same maximum power output as before, 200 watts, we get the interesting result that the archer could shoot four arrows every second! Even at a more modest 100 watts, the archer could shoot two per second. No archer then or now could reach such rates with arrows that had to be individually loaded; maximum whole-body power just doesn't limit archery.

In fact, medieval archers with longbows achieved respectable shooting rates. I've seen figures ranging from six aimed shots per minute to twelve presumably unaimed shots per minute.[4] The muzzle-loading muskets of a few centuries later fired two to six times more slowly. But even if the archers were not aerobically limited, the old accounts assure us that rapid shooting was a strenuous business. Part of the problem comes from the way archery uses the muscles of the arms and shoulders, much less massive than the muscles lower down.

Force for the Weapons

Of the weapons described in the last chapter, only the ballista appears power-limited. How about force, the other measure of what muscle can do? What limitations might it impose?

Let's look again at what those ancient artillerymen had to accomplish. Recall that the missile from a large ballista needed 160,000 joules of energy to send it on its way. Energy or work is given by the product of force (pounds or newtons) and distance (feet or meters). If we know how much force a pair of men could exert, we can figure how far they must have pulled in order to draw back the projectile and its holder. Cranks saw little use in antiquity, but a ballista (as it had to) used something analogous. A capstan with a projecting handspike, as in the illustrations (Figures 7.10 and 12.6), amounted to a more substantial version of a ship's wheel. We might estimate 110 pounds (500 newtons) as the force two men could sustain when they pulled on handspikes. That's half of what Cotterell and Kamminga measured for the best of their beer-promised Australian students, but the students pulled in optimal stance on an immovable rope and did so only briefly. Their 160,000 joules (newtons times meters) divided by 500 newtons give 320 meters. So the handspikes on their capstan had to turn at least 320 meters for each shot. If the capstan was 10 meters in circumference (radial handspikes sticking 5.2 feet or 1.6 meters outward), they would have had to turn it fully around 320 meters divided by 10 meters or 32 times per shot. If they worked at their aerobic maximum, shooting once every 6.7 minutes, they needed to turn the capstan about five times each minute, or once every twelve seconds.

Those twelve seconds had to include time to pull each handspike out of the central cyclinder and stick it in again so it pointed forward—at least twice per turn or once every six seconds. The alternative—using multiple, fixed handspikes (as on a ship's wheel)—would have required that the rest of the weapon be elevated more than five feet off the ground.

So not only must the pair have worked hard, but they must have worked fast; the ballista used them near capacity for both power and force. But the result implies nothing impossible, and it gives us an idea of how big the capstan cum handspikes had to be. A smaller one would have required unreasonably fast turning as well as limited how much of the body's muscle could be brought to bear. Could two men have worked on a single capstan? Most likely, I'd guess, the cylinder extended outward from both sides of the ballista so each man could pull against his own handspike. Some illustrations (modern guesses, really) show such an arrangement. Few, however, give the ballista a sufficiently extreme capstan—long enough handspikes and a thin enough cylinder where the rope from the slider wraps around it.

An archer, as we've seen, wasn't power-limited as a whole, although with only a small mass of muscle doing all the work, local power may have presented a problem. But the figure of 70 pounds for a typical draw weight of a bow implies force limitation. To draw with that force takes training; even a fit person can't do it right off. Still, it's not extreme and must represent not some maximum manageable force but what can be counted on when the bow is shot at a trained archer's maximum rate of discharge. Greater force gives greater speed, meaning increased distance, flatter trajectories, or greater penetration. Or it permits the use of heavier arrows. A 70-pound longbow can shoot a one-ounce arrow 250 yards, a 200-pound longbow can achieve as much as 390.[5]

For that matter, bows can do still better, but they have to be more sophisticated than the otherwise excellent pieces of yew wood bent by traditional military longbowmen. The trouble with a simple bow comes from the way draw force increases with draw distance. The archer must pull not with a uniform force but with a force that increases as the arrow is drawn. Worse, upon release the force of the string on the arrow decreases as the arrow moves forward. Recall that we took for the work of the archer only half the draw force times the draw distance; that tacitly assumed that increasing draw force. Various tricks can force the archer to exert almost the full force of the draw over the entire length of the draw, thus raising that figure we took as one-half up toward one. Perhaps the most ancient of such devices is the reflex bow, which is bent around in the "wrong" direction (tips forward) until strung. Ancient Greek vases and sculptures most often show reflex bows. These days we can purchase complex compound bows, whose strings run over eccentric pulleys and which make use of other bits of high technology.

The crossbow provided one solution to the limited draw weight an archer could manage. Winching, cranking, or levering back the arrow permitted a more forceful (and, if need be, a slower) draw. A crossbow represents a more complex technology than an ordinary bow, and it arrived later, by at least thirty thousand years. It originated in China and seems to have appeared in the West sometime during the latter stages of the Roman Empire, but it looks too much like the gastraphetes to be sure from old accounts. The crossbow was well established in continental Europe by time of the Norman conquest of England, in 1066. So severely did it wound that the Lateran Council of 1139 prohibited its use—with as much effect as most pronouncements against sin—at least against Christians.

Much ingenuity went into putting its basic advantage to practical use; Figure 13.1 shows a few arrangements for pulling back the string and

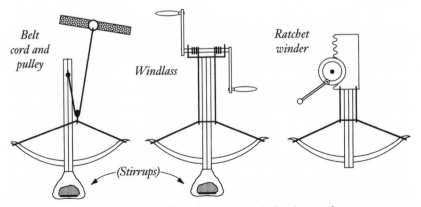

FIGURE 13.1. Several ways of drawing back the bolt of a crossbow.

bending the bow. In a foot-braced crossbow, lots more muscle can be applied to the task of pulling back the slider. In a bow with a pulley, that force can be multiplied by a factor of two. In a lever-operated crossbow, still greater force amplification can be achieved. In a cranking crossbow, energy can be put in more slowly as well as with still-greater force amplification but with what appears an awkward motion for a hand-held device. One way or another, a weapon much shorter than an efficient bow could be pulled with much greater force. Records from the Han dynasty (206 B.C.E.–220 C.E.) suggest 190 to 380 pounds.[6] Modern crossbows have been built and used with draws of up to 1,200 pounds.[7]

While we have lots of comparison between longbows and crossbows, the large number of factors involved complicates drawing conclusions. Balancing to some extent their more forceful draws, crossbows drew their bolts less far than longbows drew their arrows—often less than a foot, compared with up to thirty inches. Offsetting that, bolts weighed more than arrows, as they had to be in order to withstand the higher accelerations during launch. Thus the crossbow shot only a little faster and farther, but it shot something heavier. It was, in short, a lower-impedance weapon than the longbow, good, for instance, at penetrating armor. It also could be left cocked without needing continuous force from the archer's muscles, and its compact shape permitted its use from horseback. Conversely, the longbow could shoot more frequently, two to six times as often. The specific factor depends on the particulars of the comparison and on the particular ax ground by the authority one chooses to cite. A heavy odor of advocacy pervades the literature on weaponry.

The Chinese devised and used repeating crossbows. In these, a magazine of around a dozen arrows was attached to the top of the stock. But they shot only light arrows, ones that often used poisoned tips to compensate for their limited powers of penetration.

Mention of poison-tipped projectiles brings us to a different weapon that often used them, the blowgun, and to the forces involved in its use. In mass efficiency, the blowgun wins, hands down, over any other weapon we've considered. Only the air in the tube moves with the projectile, and that air should amount to little more than 10 percent of the projectile's weight. Blowguns come in a wide range of sizes, from a foot or so to six or eight feet in length, but except for the smallest peashooters, they all work the same way. (In tiny ones the tongue acts as a valve to admit air that has been previously compressed in the lungs.) The user must exhale as forcefully as possible some fraction of the air in the lungs. Vital capacity, the maximum volume a person can exhale, matters only in blowguns of the largest contained volume. Always a factor, however, is maximum expiratory pressure. But that maximum occurs when someone exhales against an infinite load and thus does not discharge any air at all. So the greater the internal volume, the lower the pressure the blower can develop. Pressure times flow rate gives power output, but enough other factors are in play so that simple product gives no reliable guide to design.

Consider a hypothetical large blowgun, one with a volume of two liters, about half a person's vital capacity. We'll make it 2 meters long, so its cross-sectional area must be 1.5 square inches, or 10 square centimeters, and its diameter 1.4 inches, or 3.6 centimeters. A person can achieve a maximum expiratory pressure of about three-tenths of an atmosphere; in metric units, that's 30,000 pascals. Force is pressure times cross section (pascals times square meters), so the blower might generate 30 newtons. A one-ounce projectile (to match an arrow) that fitted the bore would suffer impractically high drag relative to its weight, so let's assume instead a 4-ounce (100-gram) projectile: heavier, thus slower. From force and mass we can get acceleration ($F = ma$, recall), which comes to about 300 meters per second squared or about 30 times gravity. From acceleration and tube length, we get launch speed, about 80 miles per hour or 35 meters per second. From launch speed, we can calculate a dragless maximum range of 134 yards or 122 meters.

That's not bad. The projectile goes a shorter distance but represents a heavier load than the arrow of a longbow. At the same time, it corresponds to a higher level of momentum: 3.5 against perhaps 1.2 for a traditional arrow.[8] (The units, kilogram-meters per second, aren't familiar ones.) That level of momentum compares favorably with what a crossbow achieves. Why not use blowguns against armored knights? At least three factors apply discounts to this straightforward estimate of what a blowgun can do. First comes the use of maximum expiratory pressure. Since blowing this hypothetical blowgun takes only a little over a tenth of a second, the blower's pressure will be lower. How much, though, depends on training, and we have respiratory data on precious few trained operators of large blowguns! Second are the various frictional losses in the tube. Even a loose-fitting projectile incurs losses, and the feather and plant fibers used to prevent leakage of air around the projectile will make things worse. In addition, airspeeds up to eighty miles per hour in a tube imply substantial fluid mechanical friction. Third, long tubes of constant internal diameter can't be easy to make with primitive techniques, even from naturally tubular stock, such as bamboo.

My colleague Knut Schmidt-Nielsen has kindly given me access to his Amazonian blowgun and its darts. The gun's length is 2.25 meters (7.4 feet), and its bore diameter is 10 millimeters (0.4 inches), with a fairly rough internal surface. The typical blowgun of the Waorani of eastern Ecuador differs little, reportedly 2.75 meters long and 12 millimeters (0.5 inches) in diameter, with a volume less than 10 percent of vital capacity, one's maximum exhalation.[9] The darts of Schmidt-Nielsen's blowgun average a remarkably light 1.5 grams (about 20 per ounce) and use some kind of plant fiber (probably kapok) on their trailing ends to provide air seals in the tube. The Waorani darts weigh about 2.5 grams. Even with darts that light, the mass tax doesn't amount to much, about 10 percent. But that's the only nontroublesome feature.

If we assume our generous third of an atmosphere blowing pressure, muzzle speed comes to almost 200 miles per hour, or 84 meters per second, with a projected maximum (dragless) distance of over 700 meters. That sounds great until we follow up with the calculation that far more than a third of an atmosphere would be needed to blow that fast in such a narrow tube—without any projectile at all. Making an allowance for the fluid mechanical resistance of the tube drops the speed estimate to about 56 miles per hour, or 25 meters per second. Momentum at launch amounts to only a hundredth of that of the hypothetical large blowgun. After applying a reasonable estimate of drag, we find that such a launch speed will give a range of about 90 feet (28 meters) and, worse, a speed at impact of only 30 miles per hour, or 13 meters per second. Moreover, that assumes shooting at about 39 degrees upward, going for range and giving up much hope of accuracy. Momentum on impact? It's negligible: about 2 percent of that of an arrow.

After I did these calculations, based on accepted figures for the pressure and volume that lungs can supply and my own approximations for drag, my anthropological colleague Steve Churchill supplied me with data his professional kinfolk had obtained in the field. One source gives a blowgun's maximum effective range as 56 feet (17 meters), another as about 100 feet,[10] nice confirmation of my estimate of 90 feet.

So blowguns end up as short-distance, low-penetration weapons, dependent on poison-tipped darts and (I presume) thin-skinned prey. Their limit, once again, comes from muscular force, in this case that of diaphragmatic and intercostal muscles in the form of maximum expiratory pressure. Nonetheless, blowguns can be accurate, and even if long, they're convenient for carrying through dense vegetation.

Reenter the Horse

Back when I talked about animals as power sources, I put off the history of horses as components of what we'd now call weapons systems. The horses of classical antiquity could pull light chariots, although not heavy vehicles, and they could be ridden, if somewhat precariously. Not that they lacked military appeal, merely effectiveness. As Xenophon put it to his Greek compatriots retreating from Persia:

> . . . you must remember that ten thousand cavalry only amount to ten thousand men. No one has ever died in battle through being bitten or kicked by a horse; it is men who do whatever gets done in battle. And we are on a much more solid foundation than cavalrymen, who are up in the air on horseback, and afraid not only of us but of falling off their horses; we, on the other hand, with our feet planted on the earth, can give much harder blows to those who attack us and are much more likely to hit what we aim at. There is only one way in which cavalry have an advantage over us, and this is that it is safer for them to run away than it is for us.[11]

Only a short time thereafter Philip, Alexander, and their Macedonians transformed cavalry into an effective force. Hannibal made significant use of cavalry against Rome, although the Roman armies themselves didn't depend much on horses then or later. But things changed in the early Middle Ages. Just as harness and whippletree made horses into fine low-impedance power sources for plowing and hauling, stirrups and saddles did the same for high-impedance equestrian applications. As Lynn White put a point made in the last chapter:

> During the 730's, there occurred a sharp discontinuity in the history of European warfare, caused by the introduction, from India

by way of China, of the stirrup. The stirrup is a curious item in the history of technology because it is both cheap and easy to make, yet it makes a vast difference in what a warrior can do on a horse. As long as a man is clinging to his horse by pressure of his knees, he can wield a spear only with the strength of his arms. But when the lateral support of stirrups is added to the fore-and-aft buttressing of the pommel and cantle of a heavy saddle, horse and rider become one. . . . The blow is struck no longer with the strength of a man's muscles but rather by the impetus of a charging stallion and a rider. The stirrup thus made possible the substitution of animal power for human power. It was the technological basis for mounted shock combat, the typical Western medieval mode of fighting.[12]

To this one must add a few words about how saddles and stirrups facilitated mounted archery. Since a bow recoils negligibly, a person can shoot an arrow while riding bareback. Longbows may be impractical, but ancient art shows riders and even centaurs using short, recurved bows. With less attention needed to stay mounted, both the speed and accuracy of shooting must have improved. Saddle and stirrups must have had their greatest impact, though, on shooting posture. We retain an expression, the "Parthian shot" (commonly corrupted into "parting shot"), for a blow (now usually verbal) struck after an issue has been decided. The Parthians, who lived in what is now Iran, could shoot arrows intended to keep pursuers safely distant as they retreated or feigned retreat. Only a firmly situated rider could risk turning around and then shooting rearward from a running horse.

Historians differ on whether saddles and stirrups provided the crucial support not just for riders but for the whole feudal system of Europe. White, in particular, advocated the latter view. I'm more curious about whether the mounted knight in full armor constituted a reasonable use of the continent's potential military resource. Impressive, surely, but just as surely cumbersome, vulnerable, and expensive. Could a charging stallion and rider, both heavily armored, have been able to change direction if the target moved? Mass and maneuverability are antithetical mechanical attributes. Could a line of knights have charged against a line of much more numerous (because much cheaper) archers, each with a heavy bolt in a cocked crossbow? A suit of armor that could withstand a bolt at short range would have been too heavy to wear or for a horse to bear, although chain mail was somewhat effective against crossbows. Also, were knights effective enough to be worth the extra baggage and ancillary personnel they imposed on a military force? After all, armored knights couldn't even mount their horses without a lot of help.

Horsed cavalry persisted into the nineteenth century, and remnants lasted into the twentieth. A person on a horse can get from one place to another faster than one on foot, and that matters for command, reconnaissance, patrolling, and so forth. But heavily armored cavalry—knights—disappeared centuries earlier, a demise most often attributed to firearms. I (not alone) wonder about that explanation. Since muscle-powered weapons did as well against armored riders as early firearms, perhaps no strictly military explanation will do. The alternative? Maybe we're viewing one consequence of a general shift in social organization.

What Killed Muscle-Powered Weapons?

That brings up the larger issues of when, where, and why firearms replaced muscle-powered weaponry. The "when" gives no trouble: late medieval times. Cannon first saw use in battle about the middle of the fourteenth century, and portable firearms became common less than two centuries later. The introduction of these combustion weapons paralleled that of combustion engines, which made their appearance four centuries later. Both combustion weapons and combustion engines replaced muscle, and in both cases large, inefficient devices preceded lighter and more portable ones. Cannon preceded muskets; Newcomen atmospheric pumps preceded engines suitable for steamboats and railroads.

But the four-century interval between the two presents a puzzle. Did steam engines pose greater technological challenges than did cannon? I have grounds for doubt. In the thirteenth century the British Isles began large-scale distillation for making whiskey; the product enjoyed sufficient appeal that by 1300 laws were attempting to limit its abuse. Moreover, the basic distillation technique was available many centuries earlier, even if it wasn't put to substantial use.[13] The equipment for distilling resembles that of a simple steam engine, and the process provides a view of steam as something that can be piped and something capable of developing force; stills still explode. The puzzle persists.

Perhaps economics explains the four-century interval. The development of industrial engines enjoyed no government subsidization, and their adoption depended on rational economic comparisons. By contrast, military irrationality is the stuff of legend. For instance, in the middle of the fifteenth century the Scots built a huge cannon, Mons Meg, shown in Figure 13.2, that could fire balls weighing as much as one thousand pounds. Moving it to any battle site turned out to be impractical, and it blew up when first tested and had to be rebuilt. It was therefore mounted on the battlements of Edinburgh Castle, where its valuable mass of metal could face nothing worth hitting with such large cannonballs. (Aside from a century spent at the Tower of London, the cannon has remained at the castle, where it can be seen to this day.[14]) More recent examples of military technophilia need be mentioned only by name: zeppelins, Maginot Line, antiballistic missiles. . . .

The "where" is as easy—Europe—although explaining it isn't. China provided the means, gunpowder, to Europe but not to the Americas. China, though, had much earlier achieved substantial political unity, and it enjoyed adequate barriers to invasion from most (although not all) directions. Its military could safely grow conservative, as internal insurrections would have had neither the duration nor the resources to innovate. Western Europe, by contrast, was nothing if not politically fragmented and fractious, with frequent warfare among its constituent entities and with intermittent invasions from eastern Europe, central Asia, the Middle East, and northwestern Africa. The appeal of better weaponry, indigenously invented or expropriated, needs little further rationalization.

FIGURE 13.2. Mons Meg, on Edinburgh Castle, as it looked around 1880.

A parallel instance supports the case. The Chinese invented but then made little military use of rockets. But in the late eighteenth century the regime of Hyder Ali of Mysore, in southern India, another politically and militarily volatile region, developed gunpowder-fueled rockets, a technology alternative to gunpowder-fired cannon. They were used with great effect against the British at the end of the century. The British then hastened to develop similar weapons; "the rockets' red glare" in the national anthem of the United States refers to the rockets with which the British assaulted Fort McHenry (at Baltimore) in 1814.

The "why" presents greater difficulty. The obvious explanation boils down to better killing through chemistry. (In perverting the Du Pont company's motto, I'm mindful that its first product was gunpowder.) As the British historian of technology Donald Cardwell puts it, "The decisive advantage of the musket was that it did not require a strong and highly skilled man to use it effectively. . . . Virtually anyone could be a musketeer. Even armored horsemen and mounted knights stood no chance against a clodhopper of a musketeer with a cheap musket."[15] Or, in Lynn White's words, "When in 1595 the longbow was officially discarded by Elizabeth's army in favor of the musket, it was still technically the superior weapon. The musket, however, was reasonably effective in the hands of less well-trained soldiers, and that fact was decisive."[16]

But they may exaggerate. Anyone could shoot a musket, but to load powder and ball into a muzzle-fed weapon and then ignite some priming powder in a firing pan with a smoldering slow match—and do it in battle again and again—required extensive drill. Muskets may have been more suitable than longbows for use on horseback, but either the short, recurved bows known to antiquity or the newer crossbows should have been at least as good. For that matter, smoothbore muskets may never have achieved decisive technological superiority.[17]

So Lynn White's case against early cannon may apply to personal firearms up to five hundred years later:

Had an Office of Technology Assessment been asked to present a report on gunpowder artillery when it first appeared at Florence in 1326, the measuring rod of effectiveness would have been the trebuchet. The earliest cannon were crude, cumbersome, and inefficient. They were costly to make and costly to supply with their chemical fuel. They could not be aimed with any great exactness; they were slow to load and to fire; they could rarely hit the same spot on a fortification twice because of irregular composition and combustion of the powder. . . . Any rational technology assessment of the cannon in 1326 or for a hundred years later would have concluded: "Stick to the trebuchet."[18]

I think a less rational factor better explains the history. Firearms, especially cannon, impress one's own side and intimidate one's opponents. Noise plays a role in battle, even if artificially supplied with horns (recall Joshua at Jericho) or yelling (the Rebel yell of the Confederates in the Civil War). Cannon do especially well since the human ear interprets sounds with sharp onsets as louder; percussive music played backward sounds strangely muted. Firearms make smoke and odor as well, impressing several other senses in a way that no trebuchet or crossbow can match. Eventually, of course, firearms outperformed muscle-powered arms. But the rulers and commanders who first used them could not have been thinking of posterity any more than evolution can generate hopeful monsters with structures that merely portend their ultimate utility.

As correlative evidence for the argument, consider the preference, from the start, for cannon over rockets. The basic idea of each held no mystery, and rockets did see sporadic use during the Middle Ages. As we know from fireworks displays, gunpowder makes an adequate propellant as well as an explosive. Both cannon and rockets launch projectiles from tubes, but rockets put less stress on those tubes. For one thing, the fuel burns steadily, so the rocket generates a less sudden and less drastic increase in internal pressure. For another, the rocket continues to propel itself after leaving the launching tube, so the muzzle velocity of the projectile needn't be its maximum velocity, and the muzzle velocity doesn't determine the range. Rocket launchers can thus be much lighter than cannon; in recent wars individual soldiers have used tank-destroying rocketry. Even for shipboard use, preference went to cannon, despite the fact that they weighed as much as fifty times as much as their projectiles. Heavy cannon well above the waterline made warships less seaworthy than merchant ships with heavy cargo stowed well down in the hold; in some infamous instances, warships capsized when launched.[19] Not that rocket technology presents no problems. But rockets go whoosh, not bang.

Native Americans found firearms impressive, even though they practiced stealthy hunting and a type of warfare for which their own excellent archery equipment may have been better. And the Europeans
imported muskets and cannon rather than readopt archery—even when
they adopted some native methods of hunting and warfare. The Americans laboriously hauled cannon from Fort Ticonderoga, in New York
State, across New England to the hills outside Boston during the winter
of 1775–1776. The British, ringed by cannon, evacuated the city, to
which they have never returned.

The Americans couldn't make large numbers of firearms with their
limited metal-producing industry, and local powder mills weren't up to
the wartime demand for explosive. During that same first winter of the
war, Benjamin Franklin, ever innovative, suggested reconsideration of
bows and arrows:[20]

> 1st Because a man may shoot as truly with a bow as with a
> common musket.
>
> 2dly He can discharge four arrows in the time of charging and
> discharging one bullet.
>
> 3dly His object is not taken from his view by the smoke of his
> own side.
>
> 4thly A flight of arrows, seen coming upon them, terrifies and
> disturbs the enemies' attention to their business.
>
> 5thly An arrow striking in any part of a man puts him hors de
> combat till it is extracted.
>
> 6thly Bows and arrows are more easily provided everywhere
> than muskets and ammunition.

Nonetheless, Franklin then exerted every bit of charm and influence
to obtain a French supply of firearms and powder and to keep it moving
westward across the Atlantic.[21]

Since various isolated human populations have only recently gained
access to firearms, the issue remains alive. The accounts of anthropologists leave no doubt that firearms hold an attractiveness out of proportion to their cost-effectiveness, which, since ammunition cannot be
self-produced, remains high. But complications cloud comparisons of
the effectiveness of firearms with muscle-powered weaponry. A shotgun
may yield more protein per hour of hunting than bow and arrow or

blowgun, as found by one study of Amazonian natives,[22] but by requiring trade for ammunition, it extracts an additional cost. Furthermore, the savvy hunters, as that study also found, used blowguns for small prey, saving expensive ammunition for larger animals. Protein per hour as a measure of effectiveness—that food for thought leads to the issues of the next chapter.

CHAPTER 14

Muscle as Meat

. . . The wolves, as I should have remembered, ate the whole mouse; and my dissections had shown that these small rodents stored most of their fat in the abdominal cavity, adhering to the intestinal mesenteries, rather than subcutaneously or in the muscular tissue. It was an inexcusable error I had made, and I hastened to rectify it. From this time to the end of the experimental period I too ate the whole mouse, without the skin of course, and I found that my fat craving was considerably eased.[1]

FARLEY MOWAT, *Never Cry Wolf*

. . . Another purpose of collagen fibers in flesh is to put up the work of fracture. This is a good thing for the animal, but it is inconvenient for the people who want to eat its flesh. In other words, it is collagen which makes meat tough. Nature, however, does not seem to be on the side of the vegetarians, because she has arranged, in her wisdom, that collagen should break down to gelatin—a substance of low strength when wet—at a somewhat lower temperature than that which elastin or muscle can withstand. The process of cooking meat therefore consists of converting most of the collagen fibers into gelatin (which is jelly or glue) by roasting or frying or boiling. It is science of this kind which restores one's faith in the beneficence of Providence.[2]

JAMES E. GORDON, *Structures, or Why Things Don't Fall Down*

A S THE EDIBLE ENGINE MUSCLE SUPPLIES ENERGY IN YET another way. That edibility bears relevance to both its natural and its social history, so any proper account of muscle must flesh out the story of carnivory. Yes, we do eat other parts of animals, but as a look at any meat counter attests, muscle matters most.

We easily forget what benefits we've gained from that final element of versatility. An animal too old to work could be recycled through the stewpot. An expedition that ran out of food could eat its transport and proceed on foot. A self-propelled food source could be moved from place to place with only a small exertion from the herder. Over the course of human history these have not been minor advantages.

For Americans, the fame of no expedition exceeds that of the journey made between 1804 and 1806 by Meriwether Lewis, William Clark, and their associates from the Mississippi to the Pacific coast and back. They mainly walked and used boats, some made by themselves en route. But wherever possible they used horses as pack animals. While they carried some basic food, they could not possibly bring enough to sustain themselves and depended on hunting (and some barter with the natives) for most of their calories. Neither hunting nor barter proved fully reliable, but packhorses (also obtained by barter), provided a reserve, at one stage a crucial one. From Meriwether Lewis's journal entries in 1805, written as the expedition descended just west of the continental divide:

> September 14th. The whole stock of animal food being now exhausted, we therefore killed a colt, on which we made a hearty supper.
>
> September 16th. Though before setting out this morning we had seen four deer, yet we could not procure any of them, and were obliged to kill a second colt for our supper.
>
> September 17th. We had killed a few pheasants; but these being insufficient for our subsidence, we killed another of the colts.
>
> September 20th. At one o'clock we halted on a small stream, and made a hearty meal of horseflesh.
>
> October 2d. The hunters returned in the afternoon with nothing but a small prairie-wolf; so that, our provisions being exhausted, we killed one of the horses to eat and provide food for the sick.[3]

In view of the slightly desperate tone of these entries, I should mention that the expedition lost no member to either hunger or accident, and all but a single individual who died of an illness returned in good condition.

A little more than a century later Roald Amundsen and three other Norwegians made the trek from the coast of Antarctica to the South Pole, moving their supplies on sleds pulled by dogs. With no possibility of living off the land in the interior of that continent, even the dog food had to be hauled along. Their careful scheme resembled what we do when launching spacecraft with multistage rockets. As supplies diminished, the need for dogs diminished, and dog after dog provided food for the remaining dogs and, via stewpot conversion, food for the explorers as well. The scheme worked so well that the expedition, frugal in all ways, brought back its reserve food. Less satisfactory, though, was its public reception outside Norway. Somehow eating their dogs marked the explorers as unsporting, and the Briton Robert Scott, who bungled his own expedition and died on his return from the pole, became the gold medal hero. Scott's expedition used ponies, which gave much less satisfactory service, and they abandoned each as it died rather than use it for food.

Muscle as Food

Muscle makes fine food for humans. But we're trophically versatile animals—more so than dogs, much more so than cats, and almost as good as pigs. So we don't have to eat meat. Indeed most of the earth's population eats only negligible quantities of the stuff. A calculation done when there were half as many of us suggested that even then we were too numerous to sustain large-scale carnivory. Without meat, though, we have to pay more attention to dietary balance, eating not just plants but the right combination of plants, to get our essential amino acids and some other critical dietary components. Most vegetarian cultures pair legumes and grains, eating legumes to compensate for the lack of lysine in grains and grains to compensate for the lack of methionine in legumes. Thus we combine beans with corn tortillas, dal (dried legumes) or tofu (soybean curd) with rice, even peanut butter and wheat bread sandwiches, the peanut (or groundnut) being a legume, not a true nut.

The main advantage of meat for dietary balance comes from the fact that it's the largest single component of creatures much like us, and thus naturally (I might say) it contains most of what it takes to make us. We can get along for a while on nothing but meat, although, I hasten to add, not nearly as well as on a well-balanced vegetarian diet. We do better with a bit of fiber, which we get only from plants. In addition, we go deficient in calcium on meat-only diets. Muscle contraction may depend on calcium, but bones contain most of a body's calcium. Also, muscle contains almost no ascorbic acid (vitamin C), so a meat-only diet eventually leads to scurvy, the old hazard of long oceanic voyages and a major factor in the death of the Scott team in the Antarctic.

Even with some supplementation, a mainly meat diet provides an overdose of protein, which can have unpleasant consequences when its metabolic by-products accumulate to toxic levels. A recent paper has advanced the intriguing notion that vomiting early in pregnancy ("morning sickness," which doesn't happen only in the morning) may afford mother and embryo some protection from that toxicity. Its authors note, among other points supporting their hypothesis, that vegetarian cultures know nothing of morning sickness.[4]

Beyond dietary balance, meat represents an energy-dense food. Only pure fat or oil or particularly oily plant material supplies more energy per unit mass. For most of human history, that must have held considerable appeal. Famines have been common, and the fatter person can tolerate a longer period of caloric deprivation. Table 14.1 provides some representative numbers to flesh out the picture of energy density.[5]

First, the table ought to dispose of a bit of practical mythology, the often-cited datum that we get nine hundred kilocalories from each hundred grams of fat as against four hundred for carbohydrate or protein. Those numbers refer to the pure stuff, and we consume only fat in anything like pure form. Neither dry rice nor powdered soy protein makes it to the table.

Of more consequence here is the great difference a little fat makes to the energy content of a piece of meat. Muscle may be good food for humans trying to get enough energy; muscle with fat comes out far better. Albert Szent-Györgyi remarked on the inverse correlation between menu price and collagen content. Similarly, going to the meat market with tables of calorific values and government labeling standards reveals a direct correlation between fattiness and price. A carcass of beef rated "prime" has 46 percent fat; one rated "choice" has 40 percent fat; "good" has 34 percent; "standard," 27 percent; "utility," 24 percent. Fatty goose and duck cost more than lean chicken and turkey. Fattier lamb chops cost

| Material | Energy Density | | % fat |
	kilocalories 100 grams	joules kilogram	
Salad oil	880	37,000,000	100.0
Animal fat (suet, lard, etc.)	750	32,000,000	80.0
Walnut	650	27,000,000	64.0
Fatty bird (goose, with skin)	370	16,000,000	33.6
Lean bird (chicken breast, muscle)	105	4,400,000	4.4
Wild bird (duck, muscle)	140	5,900,000	5.2
Flightless bird (ostrich, muscle)	130	5,500,000	2.8
Fatty mammal (pig, edible carcass)	500	21,000,000	51.0
Leanest domestic beef (flank steak)	175	7,400,000	9.3
Small mammal (wild rabbit, muscle)	135	5,700,000	5.0
Large mammal (deer, muscle)	125	5,200,000	4.0
Large mammal (bison, muscle)	110	4,600,000	1.8
Fatty fish (mackerel, muscle)	190	8,000,000	12.2
Lean fish (flounder, muscle)	80	3,400,000	0.8
Turtle (muscle)	90	3,700,000	0.5
Frog (hind leg muscle)	75	3,100,000	0.3
Squid (mantle, tentacle muscle)	85	3,600,000	0.9
Lobster (abdominal, claw muscle)	90	3,700,000	1.9
Rice, boiled	110	4,600,000	0.1
Corn kernels, boiled	85	3,600,000	1.0
Potato, boiled	75	3,200,000	0.1
Carrot, raw	40	1,700,000	0.2
Spinach, raw	25	1,000,000	0.3

more than leaner lamb legs. The relationship has at least one odd aspect. Fat contributes little but its calories, so it dilutes all other nutritive components. The value we put on meat thus varies inversely with everything except its energy content.

Apparently we're wired to like fatty meat. We often deal with the toughness of collagenous meat by grinding it up and selling it as sausage. But none of the world's great sausages lacks a large dose of fat. Our expression "bringing home the bacon" alludes, perhaps accidentally, to the fattiest of all commonly consumed meats. Whenever we've raised mammals for food, we've presided over an increase in fat content. Wild boar and peccary have less fat than commercial pork. Horses, raised as power sources, have meat that's far less fatty than that of cows or pigs. Since raising the energy content relative to weight exacts a cost in feed, the farmer or rancher must pay for any extra fat; to get more energy out, you have to put more energy in. But saving on feed by selecting low-fat animal stock decreases profit in an economy where the price for fat at the slaughterhouse exceeds the cost of the extra feed that makes the fat. The fat content of lean mammalian or avian muscle runs between 2 and 5 percent, but in this range you have little choice at the butcher shop. Veal, bison, and venison among mammals and ostrich and skinless chicken breasts among birds about exhaust the choices. Of these, only veal and chicken breast appear in mass-market stores.

Could preagricultural humans have acquired meat of fat content comparable to that of our domesticates? On occasion—perhaps I should say "on special occasions"—they may have done so. Premigratory birds (and the occasional premigratory mammal) lay on extra body fat. Before overwintering, warm-blooded animals (whether or not they're true hibernators) are at their fattiest. Autumn bear or goose must have provided a fine, festive treat. Fat is not fit, we're told, and the admonition applies to fitness in the evolutionary sense as well as the nutritional or sartorial. Reproductive success pays dividends, not corporeal magnificence, except as the latter contributes to the former.

The argument that fat builds a useful energy reserve rationalizes our preference for the meat of large mammals to that of small ones. Large mammals distribute fat through their muscles, putting it where it's ultimately used. Small mammals put a larger fraction beneath the skin. (Farley Mowat, quoted at the start of the chapter, didn't give proper attention to subcutaneous fat. Perhaps it came off when he skinned the animals.) Large mammals need less insulation since they have more heat-producing volume relative to their heat-losing surface area.

That affection for the muscles of large mammals must be both ancient and (as various anthropologists have argued) a critical factor in determining the kind of creature we've become. Even if we started as scavengers, we soon became hunters. About two and a half million years ago we began making tools, and at the same time we began to take larger prey than do chimpanzees, for whom a forty-pound kill is exceptional. Early human social organization must have reflected the demands made by the cooperative hunting of large animals. Our use of language may have been jump-started by the need for coordination among hunters dealing with large and dangerous prey. Perhaps we joined forces with the first domestic dogs, which we might otherwise have eaten, because together the two species did better at bringing down large mammals.

Our taste for the muscle of large mammals must have been the main thing that initiated our further domestications. Some of us drink milk, but the majority of adult humans can't digest milk without suffering intestinal distress from its lactose. To serve as adult food (and to prolong its shelf life), milk has to be converted to something like yogurt or cheese whose production uses up its lactose. Leather lasts so well that a low rate of acquisition from hunted animals should suffice. A supply of fecal fertilizer, while useful, doesn't make or break agriculture. Finally, animal power depends on harnessing technology, which came far later than domestication itself.

We can view the issue of carnivory in a more general context. Five great apes share the planet, if a bit unequally: gorillas, orangutans, bonobos, chimpanzees, and humans. All of us are at least potentially omnivorous, but we differ in how much animal material we actually consume. Gorillas occasionally and almost accidentally eat termites but not much more. Orangs, while mainly frugivorous, consume insects, bird eggs, and a few other items. Bonobos similarly put little effort into meat acquisition, even ignoring accessible sources. Chimps eat more meat and hunt and kill other mammals for food, although they're still predominantly vegetarians, and they don't search for animals over appreciable distances. Humans practice the most extensive carnivory, as both hunting—our efficient bipedal walking and running, unique among the apes, may have been selected for its utility in hunting—and husbandry. This spectrum of increasing carnivory (implying neither historical sequence nor advocacy) runs opposite the spectrum of time invested in eating. Gorillas spend almost half their waking hours putting food in their mouths; humans spend only a little, at least if we discount the time invested in vocational activities that indirectly put food on the table. Social aggressiveness also increases in parallel with carnivory; I hope that the loose correlation reflects no inescapable cause-and-effect relationship.

In a recent book the anthropologist Craig Stanford presents this case for the formative role of hunting in making us what we've become.[6] He makes the additional case that the social value of meat to chimpanzees and humans transcends its not insignificant nutritional value. In both species males provide most of the meat. Chimps use the supply of meat for social leverage, just as we humans endow hunted meat with extensive symbolism, legends, and other cultural baggage. As Stanford puts it, "Meat, not only as a nutritionally desirable food item but also as a social currency that is controlled by males and therefore is a tool for the maintenance of patriarchal systems, plays an essential role in the social systems of both traditional human and some non-human primate societies." Or in Shakespeare's elegant metaphor, "Upon what meat doth this our Caesar feed, That he is grown so great?"[7]

Nor are we great apes unusual. Carnivorous cat and lion eat occasionally; herbivorous rabbit and gazelle graze almost continuously. Only one alternative to carnivory provides sustenance in as concentrated a form. That consists of eating the particularly energy-dense parts of plants—fruits, seeds, roots, tubers, and such special parts—rather than most leaves and stems. Even for those animals that can digest cellulose (by working jointly with some intestinal protozoa), the bulk of a plant provides food of awkwardly low value. As far as I know, no animal can both make a living on the bulk of plants and manage to fly: too dilute an energy source, too massive the requisite digestive equipment. Moths and butterflies come close, but in metamorphosis these Lepidoptera shrink down the leaf-eating caterpillar's gut and shift to an adult diet (when the adults eat at all) of nicely nutritive nectar and other such things. Birds seek out the good stuff as well: seeds, insects, worms, nectar.

A Variety of Meats

The psoas muscle—filet mignon—lies within the skeletal cage of the mammalian trunk. So it needs less protection from daily bumps and bruises than, say, rectus abdominis or latissimus dorsi, both on the outside of that cage. It thus comes to the plate with less collagenous connective tissue and with unequaled tenderness— at least unequaled by any other commonly consumed mammalian muscle. Scallop muscle may match it; like psoas, it lies inside stiff skeletal structures. Shrimp muscle has a denser texture—you can more easily thread shrimp onto skewers—but the abdomens of shrimps have flimsy exoskeletons that you peel off rather than pry open.

At the other extreme is squid. Squid jet by contracting their outer mantles, the thin sheets of muscle that surround the water-filled cavities that surround the rest of the animals. We eat the mantle (sometimes replacing the rest of the animal with wonderful stuffing), and we eat the tentacles. Both have to be resilient if they're to work. Neither mantles nor tentacles have bones, so their muscles must reexpand each other; it demands some tight interlacing. Thus squid muscle contains lots of collagen. That's culinarily bothersome. Either you cook squid very briefly—as one cookbook says, just "threatening them with heat"—so that collagen doesn't develop toughening molecular cross-links,[8] or you subject them to lengthy cooking with lemon juice, vinegar, or tomato sauce, doing to the collagen what the chemist would call an acid hydrolysis in order to break up some of those cross-links.

Horsemeat tastes sweeter than beef, pork, or venison. No great mystery why—it contains much more glycogen, the animal equivalent of starch. The glycogen breaks down into sugar with a little persuasion from the meat's enzymes while the meat is aged, or from the heat while it's being cooked, or from an enzyme in our saliva while being eaten. Animals mobilize glycogen more rapidly than fat, and having a good store right at the site of use—rather than having to pipe it in from the liver—must increase its effectiveness. That should matter most for aerobic endurance runners. Horses, as we saw earlier, win prizes for aerobic output; cows and deer do not. Bullfights, whatever their other barbarisms, don't force animals to run steadily with picadors in pursuit. But stag hunting by horse-mounted humans with their dogs turns out to be especially cruel, forcing a sprinter to run both fast and long—ten miles or so. The blood and hormone profiles of newly killed stags show them to be almost lethally stressed.[9] Few humans eat horsemeat, but some of us do acquire a taste for it. I've heard that during World War II the Faculty Club at Harvard served horse, and it became popular, perhaps because with rationing, the diners ate less meat of any sort at home. For years after the war demand persisted, and the place continued to serve it.

We've already noted that the meat of very small mammals contains less fat than that of larger ones; the leanness of Mowat's little rodents should have been expected. Fat makes good insulation, and small warm-blooded animals need more insulation than do big ones. Their surface to volume ratios are higher, as are their metabolic rates relative to their weights. Better to let fat do double duty, limiting heat loss as well as providing an energy reserve. Among aquatic mammals that live in cold water, the subcutaneous fat layer (blubber, to be sure, isn't just fat) can represent as much as half their total volume, even in animals of substantial size.[10]

In short, how muscles look and taste makes sense in terms of their biological roles, transcending even the ways we've modified animals and their muscles in the process of domestication.

Incidentally, insects show an analogous divergence in their immediate fuel, but in a different functional context. Flies, for instance, use glycogen for muscle fuel, while migratory butterflies and locusts use fat. Glycogen moves and mobilizes better—it readily breaks down into water-soluble sugar molecules—but fat can be stored with greater weight efficiency. Animals store their glycogen (as plants store their starch) in a heavily hydrated form. So stored glycogen delivers only about a hundred or so kilocalories per hundred grams rather than the eight hundred delivered by stored fat. The data given earlier for energy in carbohydrate as brought to the table provide a much better estimate of its storage weight within animals than data for the pure chemical. Flies can fly nonstop for only an hour or two without refueling; by contrast, desert locusts can fly for at least twelve hours, and monarch butterflies migrate thousands of miles with little refueling on the way. Like insects, we vertebrates use both fuels, and also like them, we tune the mix to match task and lifestyle.

Given the human fondness for meat, I'm puzzled by the wide cultural range of the acceptability of different kinds of meat. Some cultures eschew meat altogether, and I've never had the temerity to ask vegetarians whether they still find meat palatable. Anthropologists report a wide range of strong and individualistic taboos on eating meat from particular animals in small and isolated human cultures. But the variation persists even in large and interactive cultures. Traditional Chinese will eat a particularly wide range of animals and parts of animals; traditional Indians, even nonvegetarians, accept only a narrow range. Looking out my window in a small hotel in Beijing, I could see each morning a man pedaling a cart bearing the remains of the meals we had eaten the day before.

These, I was told, would be collected and delivered to the pig farms near the city. I also saw no songbirds. By contrast, in neither Israel nor southern India did I see pigs, nor was I offered pork. Despite any advantage they might offer in the way of efficient food utilization, pigs were culturally unacceptable. Songbirds, though, were a pleasant feature of both places. Consideration of the unpalatability of different meats, though, brings up a subject that can only be described by the same word, "unpalatable."

Eating Ourselves

We do it as a matter of course in sloth or starvation; we attempt it with infrequent success when dieting. Like much of the rest of our bodies, our muscle mass tracks the demand for its use and the supply of material for its construction. Keeping humans living in spacecraft from losing too much muscle has been a challenge, less well solved than the advocates of people in space prefer to admit. Long periods of enforced bedrest can be almost as detrimental. Put a limb in a cast, and it emerges, a month or so later, a shadow of its former self, unless its muscles are used by a protruding end of the appendage. At least the changes have no hereditary significance; we cannot be breeding a race of congenital couch potatoes. But such issues have come up earlier. There remains a question that might be put in facetious terms. If the Lord didn't want us to eat people, why were we made of meat? No question about it, human muscle is as edible as any other kind. So we ought to ask about cannibalism or, to use a proper classically derived term, anthropophagy.

No topic I've looked into while writing this book approaches the contentiousness of cannibalism. For one thing, no topic approaches it in underlying moral and emotional load. For another, it touches the most sensitive nerves of the anthropologists: their struggle to free themselves of racism and European ethnocentricism; their struggle to build a subject on verifiable evidence rather than the weight of accumulated anecdotes; their disparate attitudes toward Freudian and sociobiological explanations.[11] Even preference for "anthropophagous" rather than "cannibal" carries baggage; it avoids linguistic connection with native Caribs of the West Indies.

About twenty years ago a book appeared that disputed all claims of cannibalism except for isolated (and indisputably documented) cases among small groups trapped in lifeboats, on snowy mountains, and such.[12] By academic, as opposed to journalistic, standards, the book failed to make a persuasive case. But almost everyone writing during the subsequent two decades has felt compelled to take up the cudgels. It's that kind of field. Yes, many instances of cannibalism can be persuasively documented, even if many others rest on shaky evidence. To me the cannibalism deniers sound much like the evolution deniers we biologists bump up against and wish we could ignore—except that for cannibalism the main challenge comes from the sociopolitical left rather than the right.

Cannibalism comes in several guises. What has been called survival cannibalism has occurred in trapped and starving groups. A famous example involved the survivors of the wreck of the ship *Essex*, rammed by a sperm whale, in 1820, the episode that in part inspired Melville's *Moby-Dick* and forms the subject of a recent best-selling book.[13] Another happened in the Donner party, which attempted a winter crossing of the Sierra Nevada of the American West in 1847. Less famous (or infamous) but larger-scale instances have marked numerous famines and the extreme deprivations of some prisoner of war camps, concentration camps, and gulags.[14]

The consistent characteristics of survival cannibalism should surprise no one brought up in any contemporary human society. It doesn't start until all other food, including any animals the group might have had, has been exhausted. First go herded animals, then pack animals, then pets—as if in obedience to farmers' old advice not to name an animal if you mean to eat it. Groups consisting of kinfolk or groups with strong internal cohesion resort to cannibalism less often than groups thrown together by unfortunate circumstance, a point not lost on proponents of sociobiological explanations. Scavenging from recently dead or frozen corpses almost always precedes deliberate killing, and many instances never progress (or regress) to killing or assisted suicide. More germane to the subject of the present book is the sequence that ensues once hunger drives us past our normal revulsion. It follows the way we (or hunting cultures, at least) normally eat animals: skeletal muscle first, then heart muscle and liver, then other organs.

How does human taste? Anecdotal reports refer to the sweetness of the meat and suggest that it's like pork but sweeter.[15] Interesting! Like pigs, we're omnivores. Unlike pigs but like horses, whose meat is sweet, we're aerobic specialists. So a sweet, porklike taste agrees with the biologist's explanatory prejudice. One occasionally runs into suggestions that humans may be avoided by some large predators because we taste bad, but whatever the distaste of the human psyche, the human palate finds us acceptable.

Distinct from survival cannibalism are the various forms of ritual cannibalism. In these, participants expect to gain some advantage or propitiation from eating sacrificial victims or enemies killed in battle. While the Aztecs practiced the most massive ritual cannibalism on record, the practice has been convincingly documented in scores of instances. The best-known recent case occurred among the Fore people in New Guinea, where mortuary feeding on the brains of kin was considered a mark of respect. Trouble ensued because the partially cooked brains contained the pathogen of kuru (or Creutzfeldt-Jakob disease, or spongiform encephalopathy, or mad cow disease—imperfectly distinguished prion-caused maladies), a slowly but inevitably fatal degenerative disease of the nervous system. Cannibalism ended in the 1950s, and since then the incidence of the disease has diminished.[16]

More controversy surrounds what might be called routine nutritional cannibalism, in which human societies chronically short of either total calories or protein compensate by eating some of their own number. The most likely practitioners were the Aztecs. As put by one writer,[17] "availability of domesticated animal species played an important role in the prohibition of cannibalism and the development of religions of love and mercy in the Old World. Christianity, it may yet turn out, was more the gift of the lamb in the manger than the child who was born in it." On the face of it, nutritional cannibalism has a certain ghoulish logic: simultaneous nutritional supplementation and population control. But we must, as a financial analyst would put it, do the numbers.

Assume a human provides 45 pounds or 20 kilograms of relatively fat meat and other edible parts, with a nutritional value of 300 kilocalories per 100 grams, or 60,000 kilocalories in all. Assume, further, that a person needs 3,000 kilocalories a day for sustenance. A single human would provide 20 person-days of full sustenance, or 200 person-days at 10 percent sustenance. To achieve even this latter low level of supplementation, a population would have to sacrifice nearly two of its adults each year for each of its (surviving) members. That means it would decline by almost an unthinkable two-thirds each year. Even at 5 percent sustenance it would roughly halve each year. Next to this decline, natural reproduction makes no difference; even our maximal 3 or 4 percent growth pales beside that 50 percent decline. No way can a population eat enough of its own number on a steady-state basis to make a significant difference in its calorific intake.

What about humans as a protein source for an excessively granivorous population?[18] Considerable doubt exists that traditional vegetarians have a problem getting proper protein, but if we assume that they do have a problem, we can look at how much anthropophagy might help. Adults need about 50 grams of protein per day, children somewhat less, pregnant or lactating women somewhat more.[19] Water-free protein makes up about 20 percent of slightly fatty meat. At 20 kilograms of meat per human, that comes to 4,000 grams of protein, which amounts to 80 person-days of protein for full sustenance or 800 person-days at 10 percent sustenance. To look at it another way, a human would supply a little over 10 percent of the needs of two people for a year. To get that slight supplementation, the population would drop by a third each year, less than in the previous example but still far beyond any conceivable replacement rate.

Unless a culture enjoys massive and continuous immigration, it can't use humans as routine sources of either total energy or protein. Never mind what kind of society the practice might dictate; such reliance on one another has no more practicality than making a living by taking in one another's washing, that metaphor for economic impossibility. It will shrink a population far faster than total celibacy. Cannibalism among other animals does exist, but always in some special and understandable context. A female spider may eat her mate immediately after copulation. She thereby begins pregnancy well nourished—by this animal husbandry she has her cake and eats it too—but the male has at least invested his assets in his progeny. A male lion may eat his stepchildren, but he thereby brings his mate into receptivity so his own offspring may be speedily engendered. Enough, already, of this unsavory subject.

Should We Eat Meat?

Not many animals enjoy our digestive versatility. Beyond any benefit of our brains or culture, we can survive and even thrive on a wide range of diets. All of us apes descend from plant-eating ancestors, although perhaps not of the specialized cellulose-processing kind, such as cows and horses. For instance, we can't make our own vital ascorbic acid, so it's a vitamin, C to be specific; that's a characteristic of herbivorous lineages. On the other hand, the proportions of the human digestive tract differ from those of other apes. Our larger small intestines and smaller large intestines imply a diet of nutritionally more concentrated food, rather than one dominated by fruits, roots, and leafy vegetables. Among herbivores (and the predominantly herbivorous), larger size tends to be associated with lower-value food. An elephant (or a gorilla) can't get enough seeds to make ends meet and must make do with a larger volume of more ordinary herbage. But we have intestinal proportions more typical of smaller herbivores. On this basis, Katherine Milton, of the University of California at Berkeley, suggests that meat eating, meat that enriched an otherwise bulky, low-value diet, has been a basic habit of our species since its initial divergence.[20] That squares nicely with Craig Packer's view about the critical role of hunting in instigating the special features of human social life. Eating vegetables during the week supplemented by a piece of meat on the Sabbath may have roots as deep as any recent social practice.

Nonetheless, the mix of foods and the relative reliance on meat vary greatly, both culturally and historically. Again, a large fraction of us eat no meat at all, and another large fraction eat meat in such small quantities that it can contribute only minor amounts of micronutrients. That's the long-term consequence of the Agricultural Revolution. Its inception, ten thousand years ago, may mark it as prehistoric, but that's only around four hundred generations—yesterday, in evolutionary terms. We have a choice, one that few other species could enjoy even if they had sufficient volition to choose. Should we, then, eat meat? Three distinct physiological issues bear on the choice.

First, consider protein. Meat has all the essential amino acids we need, the particular amino acids that we can't synthesize, the reason why it's called high-quality protein. If you eat a bit of meat, you needn't worry about the balance of amino acids in the rest of your diet. We don't need much, since we require only fifty grams, less than two ounces, of protein in all. Only the worst of diets, all fat and sugar, will fail to provide a good fraction of those fifty grams. A few bites of meat make you far more tolerant of variations in the composition of your fruits, grains, and vegetables. (A few bites more will make a difference in the volume of low-quality vegetation you need to provide an adequate energetic intake.)

Second comes the matter of micronutrients. Meat provides a good source of a bunch of vitamins and minerals that, while present in food from plants, are present in much lower concentrations. That may make little difference to adults, but it may well be a factor in childhood growth. Some evidence suggests that children in third world vegetarian societies suffer more from inadequate supplies of micronutrients than from amino acid shortage or imbalance. We're talking about iron, zinc, calcium, and vitamin B_{12} in particular. If you choose to eat a completely vegetarian diet, you ought to consider giving your weaned children some multivitamin and mineral supplement—or milk and eggs.

Finally, and most interesting, is fat. As we've seen, all meat has some fat, but the amount varies widely. That fat represents as concentrated a chemical energy source as anything to which humans have access. Of course we'd do as well consuming butter or vegetable oil or, if we could digest it, gasoline. We need only trace amounts of fat, just enough to supply a couple of fatty acids that we can't synthesize. But to the extent that we consume fat, we can reduce the volume of all else we eat. Nothing compensates for energy-dilute vegetables as does fat.

Does that matter? Eating an almost fat-free diet while doing hard work puts a heavy burden on the digestive system. To view it in quantitative terms, consider a laborer who requires a daily input of 4,000 kilocalories and satisfies it by eating boiled rice. At 110 kilocalories per 100 grams, the person has to eat more than 8 pounds, or 3.6 kilograms, per day. That's a lot of rice. A quick trip to the kitchen (we have convenient leftovers and a good scale) puts that at thirty cups, or over seven liters. The energy equivalent in animal fat weighs about 1 pound—less than 500 grams—and occupies only a little over two cups. Potatoes look a little better than rice, but only if we ignore any air space between individual spuds. Thus fat spares volume. Over the long haul, a low-fat diet must be a high-bulk diet and, of course, vice versa unless you want to gain weight. But a bit of lean meat—muscle per se—makes little difference. Pound for pound, as you can see from a look back at the table early in the chapter, as served lean meat has about the energy content of rice or potatoes.

Bottom line: We don't need to eat muscle, but it does provide a useful and historically significant component of a human diet. That we don't need to eat it has permitted us to increase our population to its present grotesque size. Beyond these points lie culture and attitude, grounds upon which I lack the temerity to tread.

CHAPTER 15

Pulling Things Together

That's how an engineering handbook might describe muscle. The quotation comes (slightly abridged) from the notice of a lecture to the Institute of Electrical Engineers, in London.[1] Wilkie covers the territory properly. Item (1) refers to the "skeletal" muscles that attach to our bones, the kind we've mainly been talking about. Item (2) describes (a) the "smooth" muscle that propels food through our intestines, (b) heart ("cardiac") muscle, and (c) the flight muscle of many insects.

More Sarcocentric Matters

We've touched a lot of bases in these pages. We've moved from the structure and function of muscle through the way we supply it with fuel and control its contractions, the way it forces our small tasks and powers our larger ones, the limits it puts on our activities, the tuning of traditional muscle-driven devices to its capabilities, the changing ways we've harnessed the muscle of our domesticated animals, the weapons with which we've muscled our way up the food chain and brought low one another's empires, and we've concluded by eating the very subject whose saga we recounted. But I cannot end without noting bases left untouched lest the reader, perhaps dazed by the diversity of things mentioned, imagine that nothing remains to be said.

We shiver when cold, working our muscles at no specific mechanical task but simply taking advantage of the heat generated in their operation. On cool mornings, large insects can be seen doing the same thing, as heat-generating tremors of their flight muscles vibrate their wings and bring their flight motors up to takeoff temperature. Some animals take the process a step farther, with what's called thermogenic muscle—that is, muscle with no other function than generating heat.[2]

Fish of a number of lineages modify muscle for yet another function. Recall that the cell membranes within a muscle carry an electrical potential of the order of a tenth of a volt. Line them up in series (as individual electroplaques), and higher voltage will be reached, the way we line up six individual cells in the battery of an automobile to get twelve volts rather than two. Some of these fish use relatively modest overall voltages to create around themselves electric fields, the distortions of which can be sensed to reveal objects that might be eaten or avoided—a lovely scheme for dark and murky waters.[3] Some fish, electric eels being the best known, take things further, generating voltages in the hundreds, sufficient to stun prey and deter predators.[4] That takes heroic alteration; the electroplaques of an electric eel run almost the full length of each side of the body and account for more than half its mass.

We've not touched on such important subjects as how muscle develops during embryonic life and how muscle can undevelop, part of a subject, programmed cell death, that goes by the name apoptosis. Breaking down muscle for use as a source of energy in starvation, sloth, or spacecraft isn't the only situation in which we unmake our motors. Muscles may be dismantled as part of the normal course of events. An insect such as a silk moth emerges as an adult from its pupal skin and immediately spreads its wings, spreading them in area, not just in postural position. Insect wings contain no muscles, but they do have tubes that (initially at least) are filled with fluid. Compressing their abdomens, like squeezing a set of bellows, raises internal fluid pressure and does the job, whereupon the intersegmental muscles in the abdomen break down and get used as fuel.[5]

Nor have we said much at all about the muscle that propels food through our intestines—"smooth muscle" since it lacks the obvious cross-striations of the stuff that pulls on bones—or of the special character of heart, or cardiac, muscle, cross-striated but different in both structure and mode of operation. A transplanted heart needs neither nervous connections nor pacemaker; it's self-triggering. Or of a peculiar obliquely striated muscle made by squid and other cephalopod mollusks, muscle that can contract by a longer fraction of its length than other muscle. And so on. But the author reminds himself, as his editor will otherwise do, that comprehensive isn't synonymous with comprehensible.

And Some Practical Matters

Most of us do little work in the strictly mechanical sense, however stressful and frenetic our lives. Our muscular equipment, though, marching to a drummer programmed earlier in our evolutionary history, reacts in a way that's now inappropriate. It comes from a long use-it-or-lose-it tradition of dynamic biological design. Muscle-sparing technology has the downside of requiring us to work to keep the system fully functional. Were simple strength (or its lack) the only thing at issue, we might choose the easy path of technological dependency; I certainly care little about whether I'm 35 or 45 percent muscle. Unfortunately, present evidence suggests that things I do care about, such as day-to-day health and ultimate longevity, depend on keeping my cardiac and skeletal muscle occupied. So with lots of other people, I substitute cardiovascularly gainful for economically gainful work.

The cardiovascular work seems, on the whole, to be more important than the work of skeletal muscles. If I had to choose, I'd give up the weight machines before I'd quit the aerobic machines. Still, that oversimplifies the choice, since the aerobic activities do work my muscles while the muscle-strengthening machines have little cardiovascular consequence. Not that I get much immediate joy from either. They do give me time and incentive to imagine alternatives. For instance, what about sitting in a chamber in which the air contains less than its normal 21 percent oxygen? That should persuade the heart to work harder as it supplies adequate oxygen to the body's various tissues. You would be doing the equivalent of moving to a much higher altitude. No, the scheme won't work, at least in this simple fashion. The primary stimulus for harder heart work turns out to be not oxygen but carbon dioxide. When the tissues (such as muscle) produce more carbon dioxide, the heart and lungs increase their activity, removing the carbon dioxide and (not incidentally) supplying more oxygen. So what we need isn't less oxygen but more carbon dioxide in the chamber. Easy to do, but it sounds like a risky proposition. Even so, only the heart muscle and the muscles used for breathing would get any stimulation. Oh, well, back to jogging, rowing, treadmilling . . .

As if maintaining our corporeal muscle weren't enough of a problem, we've now begun to worry about making extracorporeal muscle. That's one of the ongoing endeavors of an emerging field that calls itself biomimetics. None of our traditional engines, as should be more than clear by this point, looks or works much like muscle. They don't shorten at constant volume and temperature; they don't come in anything like such a wide size range; they're far distant from Wilkie's bill of particulars with which the chapter began. Wouldn't it be nice to have some similar item of human technology?

Well, would it? Biomimetics has from the start been afflicted with a heavy dose of naturalistic romanticism—nature does better than we, nature defines perfection in design, and so forth. Lest we lose contact with reality, it's important to bear in mind that we build engines that do better than muscle in power output relative to weight and in efficiency of conversion of chemical to mechanical energy. At a different level of design, we persist in using that radical nonbiological device the crank when we want to make humans work machinery with the greatest power output and efficiency. On her own, nature never pedals anything, but we crank pedals in all our human-powered aircraft and our best human-powered boats. Both of these latter use not flapping wings or flipping tails but rotating propellers, yet another "unnatural" device.

Still, another engine has to expand our capabilities, and a small, short-stroke, linear motor of reasonable weight and power efficiency has much to offer. Look at the drive mechanism of, for instance, a small tape recorder. One or a few motors work a rat's nest of belts, pulleys, levers, and gears. Then consider how you move your fingers. A host of motors, the muscles of hand and forearm, run things with only simple bony levers and tendonous cables. Our muscles outnumber our bones. Proliferating active motors to replace passive elements might have been an unattractive trade half a century ago—after all, motors require control elements—but in a world where few things cost less than complex electronic controls, the objection becomes laughable.

In particular, using lots of little motors sounds like a wonderful thing if you intend to build walking machines, vehicles that can deal with terrain that's out of the question for anything on wheels. Mechanical complexity dogs designs for legs; a multiplicity of independently controlled motors ought to be handy (or put you a leg up, to repeat an earlier pun). Whether such vehicles carry drivers or work robotically becomes a minor issue, since any driver would give only general commands to a largely robotic control system. The same argument (in no way original with me) holds for robotic manipulators, already normal industrial devices. Not that these hypothetical small motors need be especially musclelike, of course.

But deliberately musclelike motors hold a special promise, one, moreover, that doesn't hinge on their being available at low cost. At present we lack a good prosthesis for muscle. We can make excellent hip joints, adequate knee joints, reasonable blood vessels, and various other substitutes for what we normally grow. Artificial hearts? Despite many years of work we still do transplants—and manage perpetual problems of immunological rejection. We also try things like cardiomyoplasty, described earlier. Artificial hearts have now come close to practicality; what's of interest here is how little they resemble the biological original. Valves are easy; the problem goes back to our lack of proper prosthetic muscle.

That promise is driving a great deal of effort—and not without remarkable progress. So-called electroactive polymers have already reached the power to weight ratios of respectable muscles, and they do so at musclelike frequencies of operation and relative shortening distances. With current effort and progress, whatever I say will be past history by the time this book is read; perhaps I'd do best to stop here with a good reference and a suggestion that given the commercial interest in biomimetic muscle, a website search (with the usual discount on self-serving claims) might be in order.[6]

Of course, nothing demands that prosthetic muscle be particularly biomimetic in any strict sense. Our ways and nature's ways of making things differ as much as any two manufacturing methods could. I don't expect to see practical devices based on synthetic actin, myosin, and so forth, and I doubt if we'll end up using protein at all; for that matter, few of the materials currently of interest involve protein. Winning the game depends on making something that works like muscle in the sense of output; whether it works like muscle in an internal, mechanistic sense has no relevance—unless you take to heart some mantra about nature's unique perfection, as I emphatically do not. If you want real muscle, why not employ animals as factories? The trick then becomes one of growing immunologically neutral stuff suitable for xenotransplantation—transplantation from one species to another.

Finally, Still More General Matters

Muscle, we saw, worked like any ordinary enzymatic reaction. A protein acts as an enzyme—that is, it goes through a reversible change in configuration that in turn makes other big molecules change shape. ATP feeds into the system and powers the configurational change, coming out as ADP and phosphate in the process. The system makes the change in a certain ionic environment, in this case when exposed to calcium ions. In that sense, all depends on molecular structure and biochemical reactions.

What's different is the way the configurational changes in a huge number of molecules arranged side by side and end to end add up to motion and force at our perceptual scale. Beyond these molecular events, the world of muscle action becomes macroscopic in action and consequence. We enter the world of physiology and biomechanics. We can assert that the molecular events explain the physiology and biomechanics, and in a certain sense, we must be correct. But in another sense, these latter exist in a separate domain, with emergent properties and behavior that depend on a lot more than muscle's molecules. Explanations must take cognizance of structures on larger scales: of bones and tendons; leverage and muscular hydrostats; distances for provision of material by diffusion and circulation. These matters too can be reduced to underlying phenomena and principles; reductionism may not characterize all science, but it defines most of what we scientists do. Still, these matters of physiology and biomechanics reduce not to biochemistry but to the mechanical world of physics and engineering science.

I mean to emphasize with the present account not just the legitimacy but the central importance of this physically rather than chemically reductionist science. Even in a world in which informational content may be specified by genes, much happens for which genes have little relevance. Losing sight of, or interest in, this world deprives us of any possibility of explaining the kinds of phenomena we've seen in the previous chapters. Beyond that, this science of how whole organisms work must be central to biology. If evolution establishes the strategy of life, then physiology, the study of function, deals with the tactics. That follows from the core mechanism of evolution, natural selection. Selection depends most directly on the value of a feature to the reproductive success of an organism and only indirectly on the characteristics of cell or ecosystem.

Never mind Copernicus; each of us stands at the center of the universe. What distance matters most? That from our location to some other place. From what point in time do we date things? The instant we now experience, the dividing line between past and future. The prejudice of the academic is that his or her own discipline has more to say about what truly matters than does any other. I'm a biologist. My professional biases start with the belief that we just can't understand history, literature, economics, art, and so forth without taking biology into consideration. The suggestion that particular plagues and the domestication of plants and animals have played an important role in human history raises no eyebrows. More controversial, but, in my view, undeniable, is the idea that our behavior and even our aesthetic preferences reflect our evolutionary antecedents.

We cannot escape all the baggage that we inherit as organisms, animals, mammals, primates, hominids, humans—any more than we can escape the gravitational attraction between our bodies and the earth. That goes against the grain of brave statements declaring all to be possible if only we apply sufficient intelligence, audacity, determination, the choice depending on the source of the inspirational exhortation.

That declaration of the primacy of biology offends our liberal sentiments. Some offense traces to our conflation of laudable attempts to improve the human condition with ill-starred attempts to create a perfect society. Some traces to grotesque misapplication of notions about biological imperatives past and present: social Darwinism, eugenics, and more recent declarations of genetic and behavioral predestination. Luckily, a decent distance separates my particular brand of biology from such contentious things as personal behavior and morality, so I don't have to duck, say, antisociobiological brickbats. But no mistake, I remain persuaded that biology underpins the human world. That view provides one of the chief motivations for writing this book.

Is the conviction that biological constraints lurk beneath every bush of civility and culture a paradoxical base for socially liberal views? The apparent paradox involves no real contradiction. The problems of social improvement must simply encompass an additional complication, one that cannot be swept under the rug by wishful thinking. We just have to include that additional component, which might best be put as a question. How might we transcend our biological limitations? In answering the question, we have to recognize that fitness in the evolutionary sense constitutes no key to social desirability.

I see three main avenues for getting around the bits and pieces of behavior and physiology that, however advantageous to reproductive success at some stage in our evolution, no longer fit our present best intentions. We can deliberately adjust our behavior, extending the process in which every child discovers that he or she is just one of many humans with analogous sensitivities and aspirations. We can develop technology, continuing what began when a hominid first modified some stick, bone, or rock to improve its suitability for an intended task. Although no one can give a specific formula, everything that has been said in these sarcocentric pages suggests some general advice, a kind of step number three. That consists of understanding our biological nature, including, of course, our prime mover.

Notes

Chapter 1: Body Work

1. For undocumented statements such as the preceding, the reader may assume standard textbooks or encyclopedias.
2. Bennet-Clark (1977).
3. Thomas Kuhn (1977, pp. 66–104) gives a particularly insightful account of the origin of the energy conservation principle. See also Kevles (1971, pp. 15–16).
4. These beautiful drawings, mostly done by the great Renaissance painter Titian, are available in a variety of editions. In a library, start by looking up Saunders and O'Malley (1950). Or, still in print, is Vesalius (1973) as a Dover paperback. Or (Vesalius, 1998), there's a CD-ROM. The original wooden blocks survived until the bombing of Munich during World War II. A few full-size printings done from those blocks were published in 1934; they're of course rare and valuable. Finally, all seven books are now being published (at a high price) by the Norman Publishing Company.
5. The extreme value exceeds most published figures; it comes from Graeme Taylor's (2000) measurements on crab claw closers, about which more later.
6. Act V, Scene 2, line 10, in case you've forgotten.
7. Good sources for data such as in this paragraph and the next are Weis-Fogh and Alexander (1977), McMahon (1984), Alexander (1985), and Ellington (1985).
8. Data from various sources, as collected in an earlier book (Vogel, 1998).

Chapter 2: How Muscle Works

1. Hill (1960).
2. Szent-Györgyi (1951), pp. 8–9.
3. See Szent-Györgyi (1963) or (1965) for a bit of autobiography that gives a sense of his personality; the note about Apollonians and Dionysians is Szent-Györgyi (1972).
4. The numbers for fibers, fibrils, and sarcomeres in calf muscle come from McComas (1996).
5. The original presentation, in somewhat different format, was by Gordon et al. (1966).

6. A recent issue of the most ancient English-language scientific journal, the *Philosophical Transactions of the Royal Society of London*, is devoted to such molecular motors; on muscle's cross-bridges, see in particular Holmes and Geeves (2000), A. F. Huxley (2000), and H. E. Huxley (2000).

7. This datum and those that follow come from the excellent physiology textbook of my colleague Knut Schmidt-Nielsen, who got the numbers from one of his former students, James H. Jones.

8. Sometimes they're referred to as red blood corpuscles to remind us that although they are cell size and membrane bound, they lack nuclei and can't do what cells otherwise do.

9. Wilkie (1950).

10. Mention of the event and a general account of the recognition of negative work are given by Hill (1965). The specific reference is Abbott et al. (1952).

11. A bit of jargon, useful when one queries indexes. Contraction with shortening is called concentric contraction; contraction with lengthening (as here), eccentric contraction. To me the names seem neither graceful nor informative. In addition, they draw an incompletely separate distinction from isotonic contraction, which is shortening against a constant load, and isometric contraction, contraction against an immovable load.

12. The figures come from the classic paper by Rudolpho Margaria (1938); Wallace Fenn (1923) made the original observations on the energetics of contraction during lengthening in Hill's laboratory. Hill (1965) refers to his report as "one of the most notable papers in muscle physiology."

13. McComas (1996); see also McMahon (1984).

Chapter 3: And How We Found Out

1. Szent-Györgyi (1957).

2. A. F. Huxley and R. Niedergerke (1954) and H. E. Huxley and J. Hanson (1954).

3. From the translation (and commentaries) of Nussbaum (1978).

4. For a little more on the pneuma, a good source is Bynum et al. (1981).

5. From Szent-Györgyi (1963 or 1965, p. 462).

6. As quoted in Dorothy Needham's (1971) splendid history of muscle, under the fine title of *Machina Carnis*. Needham in turn is quoting Michael Foster (1901).

7. On the basis of a variety of odd statements in it, I don't think the translation should be regarded as definitive. You might, to get a bit suspicious, take a quick look at the translator's footnote on p. 362.

8. Or perhaps those who value ladies' handbags. But that's a different Borelli.

9. He noted that blood pressure was simply insufficient for the task. A penis has a trick to boost the pressure, recently discovered, about which more in a later chapter.

10. According to Dorothy Needham (1971).

11. Frank (1980) gives a nice sense of seventeenth-century physiology, a mix of the familiar and the bizarre.

12. This account is drawn mainly from Needham (1971) and Huxley (1980); they differ slightly but in no way that matters here. Incidentally, *A* was originally *Q* and *I* was *J*—for the German equivalents of "anisotropic" and "isotropic."

13. A. V. Hill (1965) makes this point forcefully.
14. The lesson drawn by A. F. Huxley (1980).
15. The old gas furnace in my house used such a thermocouple to assure its control that the pilot flame yet burned, lest ignition fail and the gas be released into the house. It burned out every few years, but replacement was easy and inexpensive.
16. H. E. Huxley, and J. Hanson (1954); A. F. Huxley, and R. Niedergerke (1954).
17. H. E. Huxley (1952, 1996).
18. For a relatively recent discussion of remaining questions, see A. F. Huxley (1988).
19. A. F. Huxley (1980).

Chapter 4: Flying High, Making Noise, and Clamming Up

1. Part of a note from William Hartree, an engineer, to A. V. Hill, about 1960. Quoted in Hill (1965).
2. The extensor hallucis brevis—location and action of this and the soleus, below, from Warwick and Williams (1973).
3. These figures come from McComas (1996).
4. Kooyman and Ponganis (1998).
5. The common demonstrations of diffusion in air, such as releasing perfume in a room, are (not to mince words) frauds; they demonstrate the ubiquity of air currents, not diffusion, which is so much slower that it defeats ordinary exposure. In water the situation is no better: Random currents are less, but diffusion coefficients are some ten thousand times less. See Vogel (1994).
6. The story of facilitated diffusion by myoglobin and the implications for hemoglobin evolution is succinctly told in his textbook by Schmidt-Nielsen (1997). Murray (1971) gives a more specific analysis of facilitated diffusion, and Kagan (1973) looks at myoglobin in general.
7. The title of Scholander's autobiography (1990), *Enjoying a Life in Science*, strikes me as singularly apt. See also Schmidt-Nielsen (1987).
8. McShea (1994) worries about inadvertent bias that might underlay putative size increases, the trend called Cope's Rule.
9. Forward speed is defined relative to the air it flies in; rearward speed, to the insect itself.
10. Gilmour and Ellington (1993) measured a shortening (strain) of 1.9 percent, at least as low as any published estimate.
11. Ellington (1985) gives a good, if slightly technical, summary of the flight muscle story.
12. See Pringle (1981), who discovered this autonomous (asynchronous) mode of operation of flight muscle back in 1949.
13. That may be the record, according to Wooton and Newman (1979), at least for flight muscles. Some sound-producing muscles, as we'll see, do better yet.
14. Embarrassingly, I have to admit that I never published these results. They didn't fit the story line of my thesis and somehow just stayed in the notebook. I ought thus to note that the material was *Drosophila virilis*, that frequency was recorded through a phonograph pickup attached to the animal, and that the long muscle,

the trochanteral depressor, seems to be a general feature of brachycerous diptera.
15. Harvey (1628), translated by Leake (1941).
16. Krogh (1929) gives the approach explicit recognition, and Schmidt-Nielsen has made it the subject of elegant general lectures.
17. I give a more elaborate development of the point elsewhere (Vogel, 1996).
18. See Fawcett and Revel (1961).
19. Josephson (1985).
20. Rosenbluth (1969).
21. Revel (1962).
22. Rome et al. (1996).
23. Josephson and Young (1981).
24. According to Muneoka and Twarog (1983), who give much more information on the operation and pharmacology of catch muscle.
25. Kier (1985).
26. Taylor (2000).

Chapter 5: Knowing What We're Doing

1. Sherrington (1951).
2. The account here follows the standard textbooks, but I've omitted several layers of complexity. Brooks (1986) has been particularly useful. Sherrington proved the existence of these postural reflexes, and early work on their mechanisms was done in the 1920s by Edgar D. Adrian, with whom Sherrington shared the 1932 Nobel Prize.
3. Figures for total number of muscle spindles and their density in various muscles can be found in McComas (1996).
4. The information that follows comes mostly from Schmidt-Nielsen (1984) and Weibel (1984). My own earlier book (Vogel, 1992) gives still more.
5. Data from Wade and Bishop (1962), quoted by Weibel (1984).
6. The operative rule, that total flow varies with the fourth power of radius or diameter, is the Hagen-Poiseuille law for laminar flow through circular pipes. See Vogel (1994).
7. Goldspink (1991) deals with the interaction of turnover and remodeling. Schoenheimer's account (1942) is still worth reading.
8. *Man on His Nature* (2d ed.; Doubleday Anchor) pp. 125–26. I've taken the liberty of making some deletions.
9. Williams and Goldspink (1978).
10. Salmons and Sréter (1976) did the implantation experiments and give general background and references. See also Eisenberg (1985).
11. See Hoppeler and Lindstedt (1985).
12. Data quoted by Weibel (1984) from Howald (1976) and Hoppeler et al. (1973).
13. McComas (1996).
14. See Hudlická (1985).
15. I give a more complete account in Vogel (1992) and explain the basic rule in Vogel (1994).
16. A good general account, from which most of the accompanying material was derived, is Dimengo (1998).
17. Cannon (1932).

Chapter 6: Connecting Up Muscles

1. Borelli (1680).
2. The information on jellyfish comes from DeMont and Gosline (1988) and that on squid from Gosline and Shadwick (1983). The remainder is reviewed by Alexander (1988).
3. The purist will argue that they never truly pull, that they just make the external atmosphere push. Yes, and that's why they don't "pull" at more than a pressure of one atmosphere.
4. Ellis (1944); Parry and Brown (1959).
5. See Seymour (1969).
6. For a general account, see Kier and Smith (1985).
7. Vogel (1988).
8. The pendulum-based micromotor, functioning as a vibrational gyroscope, was created by Burton Boxenhorn and Paul Greiff, in the Draper Laboratory at MIT, during the 1980s. Boxenhorn says (by E-mail to me) that earlier versions probably exist and that work on similar devices goes on.
9. He's also monumentally misleading. Force produces not motion but acceleration, at least in the absence of drag, as would be the case for the earth as a celestial body. Thus, even the slightest force would suffice to move the earth, although the effect of a small force wouldn't be noticeable for quite a while. But Newton had yet to straighten us out.
10. In not all cases do we use levers to decrease force. An exception involves the three tiny bones of the middle ear, which increase the force with which the eardrum vibrates as they transmit the vibrations to the window of the inner ear, where the neural machinery is located. Muscles, however, are not involved.
11. The first pair comes from Currey (1984); the others are from Tricker and Tricker (1967).
12. See, for a good general discussion, Ellington (1985).
13. Vogel (1988), pp. 267–68.
14. The operation of this muscle and its tendon has been studied in detail by Rack and Ross (1984); Alexander (1988) summarizes their work.
15. All this material on human anatomy comes from Warwick and Williams (1973).
16. Elner and Campbell (1981).
17. See Clark (1964). For a more recent picture, see Alexander (1988). The basic story on the shark is given by Wainwright et al. (1978) and on the mammalian penis by Kelly (1997).
18. Nishikawa (1999).
19. Wainwright et al. (1991).
20. A good view of the state of the art and current questions is given by Dickinson et al. (2000).
21. See Alexander and Jayes (1983), McMahon (1984), and Lee and Farley (1998).
22. I think the earliest paper to make the point is Alexander (1974), based on work on a jumping dog. He provides good reviews in the paper of 1984 and the book of 1988. See also a recent review by Biewener (1998).
23. See Ker et al. (1987) and Alexander (1988). Terminology varies. What is in the vernacular the Achilles tendon becomes the calcanean tendon or tendo calcaneus in many anatomy books. It's connected to both the gastrocnemius and

soleus muscles of the calf, which are sometimes together referred to as the triceps surae.
24. Dimery et al. (1986).
25. Resilin was discovered by the Torkel Weis-Fogh in the late 1950s (see paper of 1960). Alexander (1988) discusses its operation and role; Vincent (1990) deals with the basis of its elasticity.
26. I give a proper derivation in Vogel (1988), as does Bennet-Clark (1977).
27. The story here is taken from Bennet-Clark and Lucey (1967).
28. Bennet-Clark (1985); Evans (1973).
29. Wainwright (1985).

Chapter 7: Using Hand Tools

1. Washburn (1959).
2. Darwin made the suggestion in *The Descent of Man* (1871), a much more radical and speculative (and lesser-known) book than *Origin*.
3. More important yet, get a low bench with a substantial vise. Many tools are easier to use if the work doesn't have to be held simultaneously, and little people build with little pieces, exacerbating the problem of support.
4. The information in this and the next paragraph comes from Fraser (1980). Even smaller handles find use in specialized dissecting and surgical instruments.
5. Rolt (1965); see also Zagorskii (1960).
6. Momentum is mass times velocity; kinetic energy is half of mass times the square of speed. Both matter, but in subtly different ways that we don't need to worry about here.
7. Cotterell and Kamminga (1990) provide an especially good analysis of the physical processes involved in making and using stone tools. Clark (1965) talks about stone blades as early items of commerce. The literature on stone tools is voluminous; the article "Tools" (McCarthy and McGeough, 1990) in the *Encyclopaedia Britannica* provides a good start with some basic references.
8. McCarthy and McGeough (1990).
9. A good recent paper, with a good literature review, is Mathieu and Meyer (1997).
10. Kelly (1949).
11. Partridge (1984).
12. See Russell (1967) and Kauffman (1972) for details and Basalla (1988) for perspective.
13. Kauffman (1972) has illustrations.
14. Hindle (1981).
15. Moritz Busch, in 1851 or 1852; translation quoted by Fitchen (1986).
16. Hertzberg (1972).
17. Kelly (1949).
18. Sleeswyk (1981).
19. Vogel (1998); Fig. 9.9, p. 193.
20. Kirby et al. (1956).
21. White (1962), p. 110.
22. Ibid., p. 28.
23. Capodaglio and Bazzini (1996).
24. Hertzberg (1972).

Chapter 8: Working Hard

1. Robert Hooke, in *Micrographia* (1665), quoted by Hart (1965).
2. One joule equals 0.239 calories; 1 calorie equals 4.19 joules. The nutritional calorie ("Calorie" or, better, "kilocalorie") is a thousand times greater. One horsepower equals 746 watts (or joules/second) or 641 kilocalories per hour or 550 foot-pounds per second.
3. Wilkie (1950). The subject, given as DW, is most likely the investigator himself.
4. Data from Thomas (1975), Schmidt-Nielsen (1984), and Taylor et al. (1987). I have calculated the values for the cat, gazelles, and cattle from the data of Taylor et al. (1981) using Kleiber's equation (see Schmidt-Nielsen or Taylor et al.) for standard (resting) metabolism.
5. The information that follows in this and the next paragraph comes from Carrier (1984).
6. Bramble (1989).
7. This plus the general information comes from Åstrand and Rodahl (1970). An especially accessible recent source is Alexander (1999).
8. The old data come, second and third hand, from Cotterell and Kamminga (1990).
9. Henderson and Haggard (1925).
10. Wilkie (1960a).
11. Data from Wilt (1968).
12. Di Prampero (1986).
13. Data from the 1999 *World Almanac* and www.nbcolympics.com; the equation for the fitted curve is $y = 0.003x^2 - 11.653x + 11908$.
14. Jones and Lindstedt (1993).
15. Beers and Berkow, eds. (1999), p. 1799.
16. I speak from personal experience. One of these can be seen at the New York State Museum, at Albany.
17. Wachsmann (1995).
18. Landels (1978).
19. Morrison (1995).
20. Shaw (1995).
21. A reconstructed monster of a medieval galley can be seen at the Maritime Museum in Barcelona, Spain. It used about 260 oarsmen, 4 to an oar, on a single deck; it's far less compact than a classical galley that had the same number of rowers. In short, it looks terribly inefficient, even if magnificently reconstructed. As an alternative to a visit to the museum, you might look at paintings by Titian, Tintoretto, and Veronese of the Battle of Lepanto in 1571.
22. Figures from Morton (1995).
23. Bondioli et al. (1995).
24. A fine full-size (and shockingly small) reconstruction is on display in Oslo.
25. I've gone into the issue of lift- versus drag-based thrust more extensively and quantitatively elsewhere (Vogel, 1994).
26. See, on the subject, Abbott et al. (1995) and Brooks (1995).
27. See, for instance, Morton (1995).
28. I might argue that Macmillan lacked the roller chain, that wonderfully efficient driving device, invented later in the century. But he might have con-

nected a pedal shaft to the rear wheel with a pair of pushrods, one associated with each pedal. For more on the history of bicycles and such, see Kyle (1995).

29. It takes 2 and 4 meters per second; 15 watts per kilogram; 8.4 and 4 meters per second. Data from Whitt and Wilson (1982). Bicycle weight: eleven to fourteen kilograms.

30. In SI units, these figures are 160, 300, 175, and 1400 kilojoules per kilometer.

31. Whitt and Wilson (1982).

32. See, for instance, the elegantly illustrated book by Hart (1985).

33. Grosser's (1981) book on the epic is a real page turner.

34. Wilkie (1959, 1960a, 1960b).

35. Wilkie (1984).

36. See, for example, Tennekes (1996), pp. 104–05.

37. Dorsey (1990) tells this story in engaging fashion.

38. Physiological factors, including power output, fatigue, hydration, and energy supply, are discussed by Nadel and Bussolari (1988).

39. For a good review, see Drela and Langford (1985); for a somewhat more technical introduction, see Drela (1990).

40. See Roper (1995) and www.ihpva.org.

41. On how wheelbarrows work, see Cotterell and Kamminga (1990).

42. On their history, see Needham (1965), vol. 4, part 2, section 27; White (1962); Gies and Gies (1995); www.iias.nl/iiasn/iiasn5/eastasia/needham.html; http://scholar.chem.nyu.edu/~tekpages/wheelbarrow.html.

43. Allusions to a specific inventor, Chuko Liang, in 234 C.E. (Gies and Gies, 1994), should, according to Needham (1965), be taken with a large grain of salt. He may have used them, but they're most likely older.

Chapter 9: More Tough Tasks

1. Leopold (1949).
2. Soule and Goldman (1969).
3. Cotterell and Kamminga (1990), p. 193.
4. Ibid.
5. Fitchen (1986), p. 172.
6. Astrand and Rodahl (1970), p. 445.
7. Lothian (1921), p. 246.
8. Dan Livingstone, personal communication.
9. Maloiy et al. (1986); Taylor (1986).
10. Soule and Goldman (1969).
11. Kram (1996).
12. Soule and Goldman (1969).
13. Kram (1991).
14. Goldman and Iampietro (1962); Givoni and Goldman (1971).
15. Hughes and Goldman (1970).
16. Minetti (1995).
17. Lothian (1921, 1922).
18. "Dark Ages" = "knight time."

19. www.bodymechanics.com/safety.htm.
20. Heizer (1966) gives a very nice summary with lots of data, including the figures in the next paragraph.
21. Renfrew (1973); Phillips (1980).
22. Merritt and Thom (1980); Service and Bradbery (1993).
23. Data and formulas from Vogel (1988b), chs. 9, 10.
24. Fitchen (1986) makes a good case against use of rollers for the pyramids. He prefers dragging stones along greased tracks.
25. The horizontal pulling force equals your weight divided by the tangent of the angle you make with the ground. When a = 0°, tan a = 0, F = W/tan a = •.
26. Data from Cotterell and Kamminga (1990), who did the measurements.
27. On obelisks in particular, see Dibner (1950).
28. That's roughly equivalent to an eight-inch cube, if we make a guess for density. The datum comes from Dibner (1950).
29. Cowan, 1997.
30. Franklin (1742).
31. Defebaugh (1906–07), p. 541.
32. Pursell (1995), p. 158.
33. Jones and Simons (1961) have such an illustration. To my admittedly unprofessional eye, the teeth look like those of a bidirectional ripsaw, which cuts with rather than across the grain of the wood.
34. Forbes and Meyer (1955), p. 16: 11.
35. Åstrand and Rodahl (1970), citing Hansson et al. (1966).
36. Hughes and Goldman (1970).
37. Kelly (1948).
38. One shouldn't too blithely dismiss the adz, however. It retained use for specialized tasks, such as roughing out the keel of a wooden boat when an irregular shape was needed. Only a century ago Joshua Slocum (1900) adzed the keel of the *Spray*, which he sailed alone around the world.
39. Information on pit saws can be found in Oliver (1956) and Jones and Simons (1961). Pictures are widespread; Peterson (1975) has several nice ones, including a drawing of a horse-powered pit saw.
40. Agnoletti (1994).

Chapter 10: Bringing Animals to Bear . . .

1. Quick (1923), pp. 248–52, is the source; some sentences have been omitted. Quoted also by Nevins and Commager (1967), p. 317.
2. The largest part of the information in the list comes from Clutton-Brock (1999). General information on the various species mentioned comes also from Nowak (1991).
3. Clutton-Brock (1992).
4. Taking basal metabolic rate proportional to body mass to the power 0.75.
5. Recalling Chapter 8 and taking maximal metabolic rate proportional to body mass to the power 0.81.
6. Hill (1950); McMahon and Bonner (1983).
7. We must be extremely skeptical of published maximum speeds for animal loco-

motion beyond such obviously well-documented cases as humans, dogs, and horses. Exaggerated figures, once mentioned, get quoted again and again in ordinarily trustworthy reference works.

8. Alexander and Jayes (1983), Alexander (1984), or any contemporary book that touches on the subject of walking and running.

9. Taylor (1957); Alexander (1985); Brody (1945).

10. Taylor et al. (1972).

11. Full and Tullis (1990).

12. Yousef and Dill (1969).

13. Figures from Nowak (1991). My friend Barbara Grubb, a fine physiologist, raised llamas. She puts a good llama load at well under a hundred pounds, or less than a third of body weight.

14. *Encyclopaedia Britannica* (1990).

15. Randall (1997), p. 122.

16. Langdon (1986), p. 226.

17. The quote is borrowed from Powers (1924), who borrowed it herself from another secondary source.

18. Ewbank (1842).

19. Agricola (1556, posthumous); Ramelli (1588); Veranzio (1615–16).

20. Needham (1965).

21. Moritz (1958), Walton (1974).

22. See http://galileo.imss.firenze.it/pompei/technica/egcalc.html for a picture and animated version of the treadwheel-powered crane; the same site (natura/emacer.html) illustrates the animal- or human-powered Roman grinding mill. The published version (in English) is edited by Ciarello and de Carolis (2000).

23. D. R. Hill (1984).

24. Burstall (1963).

25. Examples from Derry and Williams (1960), Cotterell and Kamminga (1990), Gies and Gies (1994), and the sources cited above.

26. Needham (1965).

27. But paddle wheels must be as high, with only their very bottoms dipping into the water. Otherwise they waste most of their energy pushing water up or down. Some stream ecologists drive experimental streams with paddle wheels; at least one had to learn this lesson the hard way.

28. Johnson (1978).

29. Quoted in a fine new biography by Desmond (1997), p. 462.

30. Burrows and Wallace (1999).

31. Hardie (1824).

32. "Lizzie Borden took an ax / And gave her mother forty whacks; / when she saw what she had done / She gave her father forty-one."

33. Schifler (1973).

34. Postel (2000).

Chapter 11: Bos *versus* Equus

1. De Monroy (1796).

2. In Lexington, Kentucky, and as www.imh.org.

3. Premi (1979), cited by Cotterell and Kamminga (1990).

4. Clutton-Brock (1999).
5. Much of this comes from a general history by Parkinson (1963), who took horses more seriously than do most nonspecialists.
6. Drawn from a vase in the North Carolina Museum of Art, Raleigh.
7. The original is (1931); it became well known in the English-speaking world via White (1940), republished in White (1978).
8. Gordon (1978).
9. Cotterell and Kamminga (1990).
10. Lefebvre des Noëttes and Lynn White don't represent the last word on the issue. Other measurements were done by Spruytte (1983); see also Littauer and Crouwel (1979), Mokyr (1990), Cotterell and Kamminga (1990), and the excellent http://scholar.chem.nyu.edu/~tekpages/texts/.
11. White (1965), republished in White (1978). See also Langdon (1986), Hyland (1999), and the Web site above.
12. Hyland (1990, 1999), and the Web site above.
13. White (1965), republished in White (1978); the figure is consistent with other, more recent accounts.
14. An extensive literature argues about the relative economics of slavery and wage labor, particularly with respect to the South in the years before the Civil War.
15. White's classic statement is his small book of 1962.
16. Needham (1965).
17. Bloch (1931).
18. Raepsaet (1997).
19. White (1962) thinks it made a big difference; Comet (1997), for instance, thinks not.
20. Kransberg and Gies (1975).
21. White (1975), reprinted in White (1978).
22. Fussell (1967), noted also in Kransberg and Gies (1975).
23. Derry and Williams (1960).
24. Cowan (1997) for both items.
25. Government of India (1971).
26. Langdon (1986).
27. *The Life of the Silent Majority*, from R. S. Hoyt, ed., *Life and Thought in the Early Middle Ages* (University of Minnesota Press, 1967), pp. 85–100.
28. Lopez and Raymond (1955), cited by Langdon (1986).
29. Langdon (1986).
30. Gies and Gies (1994).
31. Allen (1848), pp. 142–43.
32. Bryant (1923), p. 130.
33. UN Food and Agricultural Organization (1982).
34. Edlin (1953), p. 239.
35. Ryan et al. (1981), p. 49.

Chapter 12: Killing Tools: The Big Picture

1. Shaw, *Man and Superman* (1901–03). The quote comes from the long dream sequence in the middle, sometimes performed separately as *Don Juan in Hell*; the speaker is the devil.

2. Kelly (1952). Originally in a daily comic strip, *Pogo*; reprinted in Kelly (1959). The speaker is a character called Deacon Mushrat, who speaks (sanctimoniously) in Old English letters.
3. See, for instance, Stanford (1999).
4. See www.worldatlatl.org.
5. See www.pipeline.com/~jbundine/ or www.abotech.com.
6. Kicking equids evidently worried our forebears. "Recalcitrant" comes from the Latin word and present anatomical term for heel, "calyx," and alludes to an animal's kicking up its heels when, as we'd say, it's "stubborn as a mule."
7. Marsden (1969, 1971) appears to be the most reliable source of information on ancient ballistae; the latter is a translation of what technical information has come down to us. Payne-Gallwey (1903) and Gordon (1978) are worth a look as well.
8. White (1940). Sometimes he got carried away by his medieval hubris.
9. For an especially amusing account, see Paul (1991). Or search the World Wide Web for "catapult," "ballista," or "trebuchet."
10. Hill (1973) makes the distinction clear in both qualitative and quantitative terms.
11. Except for a large, dense object that drops vertically or lands at a lower elevation than its launch.
12. A version of the program (in QuickBasic) appears in Appendix 2 of Vogel (1988).
13. Lucas (1982) reports the ant lion work; the caterpillar reference is Caveney et al. (1998).
14. Data for spears and ballistae from Cotterell and Kamminga (1990), for trebuchets, from data in Payne-Gallwey (1903), for bows and arrows, from Klopsteg (1943).
15. From the technical appendices in Hardy (1992).
16. Data taken from Brancazio (1984).
17. There's another important difference between momentum and kinetic energy, one that happens not to concern us directly at this point. The rate in momentum is velocity, speed with a specified direction. So the overall momentum of two equal masses colliding from opposite directions is zero, both before and after collision, properly (if counterintuitively) conserving momentum. The rate in kinetic energy is simply speed, with no implied or explicit direction. Those two masses lose all their kinetic energy (it appears as heat) if they stick when they collide (an inelastic collision); they retain it if they bounce back at their original speeds (a perfectly elastic collision).
18. Pope (1909), cited by Cotterell and Kamminga (1990).

Chapter 13: Wielding the Weapons

1. White (1975).
2. Most of the information on trebuchets comes from Hill (1973).
3. Data from the appendix of Hardy (1992).
4. See, for instance, Heath (1971) and Hardy (1992).
5. Hardy (1992), appendix.
6. Heath (1971).
7. Payne-Galwey (1903) and Heath (1971) give different instances.

8. Data from Baker (1992).
9. Yost and Kelley (1983).
10. Ibid. cites Hames for blowguns of the Ye'kwana, with their own data for those of the Waorani.
11. Xenophon, *The Persian Expedition*, trans. Warner (1961).
12. White (1963).
13. Derry and Williams (1960); Vallee (1997).
14. Mons Meg does not lack literature, much of it contradictory. See, for instance, Tranter (1987). The cannon now kept down in the bowels of the castle, exposed only to people rather than to the elements as well. But I wonder what we tourists now see. The present cannon looks smaller than the one in the figure, from Grant (1881–83), and rests on a different carriage. Grant gives its length as 13 feet and its bore as 27.5 inches and notes that the balls were granite. That translates into a ball of 1,050 pounds (480 kg), in agreement with at least one other source I've seen. It's also consistent with the weight of the weapon given by Masson (1931) of 56,000 pounds ("4000 stone"), using the usual 50:1 ratio of cannon to cannonball weight. A tourist-oriented website gives a much lower 330 pounds as the weight of its balls.
15. Cardwell (1995).
16. White (1974). The statements of a pair of contemporary advocates of muskets— John Smyth (1590) and Humfrey Barwick (1594)—have been reprinted in Heath (1973).
17. Townsend (1983) gives a fine exposition of the history and the issues involved.
18. White (1974) For a similar discussion of cannon versus trebuchets, see Hill (1973).
19. A notable example is the Swedish warship *Wasa*, which capsized in 1628. It has been salvaged and restored and has become one of the tourist attractions of Stockholm.
20. Franklin to General Charles Lee, February 11, 1776, quoted by Hardy (1992).
21. See, for an especially readable account, Schoenbrun (1976).
22. Yost and Kelley (1983).

Chapter 14: Muscle as Meat

1. Mowat (1963). He is a popular Canadian author of somewhat embellished non-fiction.
2. Gordon (1978). He was a British naval engineer and splendid writer for even the most numerophobic reader.
3. From the version edited by Harper (1893).
4. Flaxman and Sherman (2000).
5. Data from Watt and Merrill (1963); Higgs and Pratt (1999).
6. Stanford (1999). Nowak (1991) provides a wider range of confirmatory information and references without an obvious thesis to support.
7. *Julius Caesar*, Act I, Scene 2.
8. I recommend, for rings cut from mantle or the individual tentacles, a dip in boiling water lasting from half a minute to no more than a minute. The results can be eaten cold, with cocktail sauce or in salads.
9. Bateson and Bradshaw (1997). A move to ban stag hunting in Britain has

severely stressed the establishment as well as the House of Commons, equivalent, I suppose, to a movement advocating a ban on bullfighting in Spain.

10. Schmidt-Nielsen (1997).
11. Sanday (1986) provides an example of the former. Petrinovich (1999), in an especially readable account in a field fraught with pompous and pretentious prose, defends the latter.
12. Arens (1979).
13. Philbrick (2000). A more general popular account is Hanson (1999).
14. Petrinovich (1999) provides good descriptions of all these instances.
15. Ibid., pp. 212–14.
16. Oldstone (1998).
17. Nutritional cannibalism by the Aztecs was suggested by Harner (1977) and Harris (1978, 1979). The quotation is from p. 110 of Harris (1978).
18. Dornstreich and Morren (1974) made the case for human flesh as a protein source. I start with the early same figures for both supply and demand, but my calculation gives a different result. Still, even their result, that a human per year would supply 6 percent of the protein needs of ten people, isn't consistent with a steady-state human population or even, I think, with steady depredation of neighboring tribes.
19. The sources I consulted agree on this figure. See, for instance, Higgs and Pratt (1999) or Beers and Berkow (1999).
20. Milton (1999).

Chapter 15: Pulling Things Together

1. Cited in Lehninger (1975).
2. Block (1994) provides a comprehensive review.
3. The classic work is that of Machin and Lissman (1960).
4. Kalmijn (1978), for example, or general ichthyological works.
5. Lockshin and Williams (1965).
6. Try Full and Meijer (2001); at this writing interesting websites include:
   www.ai.mit.edu/projects/muscle/muscle.html,
   www.unm.edu/~amri/paper.htiml,
   rcs.ee.washington.edu/brl/project/aarm/,
   www.chemie.uni-freiburg.de/makro/finkelmann/finkelma.htm,
   www.shadow.org.uk/products/airmuscles.shtml,
   www.biomimetics.com.

References

Abbott, A. V.; A. N. Brooks; and D. G. Wilson. (1995) Human-powered watercraft. Pp. 49–67 in A. V. Abbott and D. G. Wilson, *Human-Powered Vehicles*. Champaign, Ill.: Human Kinetics.

Abbott, B. C.; B. Bigland; and J. M. Ritchie. (1952) The physiological cost of negative work. *J. Physiol.* 117: 380–90.

Agnoletti, M. (1994) Technology, economics, and forestry: water powered sawmills in Italy's Cadore region. *Forest and Conserv. Hist.* 38 (1): 24–32.

Agricola, G. (1556). *De re metallica*. Trans. H. C. and L. H. Hoover. New York: Dover Publications, 1950.

Alexander, R. M. (1974) The mechanics of jumping by a dog (*Canis familiaris*). *J. Zool.* 173: 549–73.

———. (1984a) Elastic energy stores in running vertebrates. *Amer. Zool.* 24: 85–94.

———. (1984b) Walking and running. *Amer. Sci.* 72: 348–54.

———. (1985) The maximum forces exerted by animals. *J. Exp. Biol.* 115: 231–38.

———. (1988) *Elastic Mechanisms in Animal Movement*. Cambridge, UK: Cambridge University Press.

———. (1999) *Energy for Animal Life*. Oxford, UK: Oxford University Press.

———, and A. S. Jayes. (1983) A dynamic similarity hypothesis for the gaits of quadupedal mammals. *J. Zool.* 201: 135–52.

Allen, R. L. (1848) *Domestic Animals*. New York: C. M. Saxton.

Arens, W. (1979) *The Man-Eating Myth*. Oxford, UK: Oxford University Press.

Åstrand, P.-O., and K. Rodahl. (1970) *Textbook of Work Physiology*. New York: McGraw-Hill.

Baker, T. (1992) Bow design and performance. Pp. 43–116 in J. Hamm, ed., *The Traditional Bowyer's Bible*. Azle, Texas: Bois d'Arc Press.

Basalla, G. (1988) *The Evolution of Technology*. Cambridge, UK: Cambridge University Press.

Bateson, P., and E. L. Bradshaw. (1997) Physiological effects of hunting red deer (*Cervus elaphus*). *Proc. Roy. Soc.* B264: 1707–14.

Beers, M. H., and R. Berkow. (1999) *The Merck Manual of Diagnosis and Therapy*, 17th ed. West Point, Pa.: Merck Research Laboratories.

Bennet-Clark, H. C. (1975) The energetics of the jump of the locust, *Schistocerca gregaria*. *J. Exp. Biol.* 63: 53–83.

———. (1977) Scale effects in jumping animals. Pp. 185–201 in T. J. Pedley, ed., *Scale Effects in Animal Locomotion*. London: Academic Press.

———, and E. C. A. Lucey. (1967) The jump of the flea: a study of the energetics and a model of the mechanism. *J. Exp. Biol.* 47: 59–76.

Bloch, M. (1931) *Les Caractères Originaux de l'Histoire Rurale Française*. Trans. J. Sondheimer (1966) as *French Rural History: An Essay on Its Basic Characteristics*. Berkeley: University of California Press.

Block, B. A. (1994) Thermogenesis in muscle. *Annu. Rev. Physiol.* 56: 535–57.

Biewener, A. A. (1998) Muscle function in vivo: a comparison of muscles used for elastic energy savings versus muscles used to generate mechanical power. *Amer. Zool.* 38: 703–17.

Bondioli, M.; R. Burlet; and A. Zylsberg. (1995) Oar mechanics and oar power in medieval and later galleys. Pp. 172–205 in R. Gardner, ed.,*The Age of the Galley*. Annapolis, Md.: Naval Institute Press.

Borelli, G. A. (1680–81) *De motu animalium*. Trans. P. Maquet (1989). Berlin: Springer-Verlag.

Bramble, D. (1989) Axial-appendicular dynamics and the integration of breathing and gait in mammals. *Amer. Zool.* 29: 171–86.

Brancazio, P. J. (1984) *Sport Science: Physical Laws and Optimal Performance*. New York: Simon and Schuster.

Brody, S. (1945) *Bioenergetics and Growth*. New York: Hafner.

Brooks, A. N. (1995) The 20-knot human-powered watercraft. Pp. 79–92 in A. V. Abbott and D. G. Wilson, eds., *Human-Powered Vehicles*. Champaign, Ill.: Human Kinetics.

Brooks, V. B. (1986) *The Neural Basis of Motor Control*. New York: Oxford University Press.

Bryant, R. C. (1923) *Logging: The Principles and General Methods of Operation in the United States*. New York: John Wiley & Sons.

Burrows, E. G., and M. Wallace. (1999) *Gotham: A History of New York City to 1898*. New York: Oxford University Press.

Burstall, A. F. (1963) *A History of Mechanical Engineering*. London: Faber and Faber.

Busch, M. (1971) *Travels between the Hudson and the Mississippi 1851–1852*. Trans. and ed. N. H. Binger. Lexington: University Press of Kentucky.

Bynum, W. F.; E. J. Browne; and R. Porter. (1981) *Dictionary of the History of Science*. Princeton, N.J.: Princeton University Press.

Cannon, W. (1932) *The Wisdom of the Body*. New York: W. W. Norton.

Cardwell, D. (1995) *The Norton History of Technology*. New York: W. W. Norton.

Capodaglio, P., and G. Bazzini. (1996) Predicting endurance limits in arm cranking exercise with a subjectively based method. *Ergonomics* 39: 924–32.

Carrier, D. R. (1984) The energetic paradox of human running and hominid evolution. *Curr. Anthropol.* 25: 483–95.

Caveney, S.; H. McLean; and D. Surry. (1998) Faecal firing in a skipper caterpillar is pressure driven. *J. Exp. Biol.* 201: 121–33.

Ciarallo, A., and E. de Carolis, eds. (2000) *Homo Faber: Pompeii—Life in a Roman Town*. Milan: Elemond Electra.

Clark, G. (1965) Traffic in stone axe and adze blades. *Economic History Review* 18: 1–28.

Clark, R. B. (1964) *Dynamics in Metazoan Evolution*. Oxford, UK: Oxford University Press.

Clutton-Brock, J. (1992) *Horse Power: A History of the Horse and the Donkey in Human Societies.* Cambridge, Mass.: Harvard University Press.

———. (1999) *A Natural History of Domesticated Animals*, 2d ed. Cambridge, UK: Cambridge University Press.

Comet, G. (1997) Technology and agricultural expansion in the Middle Ages: the example of France north of the Loire. Pp 11–39 in G. Astill and J. Langdon, eds., *Medieval Farming and Technology.* Leiden: Brill.

Cotterell, B., and J. Kamminga. (1990) *Mechanics of Pre-Industrial Technology.* Cambridge, UK: Cambridge University Press.

Cowan, R. S. (1997) *A Social History of American Technology.* New York: Oxford University Press.

Currey, J. (1984) *The Mechanical Adaptations of Bones.* Princeton, N.J.: Princeton University Press.

Darwin, C. (1871) *The Descent of Man, and Selection in Relation to Sex.* London: James Murray.

Defebaugh, J. E. (1906–07) *History of the Lumber Industry of America.* Chicago: American Lumberman.

De Monroy, P. (1796) *Foreign Agriculture, or, an Essay on the Comparative Advantages of Oxen for Tillage in Competition with Horses.* London: G. Nicol.

DeMont, M. E., and J. M. Gosline. (1988) Mechanics of jet propulsion in the hydromedusan jellyfish, *Polyorchis penicillatus.* I. Mechanical properties of the locomotor structure. *J. Exp. Biol.* 134: 313–32.

Derry, T. K., and T. I. Williams. (1960) *A Short History of Technology from the Earliest Times to A.D. 1900.* New York: Oxford University Press.

Desmond, A. (1997) *Huxley.* Reading, Mass.: Addison-Wesley Publishing.

Dibner, B. (1950) *Moving the Obelisks.* New York: Burndy Library.

Dickinson, M. H.; C. T. Farley; R. J. Full; M. A. R. Koehl; R. Kram; and S. Lehman. (2000) How animals move: an integrative view. *Science* 288: 100–06.

Dimengo, J. M. (1998) Surgical alternatives in the treatment of heart failure. *AACN Clinical Issues* 92: 192–207.

Dimery, N. J.; R. M. Alexander; and R. F. Ker. (1986) Elastic extension of leg tendons in the locomotion of horses (*Equus caballus*). *J. Zool.* 210: 415–25.

Di Prampero, P. E. (1986) The energy cost of human locomotion on land and in water. *Int. J. Sports Med.* 7: 55–72.

Dorsey, G. (1990) *The Fullness of Wings: The Making of a New Daedalus.* New York: Viking Penguin.

Dorstreich, M. D., and G. E. B. Morren. (1974) Does New Guinea cannibalism have nutritional value? *Human Ecology* 2: 1–12.

Drela, M. (1990) Aerodynamics of human-powered flight. *Annu. Rev. Fluid Mech.* 22: 93–110.

———, and J. S. Langford. (1985) Human-powered flight. *Sci. Amer.* 253 (5): 144–51.

Edlin, H. L. (1953) *The Forester's Handbook.* London: Thames & Hudson.

Eisenberg, B. R. (1985) Adaptability of ultrastructure in the mammalian muscle. *J. Exp. Biol.* 115: 55–68.

Ellington, C. P. (1985) Power and efficiency of insect flight muscle. *J. Exp. Biol.* 115: 293–304.

Ellis, C. H. (1944) The mechanism of extension of the legs of spiders. *Biol. Bull.* 86: 41–50.

Elner, R. W., and A. Campbell. (1981) Force, function and mechanical advantage in the chelae of the American lobster (*Homarus americanus* Decapoda: Crustacea). *J. Zool.* 193: 269–86.

Evans, M. E. G. (1973) The jump of the click beetle (Coleoptera, Elateridae)— energetics and mechanics. *J. Zool.* 167: 319–36.

Ewbank, T. (1842) *A Descriptive and Historical Account of Hydraulic and Other Machines of Raising Water.* London: Tilt and Bogue.

Fawcett, D. W., and J. P. Revel. (1961) The sarcoplasmic reticulum of a fast-acting fish muscle. *J. Biophys. Biochem. Cytol.* v10 n4, pt. 2: 89–110.

Febvres des Noëttes, R. (1931) *L'Attelage et le Cheval de Selle à Travers les Âges.* Paris: A. Picard.

Fenn, W. O. (1923) A quantitative comparison between the energy liberated and the work performed by the isolated sartorius of the frog. *J. Physiol.* 58: 175–203.

Fitchen, J. (1986) *Building Construction before Mechanization.* Cambridge, Mass.: MIT Press.

Flaxman, S. M., and P. W. Sherman. (2000) Morning sickness: a mechanism for protecting mother and embryo. *Quart. Rev. Biol.* 75: 113–48.

Food and Agricultural Organization of the United Nations. (1982) Basic technology in forest operations. *FAO Forestry Paper* 36: 1–132.

Forbes, R. D., and A. B. Meyer. (1955) *Forestry Handbook.* New York: Ronald Press.

Foster, M. (1901) *Lectures on the History of Physiology during the Sixteenth, Seventeenth, and Eighteenth Centuries.* Cambridge, UK: Cambridge University Press.

Franklin, B. (1742) *An Account of the Newly Invented Pennsylvanian Fire-Place.* Philadelphia (Pamphlet). Reprinted in J. Sparks. (1838) *The Works of Benjamin Franklin,* vol.6. Boston: Tappan and Whittemore. Pp. 34–64.

Full, R. J., and K. Meijer. (2001) Metrics of natural muscle function. In Y. Bar-Cohen, ed., *Electroactive Polymer (EAP) Actuators as Artificial Muscles: Reality, Potential and Challenges.* Pasadena, Calif.: SPIE Press (JPL Laboratory).

———, and A. Tullis. (1990) Energetics of ascent: insects on inclines. *J. Exp. Biol.* 149: 307–17.

Fussell, G. E. (1967) Farming systems of the classical era. *Technology and Culture* 8: 16–44.

Frazer, T. M. (1980) *Ergonomic Principles in the Design of Hand Tools.* Occupational Safety and Health Series, No. 44. Geneva, Switzerland: International Labour Office.

Gies, F., and J. Gies. (1994) *Cathedral, Forge, and Waterwheel: Technology and Invention in the Middle Ages.* New York: HarperCollins Publishers.

Gilmour, K. and C. P. Ellington. (1993) *In vivo* muscle length changes in bumble-bees and the *in vitro* effects on work and power. *J. Exp. Biol.* 183: 101–13.

Givoni, B., and R. F. Goldman. (1971) Predicting metabolic energy cost. J. *Appl. Physiol.* 30: 429–33.

Goldman, R. F., and P. F. Iampietro. (1962) Energy cost of load carriage. *J. Appl. Physiol.* 17: 675–76.

Goldspink, D. F. (1991) Exercise-related changes in protein turnover in mammalian striated muscle. *J. Exp. Biol.* 160: 127–48.

Gordon, A. M.; A. F. Huxley; and F. J. Julian. (1966) The variation in isometric tension with sarcomere length in vertebrate muscle fibers. *J. Physiol.* 184: 170–92.

Gordon, J. E. (1978) *Structures, or Why Things Don't Fall Down.* London: Penguin Books.

Gosline, J. M.; J. D. Steeves; A. D. Harman; and M. E. DeMont. (1983) Patterns of circular and radial mantle muscle activity in respiration and jetting of the squid *Loligo opalescens*. *J. Exp. Biol.* 104: 97–109.

Grant, J. (1881–83) *Cassell's Old and New Edinburgh*. London: Cassell, Petter, Galpin.

Grosser, M. (1981) *Gossamer Odyssey: The Triumph of Human-Powered Flight*. Boston: Houghton Mifflin.

Hanson, N. (1999) *The Custom of the Sea*. New York: John Wiley and Sons.

Hardie, J. (1824) *The History of the Treadmill*. New York: S. Marks.

Hardy, R. (1992) *Longbow: A Social and Military History*, enlarged ed. Sparkford, Somerset, UK: Patrick Stephens Ltd.

Harper, F. P., ed. (1893) *History of the Expedition under the Command of Lewis and Clark*. Reprint, 1965; New York: Dover Publications.

Harner, M. (1977) The ecological basis for Aztec sacrifice. *Amer. Ethnologist* 4: 117–35.

Harris, M. (1978) *Cannibals and Kings*. New York: Random House.

Harris, M., versus M. Sahlins. (1979) "Cannibals and Kings": an exchange. *New York Review of Books* (June 28). 26 11. Pp. 51–53.

Hart, C. (1985) *The Prehistory of Flight*. Berkeley: University of California Press.

Harvey, W. (1628) *Anatomical Studies on the Motion of the Heart and Blood in Animals*. Trans. C. D. Leake. Springfield, Ill.: C. C. Thomas, 1941.

Heath, E. C. (1971) *The Grey Goose Wing: A History of Archery*. Reading, UK: Osprey Publishing, Ltd.

———, (1973) *Bow vs. Gun*. Wakefield, UK: EP Publishing.

Heizer, R. F. (1966) Ancient heavy transport, methods and achievements. *Science* 153: 821–30.

Henderson, Y., and H. W. Haggard. (1925) The maximum human power and its fuel. *Amer. J. Physiol.* 72: 264–82.

Hertzberg, H. T. E. (1972) Engineering anthropology. Pp. 467–584 in H. P. Van Cott and R. Kincade, eds., *Human Engineering Guide to Equipment Design*. Washington, D.C.: U.S. Government Printing Office.

Higgs, J., and J. Pratt (1999) Meat, poultry, and meat products. Pp. 1272–82 in M. J. Sadler, J. J. Strain, and B. Caballero, eds., *Encyclopedia of Human Nutrition*. San Diego: Academic Press.

Hill, A. V. (1950) The dimensions of animals and their muscular dynamics. *Science Progress* 38: 209–30.

———. (1960) Production and absorption of work by muscle. *Science* 131: 897–903.

———. (1965) *Trails and Trials in Physiology*. London: Edward Arnold.

Hill, D. R. (1973) Trebuchets. *Viator* 4: 99–114.

———. (1984) *A History of Engineering in Classical and Medieval Times*. London: Croom Helm.

Hindle, B. (1981) The artisan during America's wooden age. Pp. 8–16 in C. W. Pursell, Jr., ed., *Technology in America: A History of Individuals and Ideas*. Cambridge, Mass.: MIT Press.

Holmes, K. C., and M. A. Greeves. (2000) The structural basis of muscle contraction. *Phil. Trans. Roy. Soc. Lond.* B355: 419–31.

Hoppeler, H.; P. Lüthi; H. Claasen; E. R. Weibel; and H. Howald. (1973) The ultrastructure of the normal human skeletal muscle: a morphometric analysis on untrained men, women, and well-trained orienteers. *Pflügers Archiv* 344: 217–32.

Hoppeler, H., and S. L. Lindstedt. (1985) Malleability of skeletal muscle in overcoming limitations: structural elements. *J. Exp. Biol.* 115: 355–64.

Howald, H. (1976) Ultrastructure and biochemical function of skeletal muscle in twins. *Ann. Human Biol.* 3: 455–62.

Hoyt, D. F., and C. R. Taylor. (1981) Gait and the energetics of locomotion in horses. *Nature* 292: 239–40.

Hudlická, O. (1985) Development and adaptability of microvasculature in skeletal muscle. *J. Exp. Biol.* 115: 215–28.

Hughes, A. L., and R. F. Goldman. (1970) Energy cost of "hard work." *J. Appl. Physiol.* 29: 570–72.

Huxley, A. F. (1980) *Reflections on Muscle*. Princeton, N.J.: Princeton University Press.

———. (1988) Prefatory chapter: muscle contraction. *Annu. Rev. Physiol.* 50: 1–16.

———. (2000) Mechanics and models of the myosin motor. *Phil. Trans. Roy. Soc. Lond.* B355: 433–40.

———, and R. Niedergerke. (1954) Structural changes in muscle during contraction. Interference microscopy of living muscle fibres. *Nature* 173: 971.

Huxley, H. (1952) X-ray analysis and the problem of muscle. *Proc. Roy. Soc. Lond.* B141: 59–62.

———. (1996) A personal view of muscle and motility mechanisms. *Annu. Rev. Physiol.* 58: 1–19.

———. (2000) Past, present and future experiments on muscle. *Phil. Trans. Roy. Soc. Lond.* B355: 539–43.

———, and J. Hanson. (1954) Changes in the cross-striations of muscle during contraction and stretch and their structural interpretation. *Nature* 173: 973.

Hyland, A. (1990) *Equus: The Horse in the Roman World*. New Haven: Yale University Press.

———. (1999) *The Horse in the Middle Ages*. Stroud, UK: Sutton Publishing Co.

India, Government of. (1971) *Indian Livestock Census 1966*, vol. 1. New Delhi: Ministry of Agriculture.

Johnson, P. C. (1978) *Farm Power in the Making of America*. Des Moines, Iowa: Wallace-Homestead Book Co.

Jones, J. H., and S. L. Lindstedt. (1993) Limits to maximal performance. *Annu. Rev. Physiol.* 55: 547–69.

Jones, P. d'A., and E. N. Simons. (1961) *Story of the Saw*. Manchester, UK: Newman Neeme, Ltd.

Josephson, R. K. (1985) The mechanical power output of a tettigoniid wing muscle during singing and flight. *J. Exp. Biol.* 117: 357–68.

———, and D. Young. (1981) Synchronous and asynchronous muscles in cicadas. *J. Exp. Biol.* 91: 219–37.

Kagen, L. J. (1973) *Myoglobin: Biochemical, Physiological, and Clinical Aspects*. New York: Columbia University Press.

Kalmijn, A. J. (1978) Electric and magnetic sensory world of sharks, skates, and rays. Pp. 507–28 in E. S. Hodgson and R. F. Mathewson, eds., *Sensory Biology of Sharks, Skates, and Rays*. Washington, D.C.: Office of Naval Research, Department of the Navy.

Kauffman, H. J. (1972) *American Axes: A Survey of Their Development and Their Makers*. Brattleboro, Vt.: Stephen Greene Press.

Kelly, D. A. (1997) Axial orthogonal fiber reinforcement in the penis of the nine-banded armadillo (*Dasypus novemcinctus*). *J. Morph.* 233: 249–55.

Kelly, J. E. (1949) "A Survey of Felling and Bucking Tools Used in the South." M. S. Thesis, School of Forestry, Duke University, Durham, N.C.

Kelly, W. (1959) *Ten Ever-Lovin' Blue-Eyed Years with Pogo*. New York: Simon and Schuster.

Ker, R. F.; M. B. Bennett; S. R. Bibby; R. C. Kester; and R. M. Alexander. (1987) The spring in the arch of the human foot. *Nature* 325: 147–49.

Kevles, D. J. (1971) *The Physicists*. New York: Random House.

Kier, W. M. (1985) The musculature of squid arms and tentacles: ultrastructural evidence for functional differences. *J. Morph.* 185: 223–39.

———, and K. K. Smith. (1985) Tongues, tentacles and trunks: the biomechanics of movement in muscular-hydrostats. *Zool. J. Linnean Soc.* 83: 307–24.

Kirby, R. S.; S. Withington; A. B. Darling; and R. G. Kilgour. (1956) *Engineering in History*. New York: McGraw-Hill Book Co.

Klopsteg, P. E. (1943) Physics of bows and arrows. *Amer. J. Physics* 11: 175–92. Also Pp. 157–89 in C. N. Hickman, F. Nagler, and P. E. Klopsteg, eds.; *Archery: The Technical Side*. Milwaukee: North American Press, 1947.

Kooyman, G. L., and P. J. Ponganis. (1998) The physiological basis of diving to depth: birds and mammals. *Annu. Rev. Physiol.* 60: 19–32.

Kram, R. (1991) Carrying loads with springy poles. *J. Appl. Physiol.* 71: 1119–22.

———. (1996) Inexpensive load carrying by rhinoceros beetles. *J. Exp. Biol.* 199: 609–12.

Kransberg, M., and J. Gies. (1975) *By the Sweat of Thy Brow: Work in the Western World*. New York: G. P. Putnam's Sons.

Krogh, A. (1929) The progress of physiology. *Amer. J. Physiol.* 90: 243–51.

Kuhn, T. S. (1977) *The Essential Tension*. Chicago: University of Chicago Press.

Kyle, C. R. (1995) A history of human-powered land vehicles and competitions. Pp. 95–111 in A. V. Abbott and D. G. Wilson, eds., *Human-Powered Vehicles*. Champaign, Ill.: Human Kinetics.

Langdon, J. (1986) *Horses, Oxen and Technological Innovation*. Cambridge, UK: Cambridge University Press.

Lee, C. R., and C. T. Farley. (1998) Determinants of the center of mass trajectory in human walking and running. *J. Exp. Biol.* 201: 2935–44.

Lehninger, A. L. (1975) *Biochemistry*, 2d ed. New York: Worth Publishers.

Leopold, A. (1949) *A Sand County Almanac*. New York: Oxford University Press.

Littauer, M. A., and J. H. Crouwel. (1979) *Wheeled Vehicles and Ridden Animals in the Ancient Near East*. Leiden: E. J. Brill.

Lockshin, R. A., and C. M. Williams. (1965) Programmed cell death. I. Cytology of degeneration in the intersegmental muscles of the pernyi silkmoth. *J. Insect Physiol.* 11: 123–33.

Lopez, R. S., and J. W. Raymond. (1955) *Medieval Trade in the Mediterranean World: Illustrative Documents*. New York: Columbia University Press.

Lothian, N. V. (1921) The load carried by the soldier. *J. Roy. Army Med. Corps.* 37: 241–63; 342–51; 448–58.

———. (1922) The load carried by the soldier, cont'd. *J. Roy. Army Med. Corps.* 38: 9–24.

Lucas, J. R. (1982) The biophysics of pit construction by antlion larvae (*Myrmeleon*, Neuroptera). *Anim. Behav.* 30: 651–64.

Machin, K. E., and H. W. Lissman. (1960) The mode of operation of the electric receptors in *Gymnarchus niloticus*. *J. Exp. Biol.* 37: 801–11.

Maloiy, G. M. O.; N. C. Heglund; L. M. Prager; G. A. Cavagna; and C. R. Taylor. (1986) Energetic cost of carrying loads: have African women discovered an economic way? *Nature* 319: 668–69.

Margaria, R. (1938) Sulla fisiologia e specialmente sul consumo energetico della marcia e della corsa a varie velocità ed inclinazioni del terreno. *Accad. Naz. Lincei Memorie*, serie VI 7: 299–368.

Marsden, E. W. (1969) *Greek and Roman Artillery: Historical Development*. Oxford, UK: Clarendon Press.

———. (1971) *Greek and Roman Artillery: Technical Treatises*. Oxford, UK: Clarendon Press.

Masson, R. (1931) *Edinburgh*. London: A. & C. Black.

Mathieu, J. R., and D. A. Meyer. (1997) Comparing axe heads of stone, bronze, and steel: studies in experimental archaeology. *J. Field Archaeol.* 24: 333–51.

McCarthy, W. J., and J. A. McGeough. (1990) Tools. Chicago: *Encyclopaedia Britannica* 28: 712–36.

McComas, A. J. (1996) *Skeletal Muscle: Form and Function*. Champaign, Ill.: Human Kinetics.

McMahon, T. A. (1984) *Muscles, Reflexes, and Locomotion*. Princeton, N.J.: Princeton University Press.

———, and J. T. Bonner. (1983) *On Size and Life*. New York: Scientific American Library.

McShea, D. W. (1994) Evolutionary trends and the salience bias (with apologies to oil tankers, Karl Marx, and others). *Technical Communication Quarterly* 3: 21–38.

Merritt, R. L., and A. S. Thom. (1980) Le Grand Menhir Brisé. *Archeol. J.* 137: 27–40.

Milton, K. (1999) A hypothesis to explain the role of meat-eating in human evolution. *Evolutionary Anthropology* 8: 11–21.

Minetti, A. E. (1995) Optimum gradient of mountain paths. *J. Appl. Physiol.* 79: 1698–1703.

Mokyr, J. (1990) *The Lever of Riches: Technological Creativity and Economic Progress*. New York: Oxford University Press.

Moritz, L. A. (1958) *Grain-mills and Flour in Classical Antiquity*. Oxford, UK: Clarendon Press.

Morrison, J. (1995) The trireme. Pp. 49–65 in R. Gardner, ed., *The Age of the Galley*. Annapolis, Md.: Naval Institute Press.

Morton, H. A. (1975) *The Wind Commands: Sailors and Sailing Ships in the Pacific*. Middletown, Conn.: Wesleyan University Press.

Mowat, F. (1963) *Never Cry Wolf*. Boston: Little, Brown & Co.

Muneoka, Y., and B. M. Twarog. (1983) Neuromuscular transmission and excitation-contraction coupling in molluscan muscle. Pp. 35–76 in A. S. M. Saleuddin and K. M. Wilbur, eds., *The Mollusca*, vol 4. New York: Academic Press.

Murray, J. D. (1991) On the molecular mechanism of facilitated oxygen diffusion by haemoglobin and myoglobin. *Proc. Roy. Soc. Lond.* B178: 95–110.

Nadel, E. R., and S. R. Bussolari. (1988) The *Daedalus* project: physiological problems and solutions. *Amer. Sci.* 76: 350–60.

Needham, D. M. (1971) *Machina Carnis: The Biochemistry of Muscular Contraction in Its Historical Development*. Cambridge, UK: Cambridge University Press.

Needham, J. (1965) *Science and Civilisation in China*, vol. 4, part 2. Cambridge, UK: Cambridge University Press.

Nevins, A., and H. S. Commager. (1967) *A Pocket History of the United States*, 5th ed. New York: Simon & Schuster.

Nishikawa, K. C. (1999) Neuromuscular control of prey capture in frogs. *Phil. Trans. Roy. Soc.* B354: 941–54.

Nowak, R. M. (1991) *Walker's Mammals of the World*, 5th ed. Baltimore: Johns Hopkins University Press.

Oldstone, M. B. A. (1998) *Viruses, Plagues, and History*. New York: Oxford University Press.

Oliver, J. W. (1956) *History of American Technology*. New York: Ronald Press.

Parkinson, C. N. (1963) *East and West*. Boston: Houghton Mifflin Co.

Parry, D. A., and R. H. J. Brown. (1959) The hydraulic mechanism of the spider leg. *J. Exp. Biol.* 36: 423–33.

Partridge, E. (1984) *A Dictionary of Slang and Unconventional English*, 8th ed. New York: Macmillan.

Payne-Gallwey, R. (1903) *The Crossbow: Mediaeval and Modern Military and Sporting*. Reprint, 1958; New York: Bramhall House.

Paul, J. (1991) *Catapult: Harry and I Build a Siege Weapon*. New York: Villard Books.

Petrinovich, L. (1999) *The Cannibal Within*. New York: Aldine de Gruyter.

Philbrick, N. (2000) *In the Heart of the Sea*. New York: Viking.

Phillips, P. (1980) *The Prehistory of Europe*. Bloomington: Indiana University Press.

Pope, S. T. (1909) *A Study of Bows and Arrows*. Berkeley: University of California Press.

Powers, E. (1924) *Medieval People*. London: Methuen & Co.

Postel, S. (2000) Troubled Waters. *The Sciences* 40/2: 19–24.

Premi, S. C. L. (1979) "Performance of bullocks under varying conditions of load and climate." Bangkok, Thailand: Asian Institute of Technology, unpublished thesis (Master of Engineering), cited by Cotterell and Kamminga.

Pringle, J. W. S. (1981) The evolution of fibrillar muscle in insects. *J. Exp. Biol.* 94: 1–14.

Pursell, C. (1995) *The Machine in America: A Social History of Technology*. Baltimore: Johns Hopkins University Press.

Quick, H. (1923) *The Hawkeye*. New York: A. L. Burt Company.

Rack, P. M. H., and H. F. Ross. (1984) The tendon of flexor pollicis longus: its effects on the muscular control of force and position at the human thumb. *J. Physiol.* 241: 795–808.

Raepsaet, G. (1997) The development of farming implements between the Seine and the Rhine from the second to the twelfth centuries. Pp. 41–68 in G. Astill and J. Langdon, eds., *Medieval Farming and Technology*. Leiden: Brill.

Ramelli, A. (1588) *The Various and Ingenious Machines of Agostino Ramelli*. Trans. M. T. Gnudi, annotated E. S. Ferguson. Baltimore: Johns Hopkins University Press, 1976.

Randall, W. S. (1997) *George Washington: A Life*. New York: Henry Holt & Co.

Renfrew, C. (1973) *Before Civilisation*. London: Jonathan Cope.

Revel, J. P. (1962) The sarcoplasmic reticulum of the bat cricothyroid muscle. *J. Cell Biol.* 12: 571–88.

Rolt, L. T. C. (1965) *A Short History of Machine Tools*. Cambridge, Mass.: MIT Press.

Rome, L. C.; D. A. Syme; S. Hollinwork; S. L. Lindstedt; and S. M. Baylor. (1996) The whistle and the rattle: the design of sound-producing muscles. *Proc. Nat. Acad. Sci. USA* 93: 8095–8100.

Roper, C. (1995) History and present status of human-powered flight. Pp. 217–38 in A. V. Abbott and D. G. Wilson, *Human-Powered Vehicles*. Champaign, Ill.: Human Kinetics.

Rosenbluth, J. (1969) Sarcoplasmic reticulum of an unusually fast-acting crustacean muscle. *J. Cell Biol.* 42: 534–47.

Russell, C. P. (1967) *Firearms, Traps, and Tools of the Mountain Men*. New York: Alfred A. Knopf.

Ryan, M.; F. Abeyratne; and J. Farrington. (1981) *Animal Draught: The Economics of Revival*. Colombo, Sri Lanka: Agrarian Research and Training Institute.

Salmons, S., and F. A. Sréter. (1976) Significance of impulse activity in the transformation of skeletal muscle type. *Nature* 263: 30–34.

Sanday, P. R. (1986) *Divine Hunger: Cannibalism as a Cultural System*. Cambridge, UK: Cambridge University Press.

Saunders, J. B. deC. M., and C. D. O'Malley. (1950) *The Illustrations from the Works of Andreas Vesalius*. Cleveland: World Publishing Co.

Schiøler, T. (1973) *Roman and Islamic Water-Lifting Wheels*. Odense, DK: Odense University Press.

Schmidt-Nielsen, K. (1984) Scaling: *Why Is Animal Size So Important?* Cambridge, UK: Cambridge University Press.

———. (1987) Per Fredrik Thorkelsson Scholander 1905–1980. *Nat. Acad. Sci. US Biographical Memoirs* 563: 87–412.

———. (1997) *Animal Physiology: Adaptation and Environment*, 5th ed. Cambridge, UK: Cambridge University Press.

Schoenbrun, D. (1976) *Triumph in Paris: The Exploits of Benjamin Franklin*. New York: Harper and Row.

Schoenheimer, R. (1942) *The Dynamic State of Body Constituents*. Cambridge, Mass.: Harvard University Press.

Scholander, P. F. (1960) Oxygen transport through hemoglobin solutions. *Science* 131: 585–90.

———. (1990) *Enjoying a Life in Science*. Fairbanks: University of Alaska Press.

Service, A., and J. Bradbery. (1993) *The Standing Stones of Europe: A Guide to the Great Megalithic Monuments*. London: J. M. Dent.

Seymour, M. K. (1969) Locomotion and coelomic pressure in *Lumbricus terrestris L. J. Exp. Biol.* 51: 47–58.

Shaw, G. B. (1903) *Man and Superman: A Comedy and Philosophy*. New York: Brentano's (1904).

Shaw, J. T. (1995) Oar mechanics and oar power in ancient galleys. Pp. 163–71 in R. Gardner, ed., *The Age of the Galley*. Annapolis, MD: Naval Institute Press.

Sherrington, C. S. (1940) *Man on His Nature*. Cambridge, UK: Cambridge University Press.

———. (1947) *The Integrative Action of the Nervous System*, 2d ed. New Haven: Yale University Press.

Sleeswyk, A. W. (1981) Hand-cranking in Egyptian antiquity. Vol. 6, pp. 23–37 in A. R. Hall and N. Smith, eds., *History of Technology*. London: Mansell Publishing.

Soule, R. G., and R. F. Goldman. (1969) Energy cost of loads carried on the head, hands, or feet. *J. Appl. Physiol.* 27: 687–90.

Spruytte, J. (1983) *Early Harness Systems*. Trans. M. A. Littauer. London: J. A. Allen.

Stanford, C. B. (1999) *The Hunting Apes: Meat Eating and the Origins of Human Behavior*. Princeton, N.J.: Princeton University Press.

Szent-Györgyi, A. (1951) *The Chemistry of Muscle Contraction*, 2d ed. New York: Academic Press.

———. (1957) *Bioenergetics*. New York: Academic Press.

———. (1963) Lost in the twentieth century. *Annu. Rev. Biochem.* Vol. 32: 1–14. Reprinted in *The Excitement and Fascination of Science*. Palo Alto, Calif.: Annual Reviews, 1965.

———. (1972) Dionysians and Apollonians. *Science* 176: 966.

Taylor, C. R. (1986) Carrying loads: the cost of generating muscular force. *News in Physiol. Sci.* 1: 153–55.

———; S. L. Caldwell, and V. J. Rowntree. (1972) Running up and down hills: some consequences of size. *Science* 178: 1096–97.

Taylor, C. R., G. M. O. Maloiy; E. R. Weibel; V. A. Langman; J. M. Z. Kamau; H. J. Seeherman; and N. C. Heglund. (1981) Design of the mammalian respiratory system. III. Scaling maximum aerobic capacity to body mass: wild and domestic mammals. *Resp. Physiol.* 44: 25–37.

Taylor, C. R., R. H. Karas; E. R. Weibel; and H. Hoppeler. (1987) Adaptive variation in the mammalian respiratory system in relation to energetic demand. II. Reaching the limits to oxygen flow. *Resp. Physiol.* 69: 7–26.

Taylor, G. M. (2000) Maximum force production: why are crabs so strong? *Proc. Roy. Soc. Lond.* B267: 1475–80.

Taylor, R. J. F. (1957) The work output of sledge dogs. *J. Physiol.* 137: 210–17.

Tennekes, H. (1996) *The Simple Science of Flight: From Insects to Jumbo Jets*. Cambridge, Mass.: MIT Press.

Thomas, S. P. (1975) Metabolism during flight in two species of bats, *Phyllostomus hastatus* and *Pteropus gouldii*. *J. Exp. Biol.* 63: 273–93.

Townsend, J. B. (1983) Firearms against native arms: a study in comparative efficiencies with an Alaskan example. *Arctic Anthropol.* 20: 1–33.

Tranter, N. (1987) *The Story of Scotland*. Glasgow, UK: Neil Wilson Publishing.

Tricker, R. A. R., and B. J. K. Tricker. (1967) *The Science of Movement*. New York: American Elsevier.

Vallee, B. (1997) Alcohol and the development of human civilization. Pp. 125–54 in P. Day, ed., *Exploring the Universe*. Oxford, UK: Oxford University Press.

Van Leeuwen, J. L., and W. M. Kier. (1997) Functional design of tentacles in squid: linking sarcomere ultrastructure to gross morphological dynamics. *Phil. Trans. Roy. Soc. Lond.* B352: 551–71.

Veranzio, F. (1615–16) *Machinae Novae*. Munich: Heinz Moos Verlag, 1965.

Vesalius, A. (1973) *The Illustrations from the Works of Andreas Vessalius of Brussels, with Annotations and Translations, a Discussion of the Plates and Their Background*. New York: Dover Publications.

———. (1998) *De Humani Corporis Fabrica*. CD-ROM, Octavo Corp.

Vincent, J. F. V. (1990) *Structural Biomaterials*, rev. ed. Princeton, N.J.: Princeton University Press.

Vogel, S. (1988a) Capitalizing on currents, how non-rigid organisms use flow-induced pressures. *Amer. Sci.* 76: 28–34.

———. (1988b) *Life's Devices: The Physical World of Animals and Plants.* Princeton, N.J.: Princeton University Press.

———. (1992) *Vital Circuits: On Pumps, Pipes, and the Workings of Circulatory Systems.* New York: Oxford University Press.

———. (1994a) Dealing honestly with diffusion. *Amer. Biol. Teacher* 56: 405–07.

———. (1994b) *Life in Moving Fluids,* 2d ed. Princeton, N.J.: Princeton University Press.

———. (1996) Diversity and convergence in the study of organismal function. *Israel J. Zool.* 42:297–305.

———. (1998) *Cats' Paws and Catapults.* New York: W. W. Norton.

Wachsmann, S. (1995) Paddled and oared ships before the Iron Age. Pp. 10–35 in R. Gardner, ed., *The Age of the Galley.* Annapolis, Md.: Naval Institute Press.

Wade, O. L., and J. M. Bishop. (1962) *Cardiac Output and Regional Blood Flow.* Oxford, UK: Blackwell Scientific.

Wainwright, P. C.; D. M. Kraklau; and A. F. Bennett. (1991) Kinematics of tongue protrusion in *Chamaeleo oustaleti. J. Exp. Biol.* 159: 109–33.

Wainwright, S. A.; D. A. Pabst; and P. F. Brodie. (1985) Form and possible function of the collagen layer underlying cetacean blubber. *Amer. Zool.* 25: 146A.

Wainwright, S. A.; F. Vosburgh; and J. H. Hebrank. (1978) Shark skin: function in locomotion. *Science* 202: 747–49.

Walton, J. (1974) *Water-mills, Windmills and Horse-mills of South Africa.* Cape Town: C. Struik Publishers.

Warwick, R., and P. L. Williams. (1973) *Gray's Anatomy,* 35th British ed. Philadelphia: W. B. Saunders.

Washburn, S. L. (1959) Speculations on the interrelationships of the history of tools and biological evolution. *Human Biology* 31: 21–31.

Watt, B. K., and A. L. Merrill. (1963) *Composition of Foods: Raw, Processed, Prepared. Agriculture Handbook No. 8.* Washington, D.C.: U.S. Department of Agriculture.

Weibel, E. R. (1984) *The Pathway for Oxygen: Structure and Function in the Mammalian Respiratory System.* Cambridge, Mass.: Harvard University Press.

Weis-Fogh, T. (1960) A rubber-like protein in insect cuticle. *J. Exp. Biol.* 37: 889–906.

Weis-Fogh, T., and R. M. Alexander. (1977) The sustained power output available from striated muscle. Pp. 511–25 in T. J. Pedley, ed., *Scale Effects in Animal Locomotion.* London: Academic Press.

White, L. T. (1940) Technology and invention in the Middle Ages. *Speculum* 14: 141–59. Also in White, 1978.

———. (1962) *Medieval Technology and Social Change.* New York: Oxford University Press.

———. (1963) Medieval roots of modern technology. Pp. 19–34 in K. F. Drew and F. S. Lear, eds., *Perspectives in Medieval History.* Chicago: University of Chicago Press. Also in White, 1978.

———. (1965) The legacy of the Middle Ages in the American Wild West. *Speculum* 40: 191–202. Also in White, 1978.

———. (1967) The life of the silent majority. Pp. 85–100 in R. S. Hoyt, ed., *Life and Thought in the Early Middle Ages.* Minneapolis: University of Minnesota Press.

————. (1974) Technology assessment from the stance of a medieval historian. *Amer. Historical Rev.* 79: 1–13. Also in White, 1978.

————. (1975) The study of medieval technology, 1924–1974: personal reflections. *Technology and Culture* 16: 519–30. Also in White, 1978.

————. (1978) *Medieval Religion and Technology: Collected Essays.* Berkeley: University of California Press.

Whitt, F. R., and D. G. Wilson. (1982) *Bicycling Science,* 2d ed. Cambridge, Mass.: MIT Press.

Williams, P. E., and G. Goldspink. (1978) Changes in sarcomere length and physiological properties in immobilized muscle. *J. Anat.* 127: 459–68.

Wilkie, D. R. (1950) The relation between force and velocity in human muscle. *J. Physiol.* 110: 249–80.

————. (1959) Work output of animals: flight by birds and by man power. *Nature* 183: 1515–16.

————. (1960a) Man as a source of mechanical power. *Ergonomics* 3: 1–8.

————. (1960b) Man as an aero engine. *J. Roy. Aero. Soc.* 64: 477–81.

————. (1984) Muscle function: a historical view. Pp. 3–13 in N. L. Jones, N. McCartney, and A. J. Comas, eds., *Human Muscle Power.* Champaign, Ill.: Human Kinetics.

Wilt, F. (1968) Training for competitive running. Pp. 395–414 in H. B. Falls, ed., *Exercise Physiology.* New York: Academic Press.

Wooton, R. J., and D. J. S. Newman. (1979) Whitefly have the highest contraction frequency yet recorded in non-fibrillar flight muscles. *Nature* 280: 402–03.

Yost, J. A., and P. M. Kelley. (1983) Shotguns, blowguns, and spears: the analysis of technological efficiency. Pp. 189–224 in R. B. Hames and W. T. Vickers, eds., *Adaptive Responses of Native Amazonians.* New York: Academic Press.

Yousef, M. K., and D. B. Dill. (1969) Energy expenditure in desert walks: man and burro *Equus asinus. J. Appl. Physiol.* 27: 681–83.

Zagorskii, F. N. (1960) *An Outline of the History of Metal Cutting Machines to the Middle of the 19th Century.* New Delhi: Amerind Publishing.

Index

Page numbers in *italics* refer to illustrations.

molecular, 51
muscle, 6, 7, 12, 34, 70, 215
vs. muscle color, 59
vs. muscle cross-section, 34
paddling vs. rowing, 165
peak pulling, 196
of pennate muscles, 109, *110*
vs. power, 24, *25*, 70, 181
per sarcomere, 12, 15, *72*
by screws, 135
vs. speed, *25*, *25*, 71–73, *72*, *154*, 271
weaponry, 285–91
whole animal, *25*, 34
vs. work, 4
Forrest, Nathan, 191
Franklin, Benjamin, 199, 297
frequency of shortening, 61, 63, 67, 69
friction, in moving loads, 196
frogs
African clawed, 53
gastrocnemius, 3, *3*, 6, 36, 42
jumping, 3, 4
muscle types, 54
sound production, 65
tongues, 119
fruit fly, 11, *14*, *48*
Full, Robert J., 221

gaits
energy storage, 120–21
relative cost, 218
transition points, 216, *217*
Galen, 32
Galileo, 4, 59, 124
galley warships, 166–69
galloping, 123
Galvani, Luigi, 6
garlic presses, *129*, 130
gastraphetes, *265*, *265*, 287
gastrocnemius, 3, *3*, 6, 7, 12, 23, 36, 42, 56, *106*, 122
gears, 147
gelatin, 299
girdling trees, 199
gliding, bicycle-assisted, 176
gluteal muscles, 28
glycerinated muscle, 44, 46
glycogen, as energy source, 17, 308
goats
domestication, 210

use with treadmills, 225, 231
golf, 276
Golgi, Camillo, 78
Gordon, James E., 299
Gossamer aircraft, 176–78
grain mills, 227
Grand Menhir Brisé, 194
grasshopper, *106*
graters, food, 131
gravitational energy storage, 189, 258, 284
in appendages, 120–21,189
in engines, 102
trebuchets, 268
walking gait, 120–21, 189
weaponry, 258, 268
gravity
center of, 134, *134*, 143, *144*, 195
vs. drag of projectiles, 273
opening doors, 133, *134*
and tool use, 146
Great Wall of China, 197
Greeks, ancient
archery, 287
ballistae, 266–68, *266*, 283
cavalry, 236, 291
chariots, 236–37, *237*
gastraphetes, *265*, 266
moving stones, 196
screws, 135
ships, 166, *167*, 169
soldiers, 192
treadmills, 227
grinding grain, 150
gripping, 142, 146, 131–32

hammers
designs, 131, 142–43
motions, 132
role of gravity, 146
role of kinetic energy, 139
weights, 131
handedness
humans, 131
screws, cranks, 136
handles, 130, 131
axes, 141–46
doors, 133, 135
hammers, 132, 142
screwdrivers, 132